J. Fulcher, L. C. Jain (Eds.)

Applied Intelligent Systems

Springer

Berlin
Heidelberg
New York
Hong Kong
London
Milano
Paris
Tokyo

Studies in Fuzziness and Soft Computing, Volume 153

Editor-in-chief
Prof. Janusz Kacprzyk
Systems Research Institute
Polish Academy of Sciences
ul. Newelska 6
01-447 Warsaw
Poland
E-mail: kacprzyk@ibspan.waw.pl

Further volumes of this series can be found on our homepage:
springeronline.com

Vol 134. V.A. Niskanen
Soft Computing Methods in Human Sciences, 2004
ISBN 3-540-00466-1

Vol. 135. J.J. Buckley
Fuzzy Probabilities and Fuzzy Sets for Web Planning, 2004
ISBN 3-540-00473-4

Vol. 136. L. Wang (Ed.)
Soft Computing in Communications, 2004
ISBN 3-540-40575-5

Vol. 137. V. Loia, M. Nikravesh, L.A. Zadeh (Eds.)
Fuzzy Logic and the Internet, 2004
ISBN 3-540-20180-7

Vol. 138. S. Sirmakessis (Ed.)
Text Mining and its Applications, 2004
ISBN 3-540-20238-2

Vol. 139. M. Nikravesh, B. Azvine, I. Yager, L.A. Zadeh (Eds.)
Enhancing the Power of the Internet, 2004
ISBN 3-540-20237-4

Vol. 140. A. Abraham, L.C. Jain, B.J. van der Zwaag (Eds.)
Innovations in Intelligent Systems, 2004
ISBN 3-540-20265-X

Vol. 141. G.C. Onwubolu, B.V. Babu
New Optimzation Techniques in Engineering, 2004
ISBN 3-540-20167-X

Vol. 142. M. Nikravesh, L.A. Zadeh, V. Korotkikh (Eds.)
Fuzzy Partial Differential Equations and Relational Equations, 2004
ISBN 3-540-20322-2

Vol. 143. L. Rutkowski
New Soft Computing Techniques for System Modelling, Pattern Classification and Image Processing, 2004
ISBN 3-540-20584-5

Vol. 144. Z. Sun, G.R. Finnie
Intelligent Techniques in E-Commerce, 2004
ISBN 3-540-20518-7

Vol. 145. J. Gil-Aluja
Fuzzy Sets in the Management of Uncertainty, 2004
ISBN 3-540-20341-9

Vol. 146. J.A. Gámez, S. Moral, A. Salmerón (Eds.)
Advances in Bayesian Networks, 2004
ISBN 3-540-20876-3

Vol. 147. K. Watanabe, M.M.A. Hashem
New Algorithms and their Applications to Evolutionary Robots, 2004
ISBN 3-540-20901-8

Vol. 148. C. Martin-Vide, V. Mitrana, G. Păun (Eds.)
Formal Languages and Applications, 2004
ISBN 3-540-20907-7

Vol. 149. J.J. Buckley
Fuzzy Statistics, 2004
ISBN 3-540-21084-9

Vol. 150. L. Bull (Ed.)
Applications of Learning Classifier Systems, 2004
ISBN 3-540-21109-8

Vol. 151. T. Kowalczyk, E. Pleszczyńska, F. Ruland (Eds.)
Grade Models and Methods for Data Analysis, 2004
ISBN 3-540-21120-9

Vol. 152. J. Rajapakse, L. Wang (Eds.)
Neural Information Processing: Research and Development, 2004
ISBN 3-540-21123-3

John Fulcher
Lakhmi C. Jain (Eds.)

Applied Intelligent Systems

New Directions

 Springer

Professor John Fulcher
University of Wollongong
School of Information
Technology & Computer Science
2522 Wollongong, NSW
Australia
E-mail: john@uow.edu.au

Professor Lakhmi C. Jain
University of South Australia
Knowledge-Based Intelligent
Engineering Systems Centre
Mawson Lakes
5095 Adelaide
Australia
E-mail: l.jain@unisa.edu.au

ISSN 1434-9922
ISBN 3-540-21153-5 Springer-Verlag Berlin Heidelberg New York

Library of Congress Cataloging-in-Publication-Data

A catalog record for this book is available from the Library of Congress.
Bibliographic information published by Die Deutsche Bibliothek.
Die Deutsche Bibliothek lists this publication in the Deutsche Nationalbibliographie;
detailed bibliographic data is available in the Internet at http://dnb.ddb.de

This work is subject to copyright. All rights are reserved, whether the whole or part of the material is concerned, specifically the rights of translation, reprinting, reuse of illustrations, recitations, broadcasting, reproduction on microfilm or in any other way, and storage in data banks. Duplication of this publication or parts thereof is permitted only under the provisions of the German copyright Law of September 9, 1965, in its current version, and permission for use must always be obtained from Springer-Verlag. Violations are liable for prosecution under the German Copyright Law.

Springer-Verlag is a part of Springer Science+Business Media
springeronline.com

© Springer-Verlag Berlin Heidelberg 2004
Printed in Germany

The use of general descriptive names, registered names trademarks, etc. in this publication does not imply, even in the absence of a specific statement, that such names are exempt from the relevant protective laws and regulations and therefore free for general use.

Typesetting: data delivered by editors
Cover design: E. Kirchner, Springer-Verlag, Heidelberg
Printed on acid free paper   62/3020/M - 5 4 3 2 1 0

Dedicated to

Ione Lewis

Preface

Humans have always been hopeless at predicting the future...most people now generally agree that the margin of viability in prophecy appears to be ten years.[1] Even sophisticated research endeavours in this arena tend to go off the rails after a decade or so.[2] The computer industry has been particularly prone to bold (and often way off the mark) predictions, for example:

- 'I think there is a world market for maybe five computers' *Thomas J. Watson, IBM Chairman (1943)*,
- 'I have traveled the length and breadth of this country and talked with the best people, and I can assure you that data processing is a fad that won't last out the year' *Prentice Hall Editor (1957)*,
- 'There is no reason why anyone would want a computer in their home' *Ken Olsen, founder of DEC (1977)* and
- '640K ought to be enough for anybody' *Bill Gates, CEO Microsoft (1981)*.

The field of Artificial Intelligence[3] – right from its inception – has been particularly plagued by 'bold prediction syndrome', and often by leading practitioners who should know better. AI has received a lot of bad press over the decades, and a lot of it deservedly so.[4] How often have we groaned in despair at the latest 'by the year-20*xx*, we will all have...(*insert your own particular 'hobby horse' here – e.g. autonomous robot vacuum cleaners that will absolve us of the need to clean our homes...etc*)' – and this is to completely ignore the reality that most of the world's population

---

[1] Davies S (1996) *Monitor: Extinguishing Privacy on the Information Superhighway*, Pan Macmillan, Sydney.
[2] Naisbitt J (1982) *Megatrends: Ten New Directions Transforming Our Lives*, Macdonald, London.
[3] whatever *that* term means (we could present another entire book debating this topic).
[4] Fulcher JA (2001) Practical (Artificial) Intelligence, *Invited Keynote Speech, 5th National Thai Conf Computer Science & Engineering*, Chiang Mai, 7-9 November: i23-i28.

does not have internet access[5], let alone own a computer[6] or telephone[7] – indeed 1.3 billion do not have access to clean drinking water[8]. Obviously the 'we' of these predictions refers only to a small minority of first-world citizens. Don't these advocates (zealots?) realize the harm that such misplaced predictions cause in the longer term? Indeed, their misplaced enthusiasm damages *all* of us who work in the field.

Decades of spectacular failures and unfulfilled promises can only serve to antagonize the public and get them offside for years to come. More specifically, some AI researchers have misjudged the difficulty of problems and oversold the prospects of short-term progress based on the initial results.[9,10] To put it another way, the methods that sufficed for demonstration on one or two simple examples turned out to fail miserably when tried out on wider selections of problems.[11]

This then leads us to the motivation for the present book. Your Editors felt it was timely to take a step back from the precipice, as it were, to pause and reflect, and rather than indulge in yet *more* bold predictions, to report on some intelligent systems that actually work – *now* (and not at some mythical time in the future).

With this in mind, we invited the authors herein – all international experts in their respective fields – to contribute Chapters on intelligent techniques that have been tried and proved to work on real-world problems. We should also point out that our use of 'intelligent' in this context reflects the intelligence of the authors who have created these various techniques, and not on some nebulous 'ghost in the machine'.

The book commences with Edelman & Davy's application of Genetic Programming, Support Vector Machines and Artificial Neural Networks to financial market predictions, and from their results they draw conclusions about the weak form of the Efficient Markets Hypothesis.

---

[5] http://unstats.un.org/unsd (ITU estimates – e.g. Iceland 60%; Spain 14%; China 1.74%; India 0.54%; Somalia 0.01% per 100 population)

[6] http://unstats.un.org/unsd (ITU estimates – e.g. USA 57%; Slovenia 27.5%; China 1.59%; India 0.45%; Niger 0.056% per 100 population)

[7] http://unstats.un.org/unsd (ITU estimates – e.g. Monaco 147%; Australia 100%; Estonia 75%; Mexico 27%; China 17.76%; India 3.56%; Afghanistan 0.13% per 100 population)

[8] http://unstats.un.org/unsd (WHO/UNICEF estimates)

[9] Allen J (1998) AI Growing Up; the Changes and Opportunities, *AI Magazine*, Winter: 13-23.

[10] Hearst M and Hirsh H (2000) AI's Greatest Trends and Controversies, *IEEE Intelligent Systems*, January/February: 8-11.

[11] Russell S and Norvig P (1995) *AI: A Modern Approach*, Prentice Hall, Englewood Cliffs, NJ.

Chapter 2 covers the Higher Order Neural Network models developed by Zhang & Fulcher. HONNs (and HONN groups) have been successfully applied to human face recognition, financial time series modeling and prediction, as well as to satellite weather prediction – it is the latter that is reported on in this volume.

Back demonstrates the power of Independent Component Analysis in Chapter 3, and cites examples drawn from biomedical signal processing (ECG), extracting speech from noise, unsupervised classification (non-invasive oil flow monitoring and banknote fraud detection), and financial market prediction.

Chapter 4 focuses on the application of AI techniques to regulatory applications in health informatics. Copland describes an innovative combination of Evolutionary Algorithms and Artificial Neural Networks which he uses as his primary Data Mining tool when investigating servicing by medical doctors.

Swarms and their collective intelligence are the subject of the next chapter. Hendtlass first describes several ant colony optimization algorithms, then proceeds to show how they can be applied both to the Travelling SalesPerson problem and sorting (of the iris data set).

In Chapter 6, McKerrow asks the question: 'Where have all the mobile robots gone?', and in the process restores some sanity to counteract some bold predictions by practitioners who should know better. The real-world commercial applications covered in this Chapter include robot couriers, vacuum cleaners, lawn mowers, pool cleaners, and people transporters.

Zeleznikow's expertise with intelligent legal decision support systems is brought to the fore in Chapter 7, and places this emerging field within an historical context. Rule-based reasoners, case-based and hybrid systems, Knowledge Discovery in Databases and web-based systems are all covered.

Chapter 8 is devoted to Human-Agent teams within hostile environments. Sioutis and his co-authors illustrate their ideas within the context of the Jack agent shell and interactive 3D games such as Unreal Tournament.

The Fuzzy Multivariate Auto-Regression method is the focus of Chapter 9. Sisman-Yilmaz and her co-authors show how Fuzzy MAR can be applied to both Gas Furnace and Interest Rate data.

In Chapter 10, Lozo and his co-authors describe an extension of ART – Selective Attention Adaptive Resonance Theory – and illustrate its usefulness when applied to distortion-invariant 2D shape recognition embedded in clutter.

We hope these invited Chapters serve two functions: firstly, presentation of tried and proven 'intelligent' techniques, and more especially the particular application niche(s) in which they have been successfully applied. Secondly, we hope to restore some much needed public confidence in a field that has become tarnished by bold, unrealistic predictions for the future.

Lastly, we would like to thank all our contributing authors who so willingly took time from their busy schedules to produce the quality Chapters contained herein.

Enjoy your reading.

University of Wollongong John Fulcher
University of South Australia Lakhmi C Jain
Spring 2004

Table of Contents

1 Adaptive Technical Analysis in the Financial Markets Using Machine Learning: a Statistical View
David Edelman and Pam Davy

1.1	'Technical Analysis' in Finance: a Brief Background	1
1.2	The 'Moving Windows' Paradigm	2
1.3	Post-Hoc Performance Assessment	3
	1.3.1 The Effect of Dividends	5
	1.3.2 Transaction Costs Approximations	6
1.4	Genetic programming	7
1.5	Support-Vector Machines	10
1.6	Neural Networks	12
1.7	Discussion	14
	References	15

2 Higher Order Neural Networks for Satellite Weather Prediction
Ming Zhang and John Fulcher

2.1	Introduction	17
2.2	Higher Order Neural Networks	18
	2.2.1 Polynomial Higher-Order Neural Networks	20
	2.2.2 Trigonometric Higher-Order Neural Networks	23
	Output Neurons in THONN Model#1	24
	Second Hidden Layer Neurons in THONN Model#1	27
	First Hidden Layer Neurons in THONN Model#1	30
	2.2.3 Neuron-Adaptive Higher-Order Neural Network	30
2.3	Artificial Neural Network Groups	33
	2.3.1 ANN Groups	33
	2.3.2 PHONN, THONN & NAHONN Groups	34
2.4	Weather Forecasting & ANNs	35
2.5	HONN Models for Half-hour Rainfall Prediction	36
	2.5.1 PT-HONN Model	36
	2.5.2 A-PHONN Model	37

		2.5.3 M-PHONN Model	38
		2.5.4 Satellite Rainfall Estimation Results	38
2.6		ANSER System for Rainfall Estimation	39
		2.6.1 ANSER Architecture	40
		2.6.2 ANSER Operation	41
		2.6.3 Reasoning Network Based on ANN Groups	43
		2.6.4 Rainfall Estimation Results	45
2.7		Summary	47
		Acknowledgements	47
		References	47
		Appendix-A Second Hidden Layer (multiply) Neurons	51
		Appendix-B First Hidden Layer Neurons	54

3 Independent Component Analysis

Andrew Back

3.1	Introduction	59
3.2	Independent Component Analysis Methods	60
	3.2.1 Basic Principles and Background	60
	3.2.2 Mutual Information Methods	62
	3.2.3 InfoMax ICA Algorithm	64
	3.2.4 Natural/Relative Gradient Methods	65
	3.2.5 Extended InfoMax	66
	3.2.6 Adaptive Mutual Information	66
	3.2.7 Fixed Point ICA Algorithm	68
	3.2.8 Decorrelation and Rotation Methods	69
	3.2.9 Comon Decorrelation and Rotation Algorithm	71
	3.2.10 Temporal Decorrelation Methods	71
	3.2.11 Molgedey and Schuster Temporal Correlation Algorithm	72
	3.2.12 Spatio-temporal ICA Methods	73
	3.2.13 Cumulant Tensor Methods	74
	3.2.14 Nonlinear Decorrelation Methods	75
3.3	Applications of ICA	75
	3.3.1 Guidelines for Applications of ICA	75
	3.3.2 Biomedical Signal Processing	76
	3.3.3 Extracting Speech from Noise	77
	3.3.4 Unsupervised Classification Using ICA	78
	3.3.5 Computational Finance	81
3.4	Open Problems for ICA Research	83
3.5	Summary	85
	References	86
	Appendix – Selected ICA Resources	95

4 Regulatory Applications of Artificial Intelligence
Howard Copland

4.1	Introduction	97
4.2	Solution Spaces, Data and Mining	98
4.3	Artificial Intelligence in Context	102
4.4	Anomaly Detection: ANNs for Prediction/Classification	104
	4.4.1 Training to Classify on Spare Data Sets	105
	4.4.2 Training to Predict on Dense Data Sets	106
	4.4.3 Feature Selection for and Performance of Anomaly Detection Suites	109
	4.4.4 Interpreting Anomalies	113
	4.4.5 Other Approaches to Anomaly Detection	115
	4.4.6 Variations of BackProp' ANNs for Use with Complex Data Sets	120
4.5	Formulating Expert Systems to Identify Common Events of Interest	121
	A Note on the Software	130
	Acknowledgements	130
	References	130

5 An Introduction to Collective Intelligence
Tim Hendtlass

5.1	Collective Intelligence	133
	5.1.1 A Simple Example of Stigmergy at Work	134
5.2	The Power of Collective Action	136
5.3	Optimisation	138
	5.3.1 Optimisation in General	138
	5.3.2 Shades of Optimisation	139
	5.3.3 Exploitation versus Exploration	139
	5.3.4 Example of Common Optimisation Problems	140
	Minimum Path Length	140
	Function Optimisation	140
	Sorting	141
	Multi-Component Optimisation	141
5.4	Ant Colony Optimisation	141
	5.4.1 Ant Systems – the Basic Algorithm	143
	The Problem with AS	143
	5.4.2 Ant Colony Systems	144
	5.4.3 Ant Multi-Tour System (AMTS)	145
	5.4.4 Limiting the Pheromone Density – the Max-Min Ant System	145

	5.4.5	An Example: Using Ants to Solve a (simple) TSP	146
	5.4.6	Practical Considerations	154
	5.4.7	Adding a Local Heuristic	155
	5.4.8	Other Uses for ACO	157
	5.4.9	Using Ants to Sort	158
		An Example of Sorting Using ACO	162
5.5		Particle Swarm Optimisation	165
	5.5.1	The Basic Particle Swarm Optimisation Algorithm	165
	5.5.2	Limitations of the Basic Algorithm	166
	5.5.3	Modifications to the Basic PSO Algorithm	167
		Choosing the Position S	167
		The Problem of a finite t	168
		Aggressively Searching Swarms	168
		Adding Memory to Each Particle	169
	5.5.4	Performance	170
	5.5.5	Solving TSP Problems Using PSO	171
		PSO Performance on a TSP	172
	5.5.6	Practical Considerations	175
	5.5.7	Scalability and Adaptability	176
		References	177

6 Where are all the Mobile Robots?

Phillip McKerrow

6.1	Introduction		179
6.2	Commercial Applications		181
	6.2.1	Robot Couriers	182
	6.2.2	Robot Vacuum Cleaners	183
	6.2.3	Robot Lawn Mowers	187
	6.2.4	Robot Pool Cleaners	189
	6.2.5	Robot People Transporter	191
	6.2.6	Robot Toys	193
	6.2.7	Other Applications	195
	6.2.8	Getting a Robot to Market	195
	6.2.9	Wheeled Mobile Robot Research	196
6.3	Research Directions		197
6.4	Conclusion		198
	A Note on the Figures		199
	References		199

7 Building Intelligent Legal Decision Support Systems: Past Practice and Future Challenges

John Zeleznikow

7.1	Introduction	201
	7.1.1 Benefits of Legal Decision Support Systems to the Legal Profession	202
	7.1.2 Current Research in AI and Law	204
7.2	Jurisprudential Principles for Developing Intelligent Legal Knowledge-Based Systems	208
	7.2.1 Reasoning with Open Texture	209
	7.2.2 The Inadequacies of Modelling Law as a Series of Rules	210
	7.2.3 Landmark and Commonplace Cases	211
7.3	Early Legal Decision Support Systems	214
	7.3.1 Rule-Based Reasoning	214
	7.3.2 Case-Based Reasoning and Hybrid Systems	219
	7.3.3 Knowledge Discovery in Legal Databases	222
	7.3.4 Evaluation of Legal Knowledge-Based Systems	222
	7.3.5 Explanation and Argumentation in Legal Knowledge-Based Systems	231
7.4	Legal Decision Support on the World Wide Web	233
	7.4.1 Legal Knowledge on the WWW	233
	7.4.2 Legal Ontologies	234
	7.4.3 Negotiation Support Systems	240
7.5	Conclusion	246
	Acknowledgements	247
	References	247

8 Forming Human-Agent Teams within Hostile Environments

Christos Sioutis, Pierre Urlings, Jeffrey Tweedale, and Nikhil Ichalkaranje

8.1	Introduction	255
8.2	Background	256
8.3	Cognitive Engineering	257
8.4	Research Challenge	258
	8.4.1 Human-Agent Teaming	258
	8.4.2 Agent Learning	260
8.5	The Research Environment	262
	8.5.1 The Concept of Situational Awareness	262
	8.5.2 The Unreal Tournament Game Platform	263
	8.5.3 The Jack Agent	263

8.6		The Research Application	264
	8.6.1	The Human Agent Team	264
	8.6.2	The Simulated World Within Unreal Tournament	265
	8.6.3	Interacting With Unreal Tournament	267
	8.6.4	The Java Extension	268
	8.6.5	The Jack Component	269
8.7		Demonstration System	270
	8.7.1	Wrapping Behaviours in Capabilities	271
	8.7.2	The Exploring Behaviours	272
	8.7.3	The Defending Behaviour	273
8.8		Conclusions	275
		Acknowledgements	276
		References	276

9 Fuzzy Multivariate Auto-Regression Method and its Application

N Arzu Sisman-Yilmaz, Ferda N Alpaslan and Lakhmi C Jain

9.1		Introduction	281
9.2		Fuzzy Data Analysis	282
	9.2.1	Fuzzy Regression	282
	9.2.2	Fuzzy Time Series Analysis	284
	9.2.3	Fuzzy Linear Regression (FLR)	285
		Basic Definitions	285
		Linear Programming Problem	285
9.3		Fuzzy Multivariate Auto-Regression Algorithm	286
		Example - Gas Furnace Data Processed by MAR	287
	9.3.1	Model Selection	288
	9.3.2	Motivation for FLR in Fuzzy MAR	289
	9.3.3	Fuzzification of Multivariate Auto-Regression	290
	9.3.4	Bayesian Information Criterion in Fuzzy MAR	291
	9.3.5	Obtaining a Linear Function for a Variable	292
	9.3.6	Processing of Multivariate Data	293
9.4		Experimental Results	295
	9.4.1	Experiments with Gas Furnace Data	295
	9.4.2	Experiments with Interest Rate Data	296
	9.4.3	Discussion of Experimental Results	298
9.5		Conclusions	299
		References	299

10 Selective Attention Adaptive Resonance theory and Object Recognition

Peter Lozo, Jason Westmacott, Quoc V Do, Lakhmi C Jain and Lai Wu

10.1	Introduction	301
10.2	Adaptive Resonance Theory (ART)	302
	10.2.1 Limitations of ART's Attentional Subsystem with Cluttered Inputs	303
10.3	Selective Attention Adaptive Resonance Theory	305
	10.3.1 Neural Network Implementation of SAART	306
	Postsynaptic Cellular Activity	309
	Excitatory Postsynaptic Potential	309
	Lateral Competition	309
	Transmitter Dynamics	310
	10.3.2 Translation-invariant 2D Shape Recognition	312
10.4	Conclusions	317
	References	318

Index 321

1 Adaptive Technical Analysis in the Financial Markets Using Machine Learning: a Statistical View

David Edelman[1] and Pam Davy[2]

1. Department of Banking and Finance, University College, Dublin, Ireland, david.edelman@ucd.ir
2. School of Mathematics & Applied Statistics, University of Wollongong NSW 2522, Australia, Pam_Davy@uow.edu.au

1.1 'Technical Analysis' in Finance: a Brief Background

In broad terms, the term 'Technical Analysis' (TA) is generally meant to describe any method for attempting to exploit a presumed violation of *weak efficiency* in a Financial Market. In the context of Investments, the so-called *Efficient Markets Hypothesis* (EMH) is probably the most widely studied of any topic, referring (in a heuristic sense) to the question of whether or not there exist strategies which consistently 'beat the market' in terms of return for a given level of risk, given a certain Information Set. Generally speaking, discussion of the EMH is usually divided into three categories:
1. Strong form (where the Information Set includes all information, Public and Private),
2. Semi-Strong Form (where the Information set includes all Publicly available information), and
3. Weak Form (where the Information Set includes only the past price history of the security in question).

Any investor applying TA in some market, then, believes there to exist a weak-form inefficiency in that market.

Traditional academic literature has tended to focus on establishing in an evidentiary manner whether or not Market inefficiencies exist. By contrast, here we focus on several methods which appear to strongly suggest that weak-form inefficiencies exist, admitting that conclusive proof remains, as ever, elusive, since scientific rigour would require that any experiment should be repeatable, a condition which can never truly be met if Markets are assumed to be (as the authors believe them to be) non-stationary.

While it has been generally noted [26] that most traders use some form of TA, such as crossing moving averages, Elliott Wave or Gann Theory, and so on, simple logic suggests that if any such fixed trading rule continued to achieve a super-efficient performance indefinitely, this fact would become widely known, and a subsequent correction would occur. On the other hand, many believe that there could in theory exist certain types of effective super-efficient rules, such as those which might vary over time, which might be too subtle for widespread contemporary practitioners of pattern-recognition technology (or other analytical processes) to fully exploit, but which a small population of individuals might be able to successfully apply for a time. The general economic framework for such a hypothetical situation falls within the category of the theory of *Bounded Rationality*, introduced by Simon in 1955 [20], (clarified and summarized in [21]).

Thus, we shall assume in the sequel that if any effective trading rules exist, they must vary over time, and hence the task of discovering them is tantamount to what is known as a *tracking* problem. This assumption is at the basis of the entire framework of the methodology to be presented in the following Sections.

In the next section, we begin by discussing general issues relating to overall methodology, philosophy, and several practical details of our approach. Following this, three independent studies will be presented which are similar, though implemented using the different methods of Genetic Programming, Support Vector Machines, and Neural Networks. Each section has its own Introduction, Literature Review, Methodology, and Results subsections, with a general comparative Discussion deferred to a single section at the end.

1.2 The 'Moving Windows' Paradigm

Imagine that at the beginning of each Financial Quarter, an entirely new forecasting model is to be determined and used for investment over the ensuing quarter. In building the model, a dilemma occurs, in that a sufficiently long backward-looking time horizon must be chosen to ensure a model can be fit using enough data to ensure statistical reliability. However, for sufficiently long time periods into the past, the relevance of the data to the current underlying dynamic will be too low for such data to be usefully employed in the model. The magnitude of the backward time horizons must, then, be chosen experimentally, where models are built using time windows of various widths (lengths of time into the past), with investment performance of these models over the subsequent quarters (periods just following those that had been used to build the model) assessed for each and compared in retrospect.

(Ironically, experimentation with these parameters was carried out by various 'moving windows' of research students over several years, yielding the best results typically between 6-month to 18-month periods (most often one-year), depending on the Market being studied.)

While these quarterly forecasts may be thought of during model building as 'out-of-sample' forecasts, if they are to be used in retrospect for selecting from among a number of models, then they are really only *quasi-* 'out-of-sample'. For this reason, if, say, 10 windows are to be analyzed, the tuning of the parameters of the model building should be carried out using as few of these periods as possible, typically only the first 3 or 4, with the remaining periods 'saved' for validation once the modeling methodology has been finalized. Then at the end, a graphical comparison can be made of the cumulative performance over all periods, which in the ideal case should show that the performance over the oft re-used data period and that over the validation period have qualitatively similar characteristics.

1.3 Post-Hoc Performance Assessment

One of the most controversial areas of investment analysis is that of post-hoc performance assessment. One obvious reason for this is simple: if a canonical form of post-hoc performance appraisal existed, then during any period of time there would be only one 'winner', which, given the huge number of Financial Service entities competing to lay claim to being 'the best' (and with so much at stake) would be commercially unacceptable. Thus, investment firms will tend to promote the most favourable characteristics about (and measures of) their recent or past performances (often changing the measure from year to year) so as to obfuscate the issue.

The distribution of returns of major world market indices has been widely studied, and while both skewness and kurtosis (measures of asymmetry and heavy-tailedness based on third and fourth moments, respectively) have been found to be present in most markets. For periods of one day or longer the distribution of returns is generally very well-approximated by the Gaussian (though such an approximation is clearly inadequate in the context of analysis of instruments such as options contracts – not studied here – where accurate valuation typically depends heavily on extreme tail behaviour). Accepting the adequacy of the Gaussian in this context, the mean and standard deviation of return are sufficient to characterize the entire distribution of return and hence the entire risk-return profile. In particular, one quantity, the *Reward-to-Variability* or 'Sharpe' Ratio [18, 19] has been widely accepted as the single most meaningful overall measure of the unit of Return per unit of risk. Conceptually, this quantity may be expressed as:

$$\frac{(\text{Expected Return}) - (\text{Zero Risk Financial Opportunity Cost})}{(\text{Standard Deviation of Return})} \quad (1.1)$$

or in symbols:

$$\frac{E(R) - r_f}{\sigma_R} \quad (1.2)$$

where $E(R)$ and σ_R are the mean and standard deviation of R the rate of return, and r_f is the risk-free rate of return (in other words, the relevant interest rate).

For readers familiar with Capital Market theory, this quantity is directly comparable with the slope of the Capital Market Line, which, according to the theory, defines the highest return available in a market for a give level of risk, as measured by standard deviation. Alternatively, one may wish to control the probability of a return which is less than that of risk-free rate, which is given by:

$$1 - N\left(\frac{E(R) - r_f}{\sigma_R}\right) \quad (1.3)$$

N denotes the cumulative distribution function of the Gaussian, the former being strictly decreasing in the Sharpe Ratio, providing another compelling reason to use this quantity as a measure of return for a given level of risk.

Finally, we mention a third justification, which is the relevance of the Sharpe Ratio to the logarithmic growth rate. It is a well-established result that if an investor is to be presented with a sequence of similar, independent investment opportunities, and wishes to minimize the expected waiting time to any fixed financial goal, then such an investor should seek an optimal logarithmic growth rate at each stage. It may be shown that for small levels of volatility [5], to first order, the logarithmic growth rate of an investment with expectation greater than the risk-free rate is approximated by one-half the square of the Sharpe Ratio:

$$\frac{1}{2}\left(\frac{E(R) - r_f}{\sigma_R}\right)^2 \quad (1.4)$$

One point which will become relevant here is that if we go short (in other words, invest a negative amount) on an investment R_i for a given period, then our post-hoc net return for that period is:

$$-(R_i - r_f) = -R_i + r_f \qquad (1.5)$$

indicating a *negative* opportunity cost, due to the short-sale, making funds available for deposit into an interest-bearing account.

Of course, in the context of futures contracts, the risk-free opportunity cost of investment is zero for both long and short positions, since no funds are actually outlaid when the Futures contract is entered into. For investments in Futures contracts, then the Sharpe Ratio reduces to merely the ratio of Expected return to Standard Deviation of return:

$$\frac{E(R)}{\sigma_R} \qquad (1.6)$$

1.3.1 The Effect of Dividends

As it happens, the most commonly quoted versions of major Market Indices are based solely on share prices, whereas the return to investors with rights to dividends in the individual shares includes the dividends as well. Thus, when evaluating performance of investment strategies on Indices, it is standard practice to use *accumulation indices*, which include growth due to dividends. Of course, it is neither possible to invest in Raw Indices nor Accumulation Indices themselves, but instead, one invests in Index Futures, which are based on the Raw Indices, but priced with expected divided streams taken into account. Thus, (perhaps paradoxically) an investor in a Futures contract on Raw Share Price Index will be able to receive total profits (including Dividends).

To see how this works, consider the following example, in which the average dividend return for shares represented in an index will be approximately 1% this quarter, with the current interest rate at 5% per annum. If the Raw Index now is at 1000, then the 3-month Futures Price must be approximately:

$$1000\,(1 - 0.01 + \frac{1}{4}(0.05)) = 1002.5 \qquad (1.7)$$

since a portfolio consisting of Long 1 Future, and Short 1 unit of (non-rights) Index equivalent, and Long 1 unit (generated from the short-sale) deposited in an interest-bearing account should exactly cancel.

Thus, if the Index rises to 1030 (a 3% increase) in three months and dividends have amounted to 1% as expected, then an individual that entered into the 3-month futures contract at 1002.5 will receive a (net) rate of return equal to:

$$\frac{1030 - 1002.5}{1002.5} \approx 2.75\% \qquad (1.8)$$

(to the nearest 0.05 percent).

Had the investment been in the Raw Index itself, the net rate of return would have been 3%-1.25% = 1.25%, the difference being the 1% dividend rate.

For the purposes of our various studies below, we have assumed approximate equivalence between the net returns from (hypothetical) investments in accumulation indices and the net returns from investment in corresponding (Raw) index futures, noting that both include total net return, including dividends.

1.3.2 Transaction Costs Approximations

When attempting to discover presumed inefficiencies, the first steps are by nature exploratory and hence details such as transaction costs are usually ignored. However, in order to establish a true market inefficiency, the costs of transactions must be accounted for. While a certain model might indeed yield results which violate the 'random walk' presumption about the lack of predictability of market returns, nevertheless if the resulting apparent excess financial gains made possible by the model would be outweighed by the magnitude of costs of executing the required transactions, market efficiency is not violated.

Next to actually executing the actual trades, which is impossible to do in retrospect, the best approach is to develop a conservative estimate, but one which is not too conservative.

Direct research into Australian All Ordinaries Share Price Futures trading costs established that for an investment of 1 futures contract, a broker's round-trip fee of AUD$ 25.00 would apply. For a contract currently worth approximately AUD$ 75,000 (approximately 3000 points by AUD$ 25.00 per point), this represents approximately 1/60 of one percent for entry or exit. While this is small, seemingly more significant is the bid-offer spread, which, in this very liquid market, is virtually always bounded by 1 point, or AUD$ 25.00. Ideally, a full series including bid-offer spreads should be analyzed, but as these values were not always readily available for our studies, we made a series of simplifying assumptions to obtain conservative bounds on the costs arising from the bid-offer spread, where we assumed that liquidity was large enough to justify neglecting price impact or 'slippage'.

Initially, it was decided to suppose as a 'worst-case' that on each trade, the spread was at the 'wrong side' of the desired trade (above in the case of 'Long' and below in the case of 'Short' signals). In some of the work quoted below, this approximation (which as will now be discussed, was unnecessarily conservative) was used.

A much better idea involves another sort of worst-case assumption, namely that the market was 'omniscient' with regard to the bid-offer spread. For instance, for a position on a day where there had been a rise, a long position would have incurred AUD$ 25.00, whereas a short position would have actually *benefited* from the bid-ask spread. On balance, then, if a good forecasting model were able to predict correct direction 52% of the time, a conservative estimate of the average transaction cost is:

$$\$25(.52) - \$25(.48) = \$1 \tag{1.9}$$

which is much smaller. This brings the average cost per trade to AUD$ 13.50, which, even assuming a trading volatility as low as 5%, would only contribute at most:

$$\frac{(13.50 / 75000)\sqrt{260}}{.05} \approx 0.058 \tag{1.10}$$

or less than 6% to the annualised Sharpe Ratio.

In most of the sequel, then, we neglect transaction costs, noting that as much as 6% need be subtracted from the Sharpe Ratio obtained neglecting transaction costs.

While readers might find this figure for average transaction cost surprisingly low, it should be pointed out that this is due to the extremely liquid markets encountered in Index Futures contracts, and it is precisely for this reason that we examine these markets. Were these methods applied to less liquid equities or other markets, the transaction costs incurred from trading on a daily cycle would be much higher, most likely leading to poor net performance even for the best of models.

1.4 Genetic Programming

The first of the methods used to search for tradable patterns in the All Ordinaries Index is that of *Genetic Programming* (GP), and discusses results presented in [2], under supervision of the Authors.

Following the brilliant inspiration and formalization by Koza [10] there have numerous attempts to apply GP to the financial markets. Briefly, GP is a technique for producing machine-written code using *Genetic Algorithms* (GA), a special case of *Evolutionary Algorithms* (EA), which are "search and optimization techniques inspired by natural evolution" [27]. The main difference between GAs and GPs is that individuals within a GP population have a variable length compared to the fixed lengths of individuals within GAs. In addition, frequently (but not always) GPs have a tree structure, as compared to the linear representation of GAs.

Inputs or 'terminals' are first combined according to common operations such as multiplication, addition, and so on to form 'nodes', which are then treated as terminals for the next layer of the 'tree', combined successively using similar operations until a single result appears at the top of the 'tree'. The structure and operations of the tree are first simulated at random to form a population (here, of 1000 individuals), which are then 'culled' according to a 'fitness' measure, after which the 'fittest' individuals are passed on to a new generation according to several operations, including Sexual Reproduction (where features from two 'fit' individuals are combined), Cloning (where 'fit' individuals are passed on to the next generation unchanged), and Mutation (where 'fit' individuals are passed on, but with either random alterations or pruning). The process is repeated many times (here, over 30 generations), generally resulting in a set of particularly 'fit' individuals. An important innovation due to Koza [11], referred to as *Automatically Defined Functions* (which are collectively defined shared functions) was also added.

There have been a number of studies attempting to achieve excess returns using GP-optimized technical trading rules. Of these, many produced models which appeared to find predictability in the series studied but with the vast majority either not accounting for transaction costs, or finding that when transaction costs were accounted for, they offset any potential gain due to the models.

One notable exception was [15], which claimed to have achieved excess returns in the Foreign Exchange Markets, above and beyond transaction costs.

Weak form inefficiency was not established in the following studies: the S&P500 Index [1], various intra-day markets [14], the Dow Jones Industrial Average index [13], and the Nikkei225 average [7, 8]. Santini and Tettamanzi placed first in a competition to forecast the Dow Jones Index [17], while Chen *et al* [3] successfully forecast high-frequency returns of the Hang-Seng index, though again, without assessing transaction cost (which is much higher for high-frequency trading rules).

The dataset used for the present study consisted of daily closing prices of the Australian All Ordinaries Index from January 1, 1990 to August 27 2001, a total of 2948 (trading) days. The inputs consisted of daily returns for 10 days previous to the return being forecast. The decision making process of whether to buy +1 units of (in other words, to go 'long') the index or −1 units of (i.e., go 'short', which is achieved by temporarily lending an instrument) the index consists of a '7 member majority' result of 10-member voting of committees consisting of the 'fittest' of the individuals produced by the GP regime.

The resulting decisions were found to have notable characteristics, which appear to be fairly stable from quarter to quarter, consisting of an approximate 66% strike rate on the approx. 75% of decisions taken (with there being no clear majority decision the remaining 25% of the time).

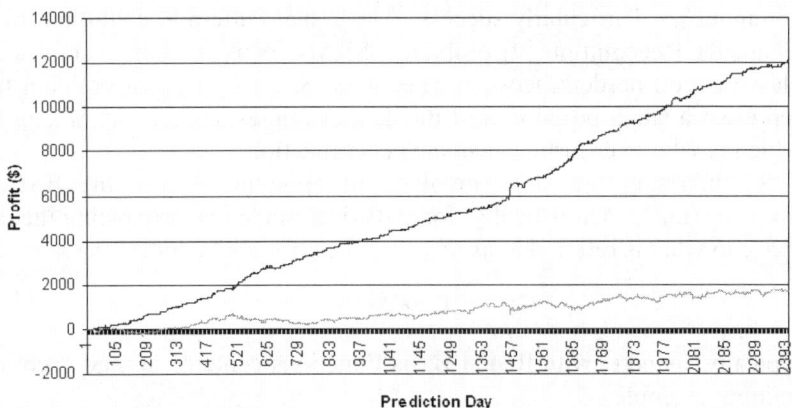

Fig. 1.1. Out-of-sample cumulative return for trading strategy based on Genetic Programming (dark gray) compared to Buy-and-Hold strategy (light gray) for Australian All Ordinaries Index, based on moving windows from January 1, 1990 to August 27, 2001

Even though the Index itself is not a tradable entity, Share Price Index (SPI) Futures contracts *are* directly tradable, with the *near* month returns being very closely correlated with the returns of Index itself. Hence, in order to (approximately) simulate Market conditions, typical SPI-Futures Transaction costs may be assumed to be $25 roundtrip for a contract worth approximately $80,000, combined with a 'worst-case' bid-offer loss of $25 per transaction. Even with transaction costs, the net result is a strikingly tradable strategy, as indicated in Fig. 1.1, where cumulative returns net of interest are shown, for the 'Buy-and-Hold' and 'GP-Advised' Funds. The Annualized Sharpe ratios (in other words, financial 'signal-to-noise ratios') are found to be consistently nearly 10 times as high for the 'GP-Advised' fund than for the 'Buy-and-Hold' fund.

1.5 Support-Vector Machines

The material in this section contains results of research work carried out and supervised by the authors, some of which is discussed in more detail in [16]. The most recent of the Data Mining techniques to be implemented here is that of Support-Vector Machines (SVMs), a method of classification (here, for 'Ups' versus 'Downs' in market returns) introduced by Vapnik in 1979 as part of what he dubbed *Statistical Learning Theory* [23, 24]. The method has gained considerable popularity in recent years, especially among the Computer Science and Machine Learning communities. Particularly successful in Visual Pattern Recognition [6] and Character Recognition [9] problems, SVMs focus on data examples near classification borders between categories. Since in a given problem these represent a small proportion of the data examples, complexity is kept low, which is vital to ensuring adequate generalization.

Summarizing the steps involved in applying SVMs, the first step involves (uncharacteristically, for statistical modeling) expanding the input space to what is referred to as *feature* space: for a give input vector $\underline{x}$,

$$\underline{x} \rightarrow \phi(\underline{x}) \tag{1.11}$$

typically, as here, with Radial Basis Functions centered around each input training example:

$$\phi_j(\underline{x}_i) = \exp\{-\frac{1}{2}\|\underline{x}_i - \underline{x}_j\|^2, \quad j = 1, 2, \ldots, n\} \tag{1.12}$$

Next, a discriminant defined by:

$$\underline{w}'\phi + b \tag{1.13}$$

where $\underline{w}$ is chosen such that $1/2\underline{w}'\underline{w}+C\Sigma\xi_i$ is minimized subject to $\xi_i \geq 0$, where ξ_i denotes the distance (norm) to the closest point in the correct classification region, and C is a suitably chosen 'slack' constant.

While the Support Vector Machine method is fairly recent, studies involving applications to the Financial Markets are beginning to appear, such as [22], which found SVMs superior to BackPropagation Neural Network models in forecasting several Chicago-based Futures prices, including those of the S&P500 index.

Fig. 1.2. Out-of-sample cumulative return for trading strategy based on Support Vector Machines (upper curve) compared to Buy-and-Hold strategy (lower curve) for Australian All Ordinaries Index, based on moving windows from 25 September, 1992 to 6 December, 2000

The dataset used for the present study resembles that used for the previous Genetic Programming Study, daily closing prices of the Australian All Ordinaries Index, beginning on 25 September 1992 and continuing to 6 December 2000 – some 2223 values – excluding weekends and public holidays. For the study, again the logarithmic returns have been used, with the values of the 10 previous days used to predict the sign of the current day's change. A 2-member committee was trained for each of the rolling 1-year periods, where in one case the outcome was labeled as a '1' if the rise was greater than 0.2%, '–1' otherwise, and in the other case the threshold was -0.2% instead of 0.2%. Thus, an investment decision was taken to go either 'long' or 'short' if there is consensus, with a neutral position being taken otherwise.

The resulting success rate was 53%, with a position taken on approximately 80% of the days. For the 'SVM' fund, the resulting Sharpe Ratio of 80% was several times as high as that of the 'Buy-and-Hold' strategy, with the former beating the latter overall and in return in 19 out of the 30 quarters examined.

The cumulative returns for the period are graphed, along with the cumulative returns for the 'Buy-and-Hold' strategy, in Fig. 1.2.

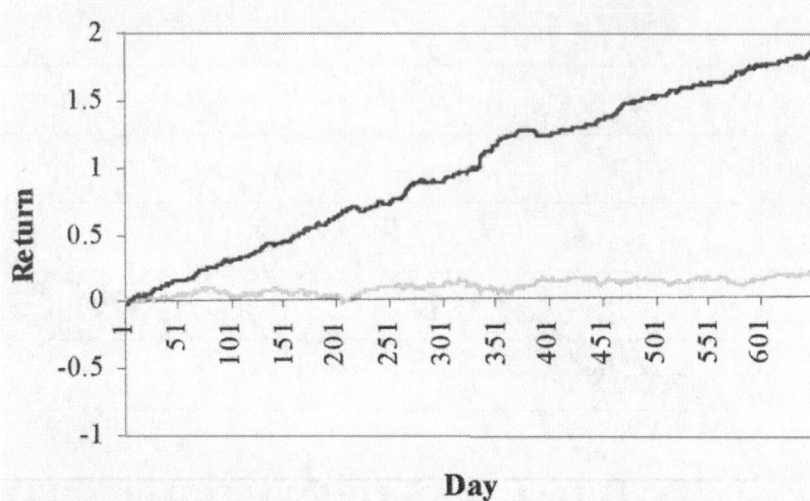

Fig. 1.3. Out-of-sample cumulative returns for open to close predictions of the Australian All Ordinaries (AORD) Index, from 4 January, 1999 to 6 July, 2001 (the Neural Network model (dark gray) used daily returns for the FTSE and Dow Jones indices in addition to previous AORD returns; the cumulative return for the Buy-and-Hold strategy is shown in light gray)

1.6 Neural Networks

The third study to be presented here, carried out under the direction and supervision of the authors and discussed in more detail in [12], involves a Neural Network analysis – a Multilayer Feedforward Perceptron with 2 hidden layers of data taken over a shorter period of time, 1998-2001, but across a wider range of series, including indices from outside Australia as well as the actual Spot Futures Market for the Australian Share Price Index (SPI), which as noted earlier is the relevant tradable instrument for those who merely wish to trade 'the index'.

Earlier work [4, 25] appears to suggest that Neural Network forecasting of the All Ordinaries Index based on same-series Adaptive Technical Analysis (ATA) is clearly successful, but only to a small degree marginal to 'Buy-and-Hold' strategies when reasonable Transaction Costs are taken into account. (This is in keeping with the results of many other similar Neural Network studies as discussed in a review paper [28]). It is for this reason that a wider range of input series has been investigated.

1 Adaptive Technical Analysis in Financial Markets using ML 13

Daily returns for the both the All Ordinaries Index and the SPI spot futures contracts were forecast with Models making use of several combinations of international series. In order to allow for different time zones, Networks were trained to predict open to close returns using previous daily close returns and differences between close and low price. The most successful of these was a model comprised of the recent time series of the Australian All Ordinaries Index, the London FTSE, and the Dow Jones Industrial Average, with the results exhibited in Fig. 1.3. Committee decision making was employed in a manner similar to the previous GP study.

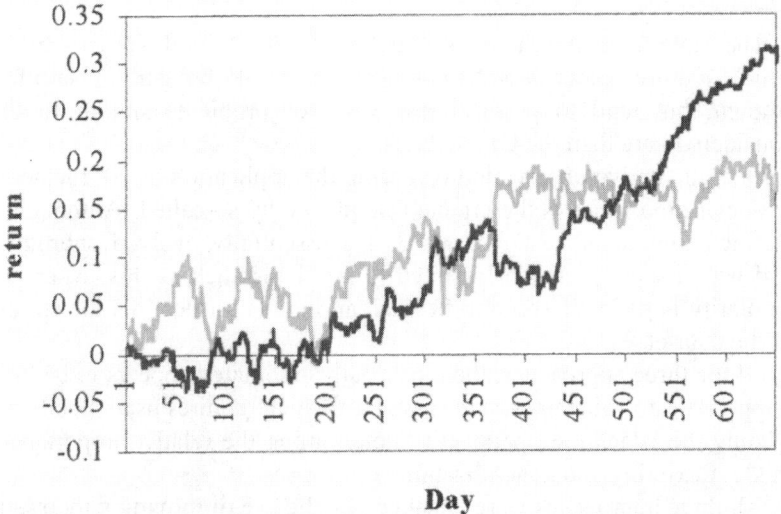

Fig. 1.4. Out-of-sample returns for open to close predictions of the Australian SPI Futures, using a Neural Network model with FTSE and Dow Jones returns as additional inputs, over the time period 12 January, 1999 to 1 August, 2001 (the corresponding Buy-and-Hold returns are shown in light gray)

Neural network prediction of SPI spot futures, using FTSE and Dow Jones indices as additional inputs, turned out to be less successful than prediction of the AORD index. As shown in Fig. 1.4, the 'Buy-and-Hold' strategy produced a greater cumulative return than the Neural Network model during the early windows, but the situation was reversed in later windows. The Sharpe Ratio for the Neural Network-Advised fund was increased by a factor of 4 relative to the 'Buy-and-Hold' strategy over the period of the study.

1.7 Discussion

While no formal statistical tests on the results of these three studies are presented here, the evidence in each case is strong enough to cast serious doubt on the Weak form of the Efficient Markets Hypothesis (in other words, that no excess risk-adjusted returns are possible using price series information alone).

Of the three methods, the GP method appears to be the most powerful, in the sense of achieving the highest return per unit of risk out-of-sample, by quite some margin, though the careful 'fine-tuning' and 'art' aspect of the model building would tend to render this method useful only to experts.

The same holds true, to a lesser extent, for the Neural Network models, which require rather less 'fine-tuning' than GP but still a significant amount, but tend to perform less well for problems similar to those considered here than the GP methods.

One qualitative observation regarding the application of GP methods to financial time series is the crucial role played by so-called 'Automatically Defined Functions' (ADFs), which are essentially evolved subroutines, and which seem particularly suited to problems such as this, where self-similarity is generally seen to be such an inherent aspect of the structure of the dataset.

Of the three approaches, the most easily automated appears to be that of Support Vector Machines, one version of which requires user-specification of only the 'slackness' parameter (determining the relative importance of model flexibility and generalizability).

All three approaches were enhanced by the use of moving windows and committees, rather than simply fitting a single model. For Genetic Programming and Neural Networks, committee members were generated via different random number seeds. In the case of Support Vector Machines, the fitting process is completely deterministic, so committee members were generated via different thresholds during the binary conversion of the response variable. The committee approach seems to be an important element when dealing with high noise problems such as financial time series prediction.

In summary, machine learning has a useful role to play in the context of Technical Analysis of Financial Markets, but informed user interaction is also a necessary ingredient for successful implementation.

References

1. Allen F and Karjalainen R (1999) Using genetic algorithms to find technical trading rules, *J Financial Economics*, 51(2): 245-271.
2. Ang L (2001) Genetic Programming: the Evolution of Forecast Equations for the All Ordinaries Index, Mathematical Finance Honours Thesis, University of Wollongong.
3. Chen SH, Wang HS and Zhang BT (1999) Forecasting High-Frequency Financial Time Series with Evolutionary Neural Trees: The Case of the Hang Seng Stock Index, *Proc 1999 Intl Conf Artificial Intelligence IC-AI'99*, 437-443.
4. Chung V (1999) Using Neural Networks to Predict the Returns of the Australian All Ordinaries Index, Honours Thesis, Dept Accounting & Finance, University of Wollongong.
5. Edelman D (2001) On the Financial Value of Information, *Annals of Operations Research*, 100: 123-132.
6. Guo G, Li SZ and Chan KL (2001) Support Vector Machines for Face Recognition, *Image & Vision Computing*, 19: 631-638.
7. Iba H and Nikolaev N (2000) Genetic Programming Polynomial Models of Financial Data Series, *Proc 2000 Congress Evolutionary Computation, 2000*, New Jersey, IEEE Press, 2: 1459-1466.
8. Iba H and Sasaki T (1999) Using Genetic Programming to Predict Financial Data, *Proc 1999 Congress Evolutionary Computation, CEC99*, New Jersey, IEEE Press, 1: 244-251.
9. Joachims T (1997) Text and Characterization with Support Vector Machines, University of Dorismund, LS VIII Technical Report no 23.
10. Koza JR (1992) *Genetic Programming: on the programming of computers by means of natural selection,* MIT Press, Cambridge, MA.
11. Koza JR (1994) *Genetic Programming II: automatic discovery of reusable programs,* MIT Press, Cambridge, MA.
12. Lawrence E (2001) Forecasting Daily returns for Equities and Equity Indices using Neural Networks: a Case Study of the Australian Market 1998-2001, Mathematical Finance Honours Thesis, University of Wollongong.
13. Li J and Tsang EPK (1999) Improving Technical Analysis Predictions: an Application of Genetic Programming, *Proc 12th Intl Florida AI Research Society Conf,* Orlando, FL, 108-112.
14. Neely CP and Weller P (1999) Intraday Technical Trading in the Foreign Exchange Market, *Working Paper Series, Federal Reserve Bank of St. Louis*, 1999-016B, http://www.stls.frb.org/research/wp/99-016.html (accessed: July, 2001)
15. Neely CP, Weller P and Dittmar R (1997) Is Technical Analysis in the Foreign Exchange Market Profitable? A Genetic Programming Approach, *J Financial & Quantitative Analysis*, 32(4).

16. Reid S (2001) The Application of Support Vector Machines for the Prediction of Stock Index Directional Movements, Mathematical Finance Honours Thesis, University of Wollongong.
17. Santini M and Tettamanzi A (2001) Genetic Programming for Financial Time series prediction, in Lanzi PL, Ryan C, Tettamanzi AGB, Langdon WB, Miller J and Tomassini M. (eds), *Proc 4th European Conf Genetic Programming (EuroGP)*, Springer Verlag, Berlin, Lecture Notes in Computer Science 2038, 361-370.
18. Sharpe WF (1966) Mutual Fund Performance, *J Business*, 39(1): 119-138.
19. Sharpe WF (1975) Adjusting for Risk in Portfolio Performance Measurement, *J Portfolio Management*, Winter: 29-34.
20. Simon HA (1955) A behavioral model of rational choice, *Quarterly J Economics*, 69: 99-118.
21. Simon HA (1987) Bounded rationality, in Eatwell J, Millgate M and Newman P (eds), *The New Palgrave: A Dictionary of Economics*, Macmillan, London & Basingstoke.
22. Tay EH and Cao LJ (2001) Application of Support Vector Machines in Financial Time Series Forecasting, *Omega* 29: 309-317.
23. Vapnik VN (1995) *The Nature of Statistical Learning Theory*, Springer Verlag, New York.
24. Vapnik VN (1999) *Statistical Learning Theory*, Wiley Interscience, New York.
25. Walmsley K (1999) Using Neural Networks to Arbitrage the Australian All Ordinaries Index, Dept. Accounting & Finance Honours Thesis, University of Wollongong .
26. Wilmott P (2001) *Paul Wilmott Introduces Quantitative Finance*, Wiley, London.
27. Wong ML and Leung KS (2000) *Data mining using grammar based genetic programming and applications*, Boston: Kluwer Academic, Boston, MA.
28. Zhang G, Patuwo BE and Hu MY (1998) Forecasting with artificial neural networks: the state of the art, *Intl J Forecasting*, 14: 35-62.

2 Higher Order Neural Networks for Satellite Weather Prediction

Ming Zhang[1] and John Fulcher[2]

1   Department of Physics, Computer Science & Engineering, Christopher Newport University, Newport News, VA 23693, USA, mzhang@pcs.cnu.edu

2   School of Information Technology & Computer Science, University of Wollongong, NSW 2522, Australia, john@uow.edu.au

2.1 Introduction

Traditional statistical approaches to modeling and prediction have met with only limited success [30]. As a result, researchers have turned to alternative approaches. In this context, Artificial Neural Networks – ANNs – have received a lot of attention in recent times. Not surprisingly, a lot of this attention has focused on MLPs [2, 7, 24, 31, 39, 43].

Traditional areas in which Artificial Neural Networks are known to excel are pattern recognition, pattern matching, mathematical function approximation, and time series modeling, simulation and prediction (the latter is the focus of the present Chapter).

ANNs, by default, employ the Standard BackPropagation – SBP – learning algorithm [35, 42], which despite being guaranteed to converge, can in practice take an unacceptably long time to converge to a solution. Because of this, numerous modifications to SBP have been proposed over the years, in order to speed up the convergence process – typical ones being:

- QuickProp [13], which assumes the error surface is locally quadratic in order to approximate second-order (or gradient) changes, and
- Resilient BackPropagation – Rprop [47], which uses only the *sign* of the derivative to affect weight changes.

Apart from long convergence times, ANNs also suffer from several other well known limitations: (i) they can often become stuck in local rather than global minima in the energy landscape, and (ii) they are unable to handle non-linear, high frequency component, discontinuous training data $\{f(x) \neq lim_{\Delta x \to 0} f(x+\Delta x)\}$, or complex mappings (associations). Moreover, since

ANNs function as 'black boxes', they are incapable of providing explanations for their behavior. This is seen as a disadvantage by users, who would rather be given a rationale for the network's decisions.

In an effort to overcome these limitations, attention has focused in recent times on using Higher Order Neural Network – HONN – models for simulation and modeling [15-17, 20, 32]. Such models are able to provide information concerning the basis of the data they are simulating, and hence can be considered as 'open box' rather than 'black box' solutions.

2.2 Higher Order Neural Networks

The aim with polynomial curve fitting is to fit a polynomial to a set of n data points by minimizing an error function [5]. This can be regarded as a non-linear mapping from input space-**x** to output space-**y**. The precise form of the function **y**(x) is determined by the values of the parameters w_0, w_1, w_2 ... and so on below (and can be likened to ANN weights). Provided there are a sufficiently large number of terms in the polynomial, we can approximate a wide class of functions to arbitrary accuracy.

Conventional polynomial curve fitting is usually restricted in practice to quadratics (cubic spline fitting is only exact over limited domains). Higher order terms enable closer fitting of short-term perturbations in the data, but at the risk of over-fitting.

The simple McCulloch & Pitts [25] neuron model is only capable of extracting first-order (linear) correlations in the input (training) data; more specifically, it produces an output (y) if the weighted sum of its inputs (x) exceeds a pre-set threshold θ:

$$y_i(x) = f[\sum_{j=1}^{n} w(i,j)x(j)] + \theta \quad (2.1)$$

The extraction of higher-order correlations in the training data requires more complex units, namely the following higher-order or polynomial form, which includes multiplicative terms [4, 33]:

$$y_i(x) = \\ f[w_0(i) + \sum_j w_1(i,j)x(j) + \sum_j \sum_k w_2(i,j,k)x(j)x(k) + ...] \quad (2.2)$$

This leads to a generalization of the input-output capability of the McCulloch & Pitts model in such polynomial (higher-order or sigma-pi) neurons. The corresponding large increase in the number of weighted terms means that complex input-output mappings normally only achievable in multi-layered architectures can be realized in a *single* layer. Moreover, since there are no hidden layers, simple training rules, such as Standard BackPropagation (generalized delta learning) can be used.

The increased power of polynomial neurons can be illustrated by considering the exclusive-OR problem. It is well known that XOR is linearly inseparable; in other words it is not possible to find a single linear discriminant which separates the two classes ('true' and 'false'). Rosenblatt's original 2-layer Perceptron was unable to solve such problems – in other words, there is no way of selecting w_1, w_2 and θ such that $w_1x_1 + w_2x_2 \geq \theta$ iff $x_1 \oplus x_2 = 1$ [34]. Nevertheless, *Multi*-Layer Perceptrons (MLPs) can successfully solve XOR (by effectively combining *two* such linear discriminants). A polynomial neuron, on the other hand, can solve XOR even more directly, since it incorporates *multiplicative* terms:

$$x_1 + x_2 - 2x_1x_2 = x_1 \oplus x_2 \qquad (2.3)$$

The first problem that needs addressing is to devise a neural network structure that will not only act as an open box to simulate modelling and prediction functions, but which will also facilitate learning algorithm convergence. Now polynomial functions are continuous by their very nature. If the data being analyzed vary in a continuous and smooth fashion with respect to time, then such functions can be effective. However, in the real world such variation is more likely to be discontinuous and non-smooth. Thus, if we use continuous and smooth approximation functions, then accuracy will obviously be a problem. We subsequently demonstrate how it is possible to simulate discontinuous functions, to any degree accuracy, using Higher Order Neural Networks (and later HONN groups), even at points of discontinuity.

The activation functions ordinarily employed in ANNs are sigmoid, generalized sigmoid, radial basis function, and the like. One characteristic of such activation functions is that they are all fixed (with no free parameters), and thus cannot be adjusted to adapt to *different* approximation problems. To date there have been few studies which emphasize the setting of free parameters in the activation function [1, 3, 6, 9, 10, 12, 18, 28, 29, 36]. ANNs which employ adaptive activation functions provide better fitting properties than classical architectures that use fixed activation function neurons. The properties of a Feedforward Neural Network (FNN) able to adapt its activation function by varying the control points of a Catmull-Rom cubic spline were presented in [40]. Their simulations confirm that this specialized learning mechanism allows the effective use of the network's free parameters. The real variables *a* (gain) and b (slope) in the generalized sigmoid activation function were adjusted during learning in [8]. Compared with classical FNNs in modelling static and dynamical systems, they showed that an *adaptive* sigmoid leads to improved data modeling. In other words, connection complexity is traded off against activation function complexity.

The development of a new HONN neural network model, which utilizes adaptive neuron activation functions, is provided in [55]. A Polynomial HONN model for data simulation has also been developed [49]. This idea was extended firstly to PHONN Group models for data simulation [50], thence to trigonometric HONNG models for data simulation and prediction [53]. Furthermore, HONN models are also capable of simulating higher frequency and higher order nonlinear data, thus producing superior data simulations, compared with those derived from ANN-based models.

2.2.1 Polynomial Higher-Order Neural Networks - PHONN

Polynomial Higher Order Neural Networks – PHONNs – were developed as part of a Fujitsu-funded project on financial time series modelling/forecasting. The starting point was the PHONN Model#0 of Fig.2.1, which uses a combination of linear, power and multiplicative neurons, rather than the sine(cosine) neurons previously employed by Fujitsu. Standard BackPropagation was used to train this ANN (as indeed it was on *all* subsequent PHONN models). Model#0 is capable of extracting coefficients a_{k1k2} of the general nth-order polynomial form:

$$z_i = \sum_{k1k2=0}^{n} a_{k1k2} x_i^{k1} y_i^{k2} \tag{2.4}$$

Adjustable weights are present in one layer only in this model. As such, Model#0 can be likened to Rosenblatt's (2-layer) Perceptron, which we have already observed is only applicable to linearly separable problems [34].

In PHONN Model#1, the general nth-order polynomial equation is reformulated by expanding each coefficient into three:

$$z_i = \sum_{k1k2=0}^{n} (a_{k1k2}^{o})[a_{k1k2}^{x} x]^{k1} [a_{k1k2}^{y} y]^{k2} \tag{2.5}$$

This results in *two* adjustable weight layers, and thus can be likened to a Multi-Layer Perceptron (MLP).

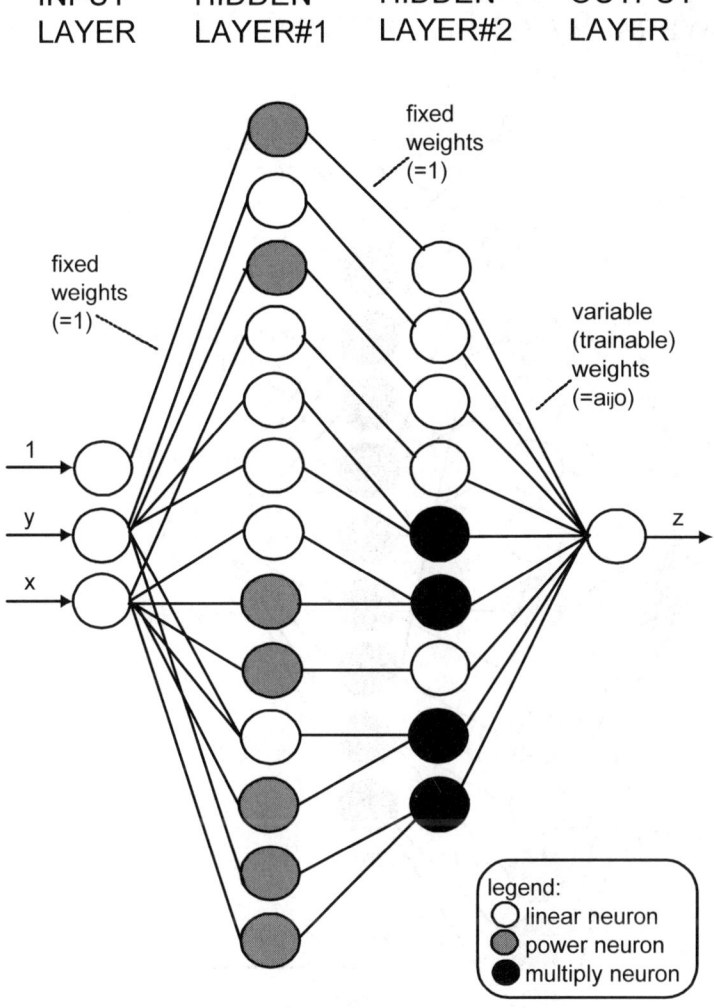

Fig. 2.1. Polynomial Higher Order Neural Network Model#0

The motivation for expanding the coefficients in Model#1 is to open up the 'black box' (closed) architecture, so that the polynomial coefficients can be correlated with network weights, and vice versa. This capability is particularly desirable for economists and others interested in time series forecasting and/or modelling.

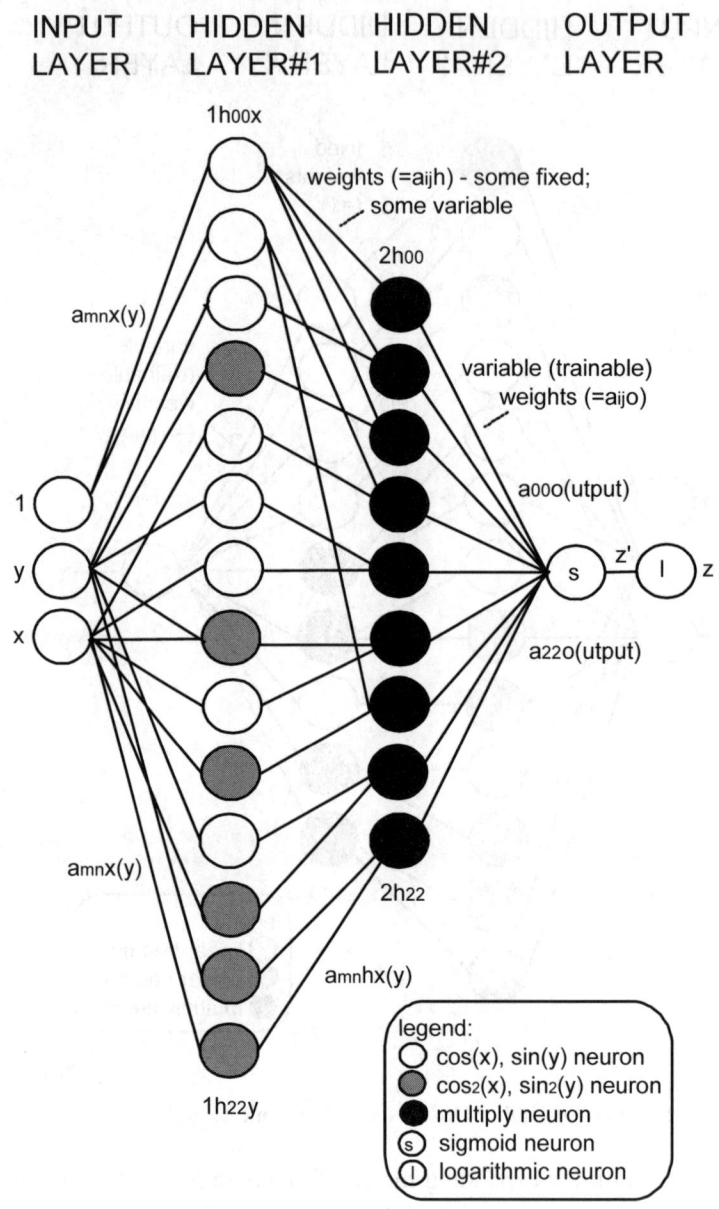

Fig. 2.2. Trigonometric Higher Order Neural Network Model#2

PHONN Model#2 is a variant in which the linear output neuron is replaced by a sigmoid-linear pair (which was found to improve convergence). Model#3 is formed from *groups* of Model#2 PHONNs, and is discussed further in Sect. 2.3.

2.2.2 Trigonometric Higher-Order Neural Networks – THONN

During a follow-up study, Trigonometric HONNs were developed, based on a *trigonometric* rather than a polynomial series expansion:

$$z = a_{00} + a_{01}\sin(y) + a_{02}\sin^2(y) + a_{10}\cos(x) + a_{11}\cos(x)\sin(y) + \quad (2.6)$$
$$a_{12}\cos(x)\sin^2(y) + a_{20}\cos^2(y) + a_{21}\cos^2(x)\sin(y) + a_{22}\cos^2(x)\sin^2(y)$$

Now since these THONN models incorporate power terms, their discrimination (correlation) ability is more powerful than Fujitsu's original sin(cos) ANNs.

THONN model#0 uses trigonometric (sine/cosine), linear, multiplicative and power neurons based on Eq. (2.6). THONN model#1 uses the same basic building blocks, but is based on the formulation given in Eq. (2.7):

$$z = \sum_{k1,k2=0}^{n} a_{k1k2} \cos^{k1}(a_{k1k2}{}^{x}x)\sin^{k2}(a_{k1k2}{}^{y}y) \quad (2.7)$$
$$= \sum_{k1,k2=0}^{n} (a_{k1k2}{}^{o})\{a_{k1k2}{}^{hx}[\cos(a_{k1k2}{}^{x}x)]^{k1}\}\{a_{k1k2}{}^{hx}[\sin(a_{k1k2}{}^{y}y)]^{k2}\}$$

THONN model#2 uses linear, multiplicative and power neurons based on the trigonometric polynomial form; it also uses a sigmoid-logarithm output neuron pair (Fig.2.2) – in other words $z=ln[(z')/(1-z')]$, where $z'=1/(1+e^{-z})$.

Training and test data from the Reserve Bank of Australia Bulletin were used to verify firstly that HONN models were capable of interpolating discontinuous, non-smooth economic data, and secondly that they converge within acceptable time frames. Both simulation and prediction experiments were performed. In the former, all of the available data were used for training; in the latter, half were used for training, and half for testing, with two previous months information used to predict a third. Errors of around 10% were observed with both PHONN and THONN [49].

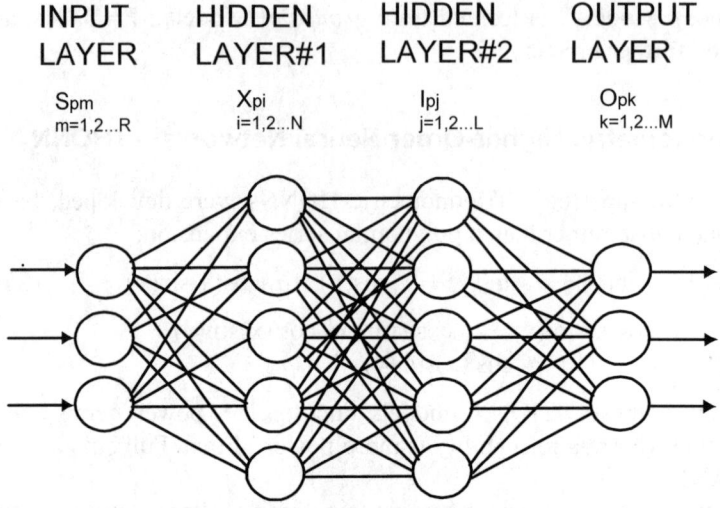

Fig. 2.3. Neuron and Weight Indices (for *p*th training vector)

Output Neurons in THONN Model#1

As is usual with ANN training (typically BackPropagation or one of its numerous variants), weight adjustment occurs in reverse order: output ➔ 2nd hidden layer ➔ 1st hidden layer ..., and so on. Accordingly, we first derive the error, derivatives, gradients and weight update equations for the output layer – this is followed by similar derivations for the 2nd, thence the 1st hidden layers.

The output layer weights are updated according to:

$$w_{kj}^{o}(t+1) = w_{kj}^{o}(t) - \eta(\partial E_p / \partial w_{kj}^{o}) \qquad (2.8)$$

where η = learning rate (positive & usually < 1)
  j = input neuron index (1…M) (ref. Fig.2.3)
  k = kth output neuron
  E = error
  t = training time
  o = output layer
  p = pth training vector
  w_{kj} = weight connecting kth neuron in output layer to jth neuron in 2nd hidden layer

The output node equations are:

$$net_{pk}^{o} = \sum_{j=1}^{L} w_{kj}^{o} i_{pj} + \theta_{k}^{o} \tag{2.9}$$

$$o_{pk} = f_{k}^{o}(net_{pk}^{o})$$

where i_{pj} = input to the output neuron (= output from 2nd hidden layer)
o_{pk} = output from the output neuron
$f^{o}{}_{k}$ = output neuron activity function
$\theta^{o}{}_{k}$ = bias term

The error at a particular output unit (neuron) will be:

$$\delta_{pk} = (y_{pk} - o_{pk}) \tag{2.10}$$

where y_{pk} = desired output value
o_{pk} = actual output from the kth unit

The total error is the sum of the squared errors across all output units, namely:

$$E_{p} = 1/2 \sum_{k=1}^{M} \delta_{pk}^{2} = 1/2 \sum_{k=1}^{M} (y_{pk} - o_{pk})^{2} \tag{2.11}$$

The derivatives $f^{o}{}_{k}'(net^{o}{}_{pk})$ are calculated as follows:
1. for linear functions ($f^{o}{}_{k}(net^{o}{}_{pk}) = net^{o}{}_{pk}$):

$$f_{k}^{o\prime}(net_{jk}^{o}) = \partial f_{k}^{o} / \partial(net_{jk}^{o}) = \partial(net_{jk}^{o}) / \partial(net_{jk}^{o}) = 1 \tag{2.12}$$

2. for sigmoid (logistic) functions ($o_{pk} = f^{o}{}_{k}(net^{o}{}_{pk}) = (1 + \exp(-net^{o}{}_{pk}))^{-1}$):

$$f_{k}^{o\prime}(net_{jk}^{o}) = \partial[(1+\exp(-net_{jk}^{o}))^{-1}] / \partial(net_{jk}^{o}) \tag{2.13}$$

$$= -(1+\exp(-net_{jk}^{o}))^{-2}(-\exp(-net_{jk}^{o}))$$

$$= (1+\exp(-net_{jk}^{o}))^{-2}((1+\exp(-net_{jk}^{o}))-1)$$

$$= f_{k}^{o}(1-f_{k}^{o}) = o_{pk}(1-o_{pk})$$

Gradients are calculated as follows:

$$\partial E_{p} / \partial w_{kj}^{o} = (\partial E_{p} / \partial o_{pk})(\partial o_{pk} / \partial(net_{pk}^{o}))(\partial(net_{pk}o) / \partial w_{kj}^{o}) \tag{2.14}$$

$$\partial E_{p} / \partial o_{pk} = \partial(1/2 \sum_{k=1}^{M}(y_{pk} - o_{pk})^{2}) / \partial o_{pk} \tag{2.15}$$

$$= 1/2(-2(y_{pk} - o_{pk})) = -(y_{pk} - o_{pk})$$

$$\partial o_{pk} / \partial(net_{jk}^{o}) = \partial f_{k}^{o} / \partial(net_{jk}^{o}) = f_{k}^{o'}(net_{jk}^{o}) \qquad (2.16)$$

$$\partial(net^{o}_{pk}) / \partial w_{kj}^{o} = \partial(\sum_{j=1}^{L} w_{kj}^{o} i_{pj} + \theta_{k}^{o}) / \partial w_{kj}^{o} = i_{pj} \qquad (2.17)$$

Combining Eqs. (2.14) through (2.17), we have for the negative gradient:

$$-\partial E_{p} / \partial w_{kj}^{o} = (y_{pk} - o_{pk}) f_{k}^{o'}(net_{pk}^{o}) i_{pj} \qquad (2.18)$$

For a linear output neuron, this becomes, by combining Eqs. (2.12) and (2.18):

$$-\partial E_{p} / \partial w_{kj}^{o} = (y_{pk} - o_{pk}) f_{k}^{o'}(net_{pk}^{o}) i_{pj} \qquad (2.19)$$
$$= (y_{pk} - o_{pk})(1) i_{pj} = (y_{pk} - o_{pk}) i_{pj}$$

Alternatively, for a sigmoid output neuron, by combining Eqs. (2.13) and (2.18), this becomes:

$$-\partial E_{p} / \partial w_{kj}^{o} = (y_{pk} - o_{pk}) f_{k}^{o'}(net_{pk}^{o}) i_{pj} \qquad (2.20)$$
$$= (y_{pk} - o_{pk}) o_{pk} (1 - o_{pk}) i_{pj}$$

The weight update equations are formulated as follows:
1. for linear output neurons, let:

$$\delta_{pk}^{ol} = (y_{pk} - o_{pk}) f_{k}^{o'}(net_{pk}^{o}) = (y_{pk} - o_{pk}) \qquad (2.21)$$

Combining Eqs. (2.8), (2.19) and (2.21), we have:

$$w_{kj}^{o}(t+1) = w_{kj}^{o}(t) - \eta(\partial E_{p} / \partial w_{kj}^{o}) \qquad (2.22)$$
$$= w_{kj}^{o}(t) + \eta(y_{pk} - o_{pk}) i_{pj}$$
$$= w_{kj}^{o}(t) + \eta \delta_{pk}^{ol} i_{pj}$$

2. for sigmoid output neurons, let:

$$\delta^{os}_{pk} = (y_{pk} - o_{pk}) f^{o}_{k}{}'(net^{o}_{pk}) = (y_{pk} - o_{pk}) o_{pk}(1 - o_{pk}) \qquad (2.23)$$

Combining Eqs. (2.8), (2.20) and (2.23) we have:

$$w_{kj}^{o}(t+1) = w_{kj}^{o}(t) - \eta(\partial E_p / \partial w_{kj}^{o}) \qquad (2.24)$$
$$= w_{kj}^{o}(t) + \eta(y_{pk} - o_{pk})o_{pk}(1 - o_{pk})i_{pj}$$
$$= w_{kj}^{o}(t) + \eta \delta_{pk}^{os} i_{pj}$$

Second-Hidden Layer Neurons in THONN Model#1

The second hidden layer weights are updated according to:

$$w_{ji}^{h}(t+1) = w_{ji}^{h}(t) - \eta(\partial E_p / \partial w_{ji}^{h}) \qquad (2.25)$$

where η = learning rate (positive & usually < 1)
  j = input neuron index (1…M)
  k = kth output neuron
  E = error
  t = training time
  h = hidden layer
  p = pth training vector
  w_{ji} = weight connecting ith neuron in 1st hidden layer to jth neuron in 2nd hidden layer

The equations for the jth hidden node are:

$$net_{pj}^{h} = \sum_{i=1}^{N} w_{ji}^{h} x_{pi} + \theta_{j}^{h} \qquad (2.26)$$

$$i_{pj} = f_{j}^{h}(net_{pj}^{h})$$

where i_{pj} = input to the output neuron (= output from 2nd hidden layer)
  x_{pi} = input to the 2nd hidden layer neuron
   (= output from the 1st hidden layer neuron)
  f^{h}_{j} = hidden neuron activation function
  θ^{h}_{j} = bias term

The error of a single hidden unit will be:

$$\delta_{pk} = (y_{pk} - o_{pk}) \qquad (2.27)$$

where y_{pk} = desired output value
  o_{pk} = actual output from the kth unit

The total error is the sum of the squared errors across all hidden units, namely:

$$E_p = 1/2 \sum_{k=1}^{M} \delta_{pk}^2 = 1/2 \sum_{k=1}^{M} (y_{pk} - o_{pk})^2 \qquad (2.28)$$

$$= 1/2 \sum_k (y_{pk} - f_k^o(net_{pk}^o))^2$$

$$= 1/2 \sum_k (y_{pk} - f_k^o(\sum_j w_{kj}^o i_{pj} + \theta_k^o))^2$$

The derivatives $f^h{}_j{}'(net^h{}_{pj})$ are calculated as follows, for a sigmoid (logistic) function:

$$i_{pj} = f_j^h(net_{pj}^h) = (1 + \exp(-net_{pj}^h))^{-1} \qquad (2.29)$$

$$f_j^{h\prime}(net_{pj}^h) = \partial[(1 + \exp(-net_{pj}^h))^{-1}]/\partial(net_{pj}^h)$$

$$= -(1 + \exp(-net_{pj}^h))^{-2}(-\exp(-net_{pj}^h))$$

$$= (1 + \exp(-net_{pj}^h))^{-2}((1 + \exp(-net_{pj}^h)) - 1)$$

$$= f_j^h(1 - f_j^h) = i_{pj}(1 - i_{pj})$$

The gradient $(\partial E_p / \partial w^h{}_{ji})$ is given by:

$$\partial E_p / \partial w_{ji}^h = \partial (1/2 \sum_k (y_{pk} - o_{pk})^2) / \partial w_{ji}^h \qquad (2.30)$$

$$= (\partial (\sum_k (y_{pk} - o_{pk})^2) / \partial o_{pk})(\partial o_{pk} / \partial (net_{pk}^o))$$

$$(\partial (net_{pk}^o) / \partial i_{pj})(\partial i_{pj} / \partial (net_{pj}^h))(\partial (net_{pj}^h) / \partial w_{ji}^h)$$

$$\partial (1/2 \sum_k (y_{pk} - o_{pk})^2) / \partial o_{pk} = -\sum_k (y_{pk} - o_{pk}) \qquad (2.31)$$

$$\partial o_{pk} / \partial (net_{pk}^o) = (\partial f_k^o / \partial (net_{pk}^o) = f_k^{o\prime}(net_{pk}^o) \qquad (2.32)$$

$$\partial (net_{pk}^o) / \partial i_{pj} = \partial (\sum_{j=1}^L (w_{kj}^o i_{pj} + \theta_k^o)) / \partial i_{pj} = w_{kj}^o \qquad (2.33)$$

$$\partial i_{pj} / \partial (net_{pj}^h) = \partial (f_j^h(net_{pj}^h)) / \partial (net_{pj}^h) = f_j^{h\prime}(net_{pj}^h) \qquad (2.34)$$

$$\partial(net_{pj}^{h})/\partial w_{ji}^{h} = \partial(\sum_{i=1}^{N}(w_{ji}^{h}x_{pi}+\theta_{j}^{h}))/\partial w_{ji}^{h} = x_{pi} \quad (2.35)$$

Combining Eqs. (2.30) through (2.35) we have, for the negative gradient:

$$-\partial E_{p}/\partial w_{ji}^{h} = \sum_{k}(y_{pk}-o_{pk})f_{k}^{o\prime}(net_{pk}^{o})w_{kj}^{o})f_{j}^{h\prime}(net_{pj}^{h})x_{pi} \quad (2.36)$$

The weight update equations are formulated as follows:
- for sigmoid neurons, let:

$$\delta^{os}{}_{pk} = (y_{pk} - o_{pk})f^{o}{}_{k}'(net^{o}{}_{pk}) = (y_{pk} - o_{pk})o_{pk}(1 - o_{pk}) \quad (2.37)$$

Combining Eqs. (2.25), (2.29), (2.36) and (2.37):

$$w_{ji}^{h}(t+1) = w_{ji}^{h}(t) - \eta(\partial E_{p}/\partial w_{ji}^{h}) \quad (2.38)$$

$$= w_{ji}^{h}(t) + \eta(\sum_{k}(y_{pk}-o_{pk})f_{k}^{o\prime}(net_{pk}^{o})w_{kj}^{o})f_{j}^{h\prime}(net_{pj}^{h})x_{pi})$$

$$= w_{ji}^{h}(t) + \eta(\sum_{k}\delta_{pk}^{os}w_{kj}^{o})f_{j}^{h\prime}(net_{pj}^{h})x_{pi})$$

$$= w_{ji}^{h}(t) + \eta(\sum_{k}\delta_{pk}^{os}w_{kj}^{o})i_{pj}(1-i_{pj})x_{pi}$$

- for linear neurons, let:

$$\delta^{ol}{}_{pk} = (y_{pk} - o_{pk})f^{o}{}_{k}'(net^{o}{}_{pk}) = (y_{pk} - o_{pk}) \quad (2.39)$$

Combining Eqs. (2.25), (2.29), (2.36) and (2.39):

$$w_{ji}^{h}(t+1) = w_{ji}^{h}(t) - \eta(\partial E_{p}/\partial w_{ji}^{h}) \quad (2.40)$$

$$= w_{ji}^{h}(t) + \eta(\sum_{k}(y_{pk}-o_{pk})f_{k}^{o\prime}(net_{pk}^{o})w_{kj}^{o})f_{j}^{h\prime}(net_{pj}^{h})\delta_{ji}^{hm}x_{pi})$$

$$= w_{ji}^{h}(t) + \eta(\sum_{k}\delta_{pk}^{ol}w_{kj}^{o})f_{j}^{h\prime}(net_{pj}^{h})\delta_{ji}^{hm}x_{pi})$$

$$= w_{ji}^{h}(t) + \eta\sum_{k}\delta_{pk}^{ol}w_{kj}^{o})\delta_{ji}^{hm}x_{pi}$$

For THONN model#1, there is only a single output neuron (i.e. k=1), and each hidden node has only two input weights. Thus for sigmoid neurons, this weight update rule becomes:

$$w_{ji}^h(t+1) = w_{ji}^h(t) + \eta(\delta_{pk}^{os} w_{kj}^o)\delta_{ji}^{hm} x_{pi} \quad (2.41)$$

where: $\delta_{ji}^{hm} = (\prod_{i'=1}^{2} w_{ji'}^h x_{pi'})/(w_{ji}^h x_{pi})$.

The above derivation was for sigmoid neurons – a similar derivation exists for multiplicative neurons, and this can be found in Appendix-A.

First Hidden Layer Neurons in THONN Model#1

The neurons in the 1st hidden layer can be either trigonometric, linear or power. Derivation of the appropriate weight update rules follows a similar pattern to that just described for 2nd hidden layer neurons, and while included here for the sake of completeness, is relegated to Appendix-B.

2.2.3 Neuron-Adaptive Higher-Order Neural Network – NAHONN

We now turn our attention to the Neuron-Adaptive HONN. The NAHONN Activation Function – NAF – is defined as follows:

$$\Psi_{i,k}(net_{i,k}) = \sum_{h=1}^{s} f_{i,k,h}(net_{i,1,k}, net_{i,2,k}, \ldots, net_{i,m,k}) \quad (2.42)$$

where i = the ith neuron in layer-k
 k = the kth neural network layer
 h = the hth term in the Neural network Activation Function (NAF)
 s = the maximum number of terms in the NAF
 x = first neural network input
 y = second neural network input
 $net_{i,j}$ = input (internal state) of the ith neuron in the kth layer
 $w_{i,j,k}$ = the weight that connects the jth neuron in layer k-1 with the ith neuron in layer-k
 $o_{i,k}$ = output of ith neuron in layer-k

Let $net_{i,1,j} = w_{i,1,k} * x_1$, $net_{i,2,j} = w_{i,2,k} x_2$ ... $net_{i,m,j} = w_{i,m,k} * x_m$. The m-dimensional NAHONN is then defined as:

$$\sum_{i=0}^{n} w_{z,i,k+1} \cdot \Psi_{i,k}(net_{i,1,k}, net_{i,2,k}, \ldots, net_{i,m,k}) \quad (2.43)$$

$$= \sum_{i=0}^{n} w_{z,i,k+1} \cdot \sum_{h=1}^{s} f_{i,k,h}(w_{i,1,k} \cdot x_1, w_{i,2,k} \cdot x_2, \ldots, w_{i,m,k} \cdot x_m)$$

2 Higher-Order Neural Networks for Satellite Weather Prediction 31

The network structure of an *m*-dimensional NAHONN is the same as that of a multi-layer feedforward ANN. In other words, it comprises an input layer of *m* input-units, an output layer of one output-unit, and a single hidden layer consisting of intermediate processing units. Again, while there is no activation function in the input layer and the output neuron is a summing unit (linear activation), the activation function for the hidden units is the *m*-dimensional NAHONN NAF defined by Eq. (2.42).

Now a *multi* m-dimensional NAHONN is defined by the following:

$$O_{j,l} = \sum_{i=0}^{n} w_{j,i,k+1} \cdot \Psi_{i,k}\left(net_{i,1,k}, net_{i,2,k}, \ldots, net_{i,m,k}\right) \quad (2.44)$$

$$= \sum_{i=0}^{n} w_{j,i,k+1} \cdot \sum_{h=1}^{s} f_{i,k,h}(w_{i,1,k} \cdot x_1, w_{i,2,k} \cdot x_2, \ldots, w_{i,m,k} \cdot x_m)$$

where l = output layer
   j = the *j*th output in the output layer
   $o_{j,l}$ = one output of the multi *m*-dimensional NAHONN

The architecture of this multi *m*-dimensional NAHONN is illustrated in Fig. 2.4.

One- and 2-dimensional NAHONN definitions, the NAHONN learning algorithm and the universal approximation capability of NAHONN are all described in [55].

The 2D Trigonometric NAHONN is defined as:

$$\sum_{i=0}^{n} w_{z,i,k+1} \cdot \Psi_{i,k}\left(net_{x,k}, net_{y,k}\right) \quad (2.45)$$

$$= \sum_{i=0}^{n} w_{z,i,k+1} \cdot \sum_{h=1}^{3} f_{i,k,h}(w_{i,x,k} \cdot x, w_{i,y,k} \cdot y)$$

$$= \sum_{i=0}^{n} w_{z,i,k+1} \cdot ((a1_{i,k} \cdot \sin^{c1_{i,k}}\left(b1_{i,k} \cdot (w_{i,x,k} \cdot x)\right) \cdot \cos^{e1_{i,k}}\left(d1_{i,k} \cdot (w_{i,y,k} \cdot y)\right))$$

where $a1_{i,k}$, $b1_{i,k}$, $c1_{i,k}$, $d1_{i,k}$, $e1_{i,k}$, are free parameters which can be adjusted (along with the weights) during training, and $net_{i,x,k} = w_{i,x,k} * x$.

Now since T-NAHONN is an open box model, it is capable of handling nonlinear data containing high frequency components. Similarly, PT-NAHONN can be used to model data which exhibits both polynomial and trigonometric features, while PES-NAHONN is suited to modeling data comprising discontinuous, piecewise functions.

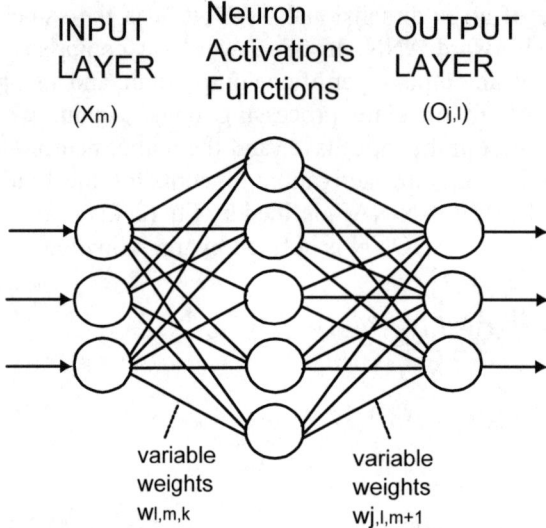

Fig. 2.4. Multi *m*-dimensional NAHONN Model

PT-NAHONN is defined by:

$$\sum_{i=0}^{n} w_{z,i,k+1} \cdot \Psi_{i,k}\left(net_{i,x,k}, net_{i,y,k}\right) \qquad (2.46)$$

$$= \sum_{i=0}^{n} w_{z,i,k+1} \cdot \sum_{h=1}^{4} f_{i,k,h}(w_{i,x,k} \cdot x, w_{i,y,k} \cdot y)$$

$$= \sum_{i=0}^{n} w_{z,i,k+1} \cdot (\ (a1_{i,k} \cdot \sin^{c1_{i,k}}\left(b1_{i,k} \cdot (w_{i,x,k} \cdot x)\right)) \cdot \cos^{e1_{i,k}}$$

$$\left(d1_{i,k} \cdot (w_{i,y,k} \cdot y)\right) + a4_{i,k} \cdot (net_{i,x,k})^{b4_{i,k}} \cdot (net_{i,y,k})^{d4_{i,k}}\ )$$

whereas TES is defined as follows:

$$\sum_{i=0}^{n} w_{z,i,k+1} \cdot \Psi_{i,k}\left(net_{i,k}\right) = \sum_{i=0}^{n} w_{z,i,k+1} \cdot \sum_{h=1}^{4} f_{i,k,h}(w_{i,x,k} \cdot x) \qquad (2.47)$$

$$= \sum_{i=0}^{n} w_{z,i,k+1} \cdot (a1_{i,k} \cdot \sin\left(b1_{i,k} \cdot (w_{i,x,k} \cdot x)\right) + a2_{i,k} \cdot e^{-b2_{i,k} \cdot (w_{i,x,k} \cdot x)}$$

$$+ a3_{i,k} \cdot \frac{1}{1+e^{-b3_{i,k} \cdot (w_{i,x,k} \cdot x)}})$$

2.3 Artificial Neural Network Groups

Neural network-based models are not yet sufficiently powerful to characterize complex, nonlinear and discontinuous systems, nor do they always provide appropriate reasoning [22]. Moreover, a gap exists in the research literature between complex systems and general systems. As it turns out however, it is a straightforward step to bridge this gap using neural network group theory.

In the real world, data can often vary in a discontinuous and non-smooth manner. In such cases, neural network groups are capable of performing much better than ANNs. Indeed, it is possible to simulate discontinuous functions (and to any degree of accuracy) using neural network group theory, even at points of discontinuity. As a result, neural network groups are capable of yielding appropriate reasoning – a function that simple ANNs cannot perform [18, 45, 51, 54].

Earlier work on ANN groups include that of [45], in which a hierarchical model of binary neuron clusters was analyzed in terms of renormalization groups. A level-by-level learning scheme for neural groups was said to closely resemble human knowledge acquisition, and lead to improvements in both learning efficiency and generalization ability [18]. Now whereas both these groups comprise discrete neurons, the groups under discussion in this Chapter comprise entire neural *networks*.

2.3.1 ANN Groups

Applying standard set theory to ANNs, we can define a specific ANN – **MLP**$_{20:15:10}$ say – as belonging to the **MLP** ANN set, which in turn is a subset of *all* ANN models:

$$\textbf{MLP}_{20:15:10} \subset \textbf{MLP} \subset \textbf{ANN}, \text{ where} \qquad (2.48)$$

$$\textbf{ANN} = \textbf{MLP} \cup \textbf{LVQ} \cup \textbf{RBF} \cup \textbf{SOM} \cup \textbf{HOPF} \cup \textbf{ART} \cup \textbf{SVM} \ldots \text{etc}$$

Further, we call the non-empty set N a 'neural network group' if $N \subset \textbf{ANN}$ (the generalized neural network set), and either the product $n_i * n_j$ or sum $n_i + n_j$ can be defined for every two elements $n_i, n_j \ni N$ [19, 27].

This enables us to define Piecewise Function Groups, namely those in which each addition $n_i + n_j$ is a piecewise function for every two elements $n_i, n_j \ni N$:

$$\begin{aligned} O_i + O_j &= O_i \quad (A < I < B) \\ &= O_j \quad (B < I < C) \quad \text{where } i = \text{neural network input} \end{aligned} \qquad (2.49)$$

Following the earlier findings of [15] and [22], it is possible to prove the following general result:

"Consider a neural network piecewise function group, in which each member is a standard MLP, and which has a locally bounded, piecewise continuous (rather than polynomial) activation function and threshold. Each such group can approximate *any* kind of piecewise continuous function, and to *any* degree of accuracy." [52]

ANN groups have been successfully applied to satellite rainfall estimation [50, 52], human face recognition [51] and financial time series simulation [53].

Results of the first study are presented in Sect. 2.6. In the latter study, experimental results obtained on All Banks Lending Data (supplied by the Reserve Bank of Australia Bulletin) showed that ANN group simulations were around 10% more accurate than ANNs.

2.3.2 PHONN, THONN & NAHONN Groups

The PHONN models of Sect. 2.2 can be extended to incorporate groups, indeed Model#3 is formed from groups of Model#2 PHONNs. PHONNGs have been subsequently applied to financial time series prediction and found to offer order of magnitude performance improvement, namely ~1.2% error for simulation (compared with ~11% for PHONN), and ~5.6% for prediction (as opposed to ~12.7% error for simulation). Similar improvements were forthcoming for THONNGs, as compared with THONNS and especially ordinary ANNs [54].

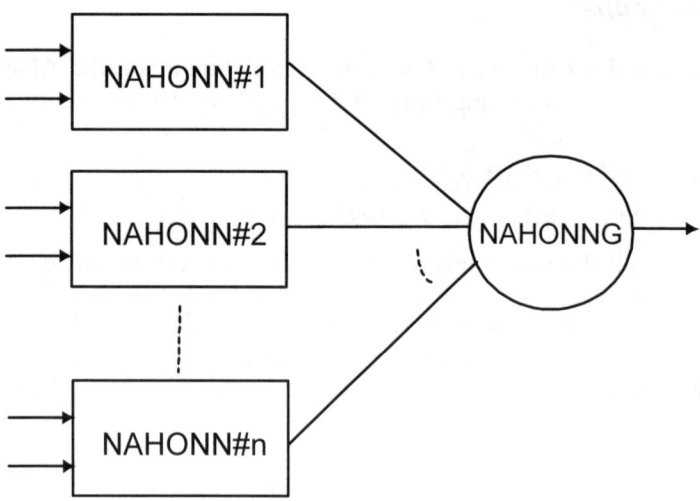

Fig. 2.5. Neuron-Adaptive Higher-Order Neural Network Group Model

A Neuron-Adaptive Higher Order Neural network Group – NAHONG – is a type of neural network group in which each element is a Neuron-Adaptive Higher Order Neural Network (Z_i):

$$NAHONG = \{ Z_1, Z_2, Z_3, \ldots, Z_i, \ldots, Z_n \} \qquad (2.50)$$

Now since NAHONG comprises Artificial Neural Networks, we can infer the following from [22]:

> "Consider a Neuron-Adaptive Higher Order Neural Network Group (NAHONG), in which each element is a standard multi-layer higher order neural network with adaptive neurons, and which has locally bounded, piecewise continuous (rather than polynomial) activation function and threshold. Each such group can approximate *any* kind of piecewise continuous function, and to *any* degree of accuracy".

A detailed proof is provided in [55].

NAHONN Groups are useful for modelling certain piecewise continuous functions – their structure is shown in Fig. 2.5.

The results of our experiments to date confirm that Neuron-Adaptive Higher-Order Neural network Groups (NAHONGs) possess superior approximation properties – namely faster learning, greatly reduced network size, and more accurate fitting of piecewise functions – compared with ANN groups. The reason for this is that NAHONGs can not only automatically find the best model and correct order, but can also model higher frequency, multi-polynomial, discontinuous, and piecewise functions [55].

2.4 Weather Forecasting & ANNs

Assessment of global climate change is very important for the future of both mankind and our environment, with rainfall estimation being a key element in this assessment [48]. Indeed, global weather prediction is often touted as one of computing's grand challenges (along with natural language understanding, human face recognition, and autonomous robots, to name but three more). It is essentially a time series prediction problem, and as such stands to benefit from techniques developed in allied areas, such as stock market prediction. Current weather prediction techniques center around grid(mesh) numerical models of the fluid and dynamical systems that prevail in the atmosphere [26], and therefore require complex differential equations and substantial computational overhead. Computations are further complicated by incomplete boundary conditions, various model assumptions and numerical instabilities [21].

During the past 20 years, there has been a substantial increase in our understanding of how satellite data can be used to estimate rainfall [14, 37, 38, 41]. Nevertheless, despite the use of interactive computer systems, the time needed to produce rainfall estimates typically exceeds half an hour. Furthermore, verification studies show the average error for rainfall events is around 30%.

Now whilst the traditional approach to weather forecasting is heavily based on numerical modeling as just described, in more recent times researchers have turned to soft computing techniques such as Expert Systems and Artificial Neural Networks. We discuss an example of the former in Sect. 2.6. As far as the latter is concerned, the earliest attempts were with ADALINE [44]; more recent efforts have focused on BackPropagation [11] and Naïve Bayesian Networks [23]. Use of such simple (naïve) ANN models is bound to produce limited results however, due to the limitations discussed earlier in Sect. 2.1, which in turn became the motivation for developing Higher-Order ANNs (Sect. 2.2) and ANN groups (Sect. 2.3). In Sect. 2.5 we develop HONNs more suited to this challenging task, and report on the success of these in predicting half-hour rainfall data.

2.5 HONN Models for Half-hour Rainfall Prediction

In this Section we introduce Higher-Order Neural Network models suitable for predicting rainfall.

2.5.1 PT-HONN Model

This network architecture combines characteristics of both PHONN and THONN (Sects. 2.2.1 and 2.2.2, respectively). PT-HONN is a multi-layer network that consists of an input layer, output layer, and two hidden layers (of intermediate processing units), and is described by the following:

$$x_{pi} = f_i^h(net_{pi}^h) = (net_{pi}^h)^n + \sin^a(net_{pi}^h) + \cos^g(net_{pi}^h) \qquad (2.51)$$

where net^h_{pi} = neuron input
x_{pi} = neuron output
f^h_{pi} = neuron input-output mapping function

2.5.2 A-PHONN Model

The network architecture of A-PHONN is based on characteristics of the PHONN, THONN, and PT-HONN models. Where it differs however is that it incorporates *adaptive* coefficients. A-PHONN is a multi-layer higher order neural network that consists of an input layer with input-units, and output layer with output-units, and two hidden layers consisting of intermediate processing units. A-PHONN is defined as:

$$z = \sum_{i,j=0}^{n} (a_{ij}(a1_{ij}x)^i (a2_{ij}y)^j \tag{2.52}$$
$$+ b_{ij} \sin^i(b1_{ij}x) \cos^j(b2_{ij}y)$$
$$+ c_{ij}/((1+\exp(-c1_{ij}x))^i (1+\exp(-c2_{ij}y))^j))$$

where the coefficients $a1_{ij}$, $a2_{ij}$, $b1_{ij}$, $b2_{ij}$, $c1_{ij}$ and $c2_{ij}$ can be adjusted during training, according to the input data. The derivatives for A-PHONN can be obtained as follows:

Let net_{pi}^h be the neuron input, x_{pi}^h the neuron output, and f_i^h be the input-output mapping function. If f_i^h is a polynomial function, then the derivative becomes:

$$f_i^{h'}(net_{pi}^h) = \partial x_{pi}/\partial(net_{pi}^h) = n((net_{pi}^h)^{n-1} \tag{2.53}$$

On the other hand, if f_i^h is a trigonometric polynomial function, then the derivative becomes:

$$f_i^{h'}(net_{pi}^h) = \partial x_{pi}/\partial(net_{pi}^h) \tag{2.54}$$
$$= \partial(\sin^n(net_{pi}^h))/\partial(net_{pi}^h)$$
$$= n\sin^{n-1}(net_{pi}^h)\cos(net_{pi}^h)$$

or

$$\partial(\cos^n(net_{pi}^h))/\partial(net_{pi}^h)$$
$$= n\cos^{n-1}(net_{pi}^h)\sin(net_{pi}^h)$$

Alternatively, for a sigmoid polynomial function:

$$x_{pi} = f_i^h(net_{pi}^h) \tag{2.55}$$

$$f_i^{h\prime}(net_{pi}^h) = \partial x_{pi} / \partial(net_{pi}^h)$$

$$= \partial(1/(1+\exp(-net_{pi}^h))^n)/\partial(net_{pi}^h)$$

$$= n(x_{pi})^n(1-x_{pi})$$

2.5.3 M-PHONN Model

The M-PHONN network architecture incorporates characteristics of the PHONN, THONN and PT-HONN models. Where it differs is that it includes a sigmoid polynomial function. It is described by the following:

$$z = \sum_{i,j=0}^{n}(a_{ij}(x)^i(y)^j + b_{ij}\sin^i(x)\cos^j(y) \tag{2.56}$$
$$+ c_{ij}/((1+\exp(-x))^i(1+\exp(-y))^j))$$

The derivatives for M-PHONN are the same as those derived for A-PHONN in Sect. 2.5.2 above.

Table 2.1. Cloud Top Temperature, Cloud Growth and Rainfall ('Expert' rule base)

Cloud Top Temp (°C)	Cloud Growth Latitude (°)	half-hour Rainfall (inches)	Cloud Top Temp (°C)	Cloud Growth Latitude (°)	half-hour Rainfall (inches)
> -32	2/3	0.05	- 70	1/3	0.85
-36	2/3	0.20	< - 80	1/3	0.95
-46	2/3	0.48	> -32	0	0.03
-55	2/3	0.79	- 36	0	0.06
-60	2/3	0.94	- 46	0	0.11
> - 32	1/3	0.05	- 55	0	0.22
- 36	1/3	0.13	- 60	0	0.36
- 46	1/3	0.24	-70	0	0.49
- 55	1/3	0.43	< -80	0	0.55
- 60	1/3	0.65			

2.5.4 Satellite Rainfall Estimation Results

For testing rainfall prediction, we relied on expert knowledge of rainfall estimation, derived from satellite observations of cloud top temperature and cloud growth [37, 38]. For example, one such rule is 'when the cloud top temperature is between −58°C and −60°C, and the cloud growth is more than 2/3 latitude, the half hour rainfall estimate is 0.94 inch'. This expert knowledge is summarized in Table 2.1.

Table 2.2 presents rainfall prediction resulting from using PHONN, PT-HONN, M-HONN and A-HONN models. For example, the average errors of PT-HONN, M-HONN and A-HONN are 5.68%, 5.42% and 5.34%, respectively – or around 10.7%, 14.8% and 16% better than PHONN.

Table 2.2. HONN Rainfall Simulation Results

Cloud Top Temp (°C)	Cloud Growth Latitude (°)	PHONN |Error| %	PT-HONN |Error| %	M-HONN |Error| %	A-PHONN |Error| %
> -32	2/3	3.22	8.80	8.78	8.76
-36	2/3	9.92	6.48	6.52	6.48
-46	2/3	25.19	23.18	22.58	21.48
-55	2/3	14.63	14.22	14.21	13.45
-60	2/3	3.66	0.09	0.12	0.14
> - 32	1/3	10.47	10.11	8.03	8.23
- 36	1/3	3.50	4.25	4.18	4.15
- 46	1/3	3.52	4.63	4.24	4.07
- 55	1/3	0.22	2.04	1.89	1.97
- 60	1/3	3.21	0.30	0.65	0.78
- 70	1/3	9.01	5.08	5.12	5.23
< - 80	1/3	3.89	1.21	1.32	1.25
> -32	0	9.81	7.86	7.24	7.17
- 36	0	2.98	3.62	3.25	3.13
- 46	0	5.69	5.68	5.67	5.54
- 55	0	5.28	4.35	4.03	4.18
- 60	0	3.32	1.33	1.43	1.35
-70	0	0.77	3.57	2.78	2.95
< -80	0	2.50	1.10	1.12	1.06
Average		6.36	5.68	5.42	5.34

2.6 ANSER System for Rainfall Estimation

An Artificial Neural network Expert System for Satellite-derived Estimation of Rainfall – ANSER – was developed earlier at the National Oceanic and Atmospheric Administration (NOAA), US Department of Commerce [50]. The satellite data used to estimate rainfall includes cloud temperature, rainburst factor, cloud merger factor, storm speed and the like. ANSER incorporates both HONNs and Neural Network groups, since simple ANN models are not sufficiently powerful to characterize complex systems such as rainfall estimation, which involves complicated, nonlinear, and discontinuous processes.

By themselves, ANN-based reasoning networks perform poorly (but nevertheless about twice as good as conventional techniques – typically ~17% test error, compared with ~30% for the latter). As already mentioned however, single ANNs cannot cope with discontinuous and non-smooth input data; ANN groups, by contrast, fare much better (which is borne out by average test errors dropping to less than 4%) [50].

In subsequent studies, we have demonstrated that HONN models can reduce rainfall estimation times by a factor of ten, compared with Decision Tree techniques. Moreover, the average rainfall estimate errors for the total precipitation event can be reduced to less than 20% (see Sect. 2.5.5).

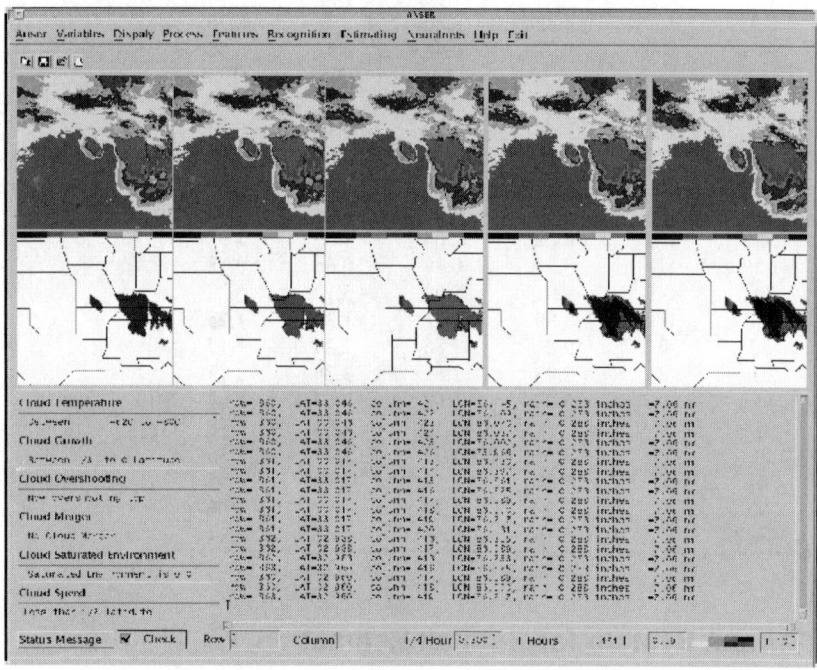

Fig. 2.6. ANSER User Interface

2.6.1 ANSER Architecture

The architecture of the ANSER system consists of three components:
- ANSER USER System
- ANSER TRAINING System

- ANSER CENTER System

The USER System consists of one or more (IBM PC-based) USER Subsystems, which provide estimates to the users. Each constituent USER Subsystem comprises a Training Subsystem, a Weight Base and an Estimate Subsystem.

The TRAINING System uses several TRAINING Subsystems for training weights. Each TRAINING Subsystem incorporates Initial Training, Re-training and Output Result functions. Input to the ANSER CENTER System comes from satellite data derived from the Expert System. The USER, TRAINING and CENTER Systems communicate with each other over an Ethernet LAN. The ANSER Graphical User Interface is illustrated in Fig. 2.6.

2.6.2 ANSER Operation

The ANSER system is shown in Fig. 2.7.

The operation of ANSER can be summarized as follows: using satellite data as input, the ANSER CENTER System firstly uses pattern recognition and image processing techniques to extract all features, rules, models, and knowledge (see Fig. 2.8). Especially significant are measurements of:
- cloud top temperature (CT),
- cloud growth factor,
- rainburst factor (RB),
- overshooting top factor (OS),
- cloud merger factor (M),
- saturated environment factor (SE),
- storm speed (S), and
- moisture correction (MC).

In order to recognize cloud merger, an Artificial Neural Network pattern recognition technique is used – details are provided in [48].

Next an ANN knowledge base technique (HONN) is used to obtain the rainfall estimate factor G ($=f$ {cloud top temperature + cloud growth}). All seven inputs – G, RB, OS, M, SE, S, MC – are fed into the ANN Reasoning Network.

There are *several* ANN reasoning network groups within the ANSER CENTER System, appropriate for different operating conditions. Based on the rule-, model- and knowledge bases, the Expert System chooses the *most suitable* ANN reasoning network group from those available, in order to produce half-hourly rainfall estimates, in real time.

The ANSER TRAINING System is used for deriving the weights of the neural networks, however since it runs on a mainframe computer, the training time is much less than for the USER System.

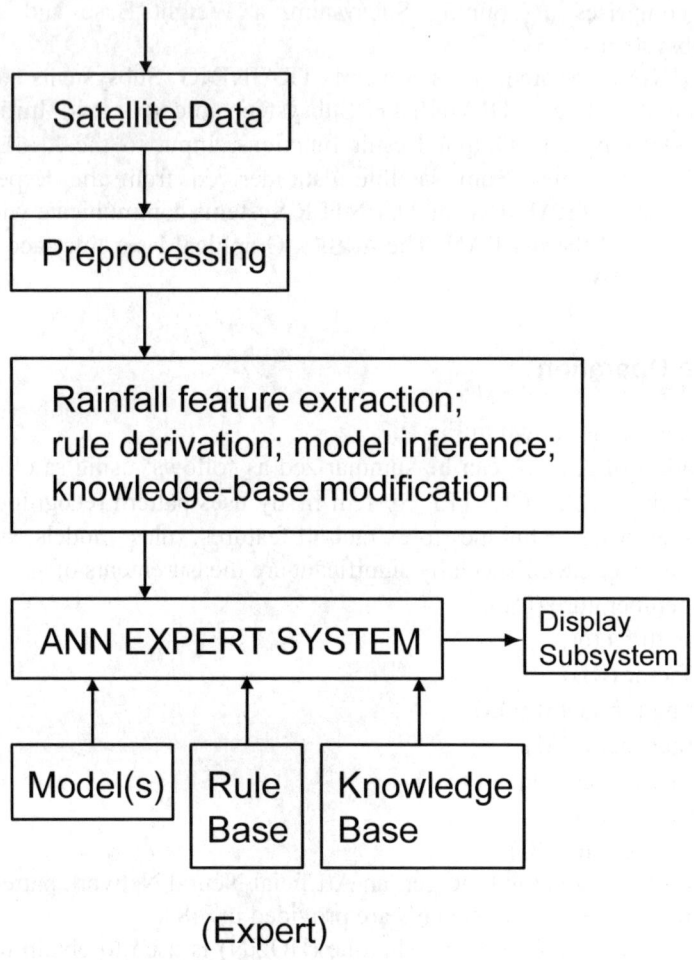

Fig. 2.7. ANSER System

Now ANSER provides *multi-user* capability, but in order to reduce training times, network weights are downloaded from the TRAINING System via Ethernet to the USER System(s). Once the input data (CT, CG, RB, OS M, SE, S, and MC) has been provided, the ANSER USER System can provide rainfall estimates in real time.

The output functions of ANSER can be either:

1. fixed-point rainfall estimates,
2. maximum isohyet area rainfall estimates, and/or
3. total aerial rainfall estimates.

All three outputs can be saved to files, as well as being displayed on a colour screen (Fig.2.6).

2.6.3 Reasoning Network Based on ANN Groups

Decision Trees – DTs – are an inherently sequential technique, due to the manner in which they search the state space. DTs cannot simultaneously compute all (complex, nonlinear) rules, models, knowledge and facts stored in their individual nodes. As a result, DTs are only capable of yielding approximate results (and moreover require longer execution times compared with parallel methods).

Artificial Neural Networks, on the other hand, are massively parallel architectures. Accordingly, rainfall estimates produced by ANN-based reasoning networks (and especially neural network *groups*), will not only be produced more quickly, but will be more accurate in comparison with DTs.

The basic architecture of the reasoning network for half-hourly satellite-derived rainfall estimation is a 3-layer BackPropagation ANN (or MLP), comprising 7 input neurons, 30 hidden neurons (divided into 2 layers) and 1 output neuron, with a total of 345 weights.

The inputs to and output from this reasoning network are as follows:
1. Inputs: G, RB, OS, M, MC, SE, MC, S
2. Output: half-hourly satellite-derived rainfall estimates

Once the input data become available, they are *simultaneously* fed into the reasoning network. The reasoning rules, models, and knowledge are stored in the weights of the (massively-parallel) neural network, so the reasoning network is capable of estimating rainfall using all rules and models simultaneously.

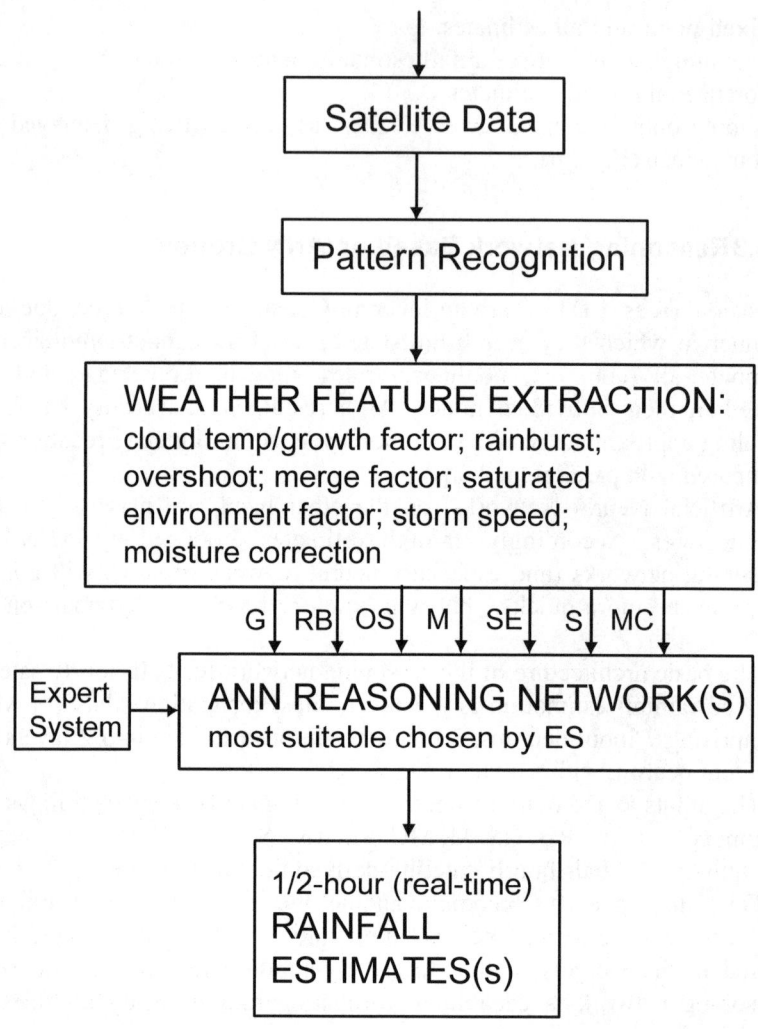

Fig. 2.8. ANSER Operation

Training of the reasoning network was performed using the NCCF HDS 9000 mainframe computer. Once trained, the reasoning network is able to estimate rainfall using rain burst, overshoot, cloud merger, saturated environment, moisture correction, storm speed and G (output from the knowledge base, inputs to which are cloud temperature and cloud growth).

The weights for the reasoning network are as follows:
1. $w1_{i,j}$ – 105 weights between the input layer and the first hidden layer,
2. $w2_{i,j}$ – 225 weights between the first and second hidden layers, and

3. $w3_{i,j}$ –15 weights between the second hidden layer and the output layer. Using these weights, estimation of the total aerial rainfall took only several seconds, once all the input data were fed into the reasoning network.

Table 2.3. Satellite-derived Precipitation Estimates

No.	Date	Location	Observation (inches)	X/S Error %	MLP Error %	NNG Error %	NAHONN Error %
1 #	07/19/85	NY	6.0	+0.83	+14.0	+6.28	+5.02
2 *	05/16/86	IL, MO	7.0	+10.0	-5.52	-3.37	-3.05
3 *	08/04/82	WI	5.3	+62.08	-8.30	-1.64	-1.82
4 *	05/03/87	MO	7.0	+47.86	-10.39	-0.94	-1.06
5 *	05/26/87	IA	5.8	-18.62	-7.08	-1.64	-1.37
6 #	08/13/87	KS	12.2	+16.62	+15.25	+1.59	+2.16
7 #	08/12/87	KS	8.7	-30.57	-21.0	-1.92	-2.01
8 #	07/19/85	IA	9.5	+96.21	+18.2	-0.74	-1.42
9 #	08/22/85	KS	4.2	-22.86	-13.3	+4.12	+3.92
10 #	07/02/83	IL	5.2	+13.65	+18.1	+8.71	+7.75
11 *	05/01/83	MO	6.0	-31.17	+4.0	+0.72	+0.70
12 #	05/27/85	MO	4.2	+2.62	+18.5	-3.88	-4.01
13 *	09/05/85	KS	6.2	-5.16	-1.90	-0.90	-0.87
14 *	07/15/84	IA	5.0	+78.4	+8.80	+1.20	+1.18
15 *	07/16/84	TN	4.2	+19.52	+6.43	-1.00	-1.01
Av					6.55 *	1.43 *	1.38*
Error %				30.41	16.91 #	3.89 #	3.75#

X/S: Xie & Scofield Study [46]; MLP basic reasoning network; NNG reasoning; NAHONN.
* training cases; # test cases.

2.6.4 Rainfall Estimation Results

A comparison of the results obtained using a conventional system [46], a basic (MLP) reasoning network, a neural network group and the NAHONN model of Sect.2.2.3 are summarized in Table 2.3. When the Xie & Scofield [46] technique was used, the average error of the operator-computed Interactive Flash Flood Analyzer (IFFA) rainfall estimates was 30.41%. For the basic (MLP) reasoning network, the training error was 6.55% and the test error 16.91%, respectively. When the ANSER technique (reasoning Neural Network Group) was used on these same fifteen cases, the average training error of rainfall estimation was 1.43%, and the average test error of rainfall estimation was 3.89%. When the NAHONN model was used on these same fifteen cases, the average training error of rainfall estimation was 1.38%, and the average test error of rainfall estimation was 3.75%, respectively.

Let us now consider a specific example (case 8 of Table 2.3): the rainfall estimation error resulting from the Xie & Scofield study is +96.21%. This falls to +18.2% with the basic (MLP) reasoning network. However, when the neural network group technique is used, the rainfall estimation test error is only –0.74%! The largest observed error using ANSER and NAHONN was only +8.71% and +7.75%, respectively (case 10). By contrast, the largest error reported in [46] was 96.21% (case 8). The largest error resulting from using the basic (MLP) reasoning network on its own was –21.0% (case 7).

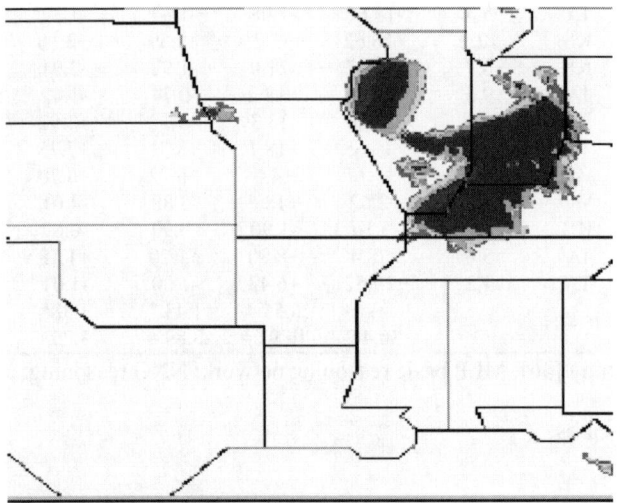

Fig. 2.9. Typical ANSER Rainfall Estimation Result

Now since rainfall estimation is a complex, nonlinear, discontinuous process, only neural network groups and HONN models are able to perform this task well; simple ANN models are inaccurate at points of discontinuity. This results in large errors with the latter technique, even though it performs better than the classical technique. In Sect. 2.3 we showed that neural network piecewise function groups are able to approximate any kind of piecewise continuous function, and to any degree of accuracy. This is the basic reason why ANN groups are able to estimate rainfall with an accuracy of around 96%. A typical rainfall estimation result for the mid-western USA states using year-2000 satellite data is shown in Fig. 2.9.

2.7 Summary

The definitions, basic models and characteristics of Higher Order Neural Networks (HONNs), Higher Order Neural Network Groups (HONNG), and Neuron-Adaptive Higher Order Neural Network (NAHONN) models have been presented. Use of such techniques has facilitated the development of an Expert System – ANSER – which is capable of estimating rainfall ten times faster than conventional techniques [38]. Moreover, the average rainfall estimation errors for the total precipitation event drop to below 10%.

Acknowledgements

The authors would like to thank Dr. Shuxiang Xu for his contribution to this work on HONNs. Heartfelt thanks are also due to the following Office of Research and Application staff at NOAA: Dr. Roderick Scofield, Dr. James Purdom, Ms. Frances Holt, Dr. Arnie Gruber, Dr. Fuzhong Weng and Mr. Clay Davenport. The initial work on HONNs was funded by Fujitsu Research Laboratories, Japan.

References

1. Arai M, Kohon R and Imai H (1991) Adaptive control of a neural network with a variable function of a unit and its application, *Trans Inst Electron Inform Communication Engineering*, J74-A: 551-559.
2. Azoff E (1994) *Neural Network Time Series Forecasting of Financial Markets* Wiley, New York.
3. Barron A (1993) Universal approximation bounds for superposition of a sigmoidal function, *IEEE Trans Information Theory*, 3: 930-945.
4. Barron R, Gilstrap L and Shrier S (1987) Polynomial and Neural Networks: Analogies and Engineering Applications, *Proc Intl Conf Neural Networks*, New York, II: 431-439.
5. Bishop C (1997) Neural Networks: a Pattern Recognition Perspective, in Fiesler E and Beale R (eds) *Handbook of Neural Computation* Oxford University Press, UK: B6.2:2.
6. Blum E and Li K (1991) Approximation theory and feed-forward networks, *Neural Networks*, 4: 511-515.
7. Chakraborty K, Mehrotra K, Mohan C and Ranka S (1992) Forecasting the behavior of multivariate time series using neural networks, *Neural Networks*, 5: 961-970.

8. Chen CT and Chang WD (1996) A feedforward neural network with function shape autotuning, *Neural Networks*, 9(4): 627-641.
9. Chen T and Chen H (1993) Approximations of continuous functionals by neural networks with application to dynamic systems, *IEEE Trans Neural Networks*, 4(6): 910-918.
10. Chen T and Chen H (1995) Approximation capability to functions of several variables, nonlinear functionals, and operators by radial basis function neural networks, *IEEE Trans Neural Networks*, 6(4): 904-910.
11. Chung CC and Kumar VR (1993) Knowledge Acquisition using a Neural Network for a Weather Forecasting Knowledge-based System, *Neural Computing & Applications*, 1: 215-223.
12. Cybenko G (1989) Approximation by superposition of a sigmoidal function, *Math Control, Signals, Systems*, 2: 303- 314.
13. Fahlman S (1988) Faster-learning Variations on Back-Propagation: an Empirical Study, *Proc 1988 Connectionist Models Summer School.*
14. Gorr WL (1994) Research prospective on neural network forecasting, *Intl J Forecasting*, 10(1): 1-4.
15. Hornik K (1991) Approximation capabilities of multi-layer feed-forward networks, *Neural Networks*, 4: 2151-2157.
16. Hornik K (1993) Some new results on neural network approximation, *Neural Networks*, 6: 1069-1072.
17. Hornik M, Stinchcombe M and White H (1989) Multi-layer feed-forward networks are universal approximators, *Neural Networks*, 2: 359-366.
18. Hu S and Yan P (1992) Level-by-level learning for artificial neural groups, *ACTA Electronica SINICA*, 20(10): 39-43.
19. Inui T, Tanabe Y and Onodera Y (1978) *Group Theory and its Application in Physics*, Springer Verlag, Berlin.
20. Karayiannis N and Venetsanopoulos A (1993) *Artificial Neural Networks: Learning Algorithms, Performance Evaluation and Applications*, Kluwer, New York (Chapter 7).
21. Lee RST and Liu JNK (2000) Teaching and Learning AI Modeling, in Jain lC (ed) *Innovative Teaching and Learning: Knowledge-Based Paradigms*, Physica Verlag, New York: 31-86 (Chapter 3).
22. Leshno M, Lin V, Pinkus A and Schoken S (1993) Multi-layer feed-forward networks with a non-polynomial activation can approximate any function, *Neural Networks*, 6: 861-867.
23. Li B, Liu J and Dai H (1998) Forecasting from Low Quality Data with Applications in Weather Forecasting, *Intl J Computing & Informatics*, 22(3): 351-358.
24. Lippmann DR (1987) An introduction to computing with neural nets, *IEEE Trans Acoustics, Speech & Signal Processing*, 4 (2): 4-22.
25. McCulloch W & Pitts W (1943) A Logical Calculus of Ideas Immanent in Nervous Activity, *Bulletin Mathematical Biophysics*, 5: 115-133.
26. McGreggor JL, Walsh KJ and Katzfey JJ (1993) Climate Simulations for Tasmania, *Proc 4th Intl Conf Southern Hemisphere Meteorology & Oceanography*, American Meteorological Society: 514-515.

27. Naimark M and Stern A (1982) *Theory of Group Representation* Springer Verlag, Berlin.
28. Park J and Sandberg IW (1991) Universal approximation using radial-basis-function networks, *Neural Computation*, 3: 246-257.
29. Park J and Sandberg IW (1993) Approximation and radial-basis-function networks, *Neural Computation*, 5: 305-316.
30. Peters E (1991) *Chaos and Order in the Capital Markets,* Wiley, New York.
31. Pham D and Liu X (1995) *Neural Networks for Identification, Prediction and Control*, Springer Verlag, Berlin.
32. Redding N, Kowalczyk A and Downs T (1993) Constructive high-order network algorithm that is polynomial time, *Neural Networks*, 6: 997-1010.
33. Psaltis D, Park C and Hong J (1988) Higher Order Associative Memories and their Optical Implementations, *Neural Networks*, I: 149-163.
34. Rosenbaltt F (1962) *Principles of Neurodynamics,* Spartan, New York.
35. Rumelhart DG, Hinton G and Williams R (1986) Learning Representations by Back-Propagating Errors, in Rumelhart D and McClelland J (eds) *Parallel Distributed Processing: Explorations in the Microstructure of Cognition, 1*, MIT Press, Cambridge, MA (Chapter 8).
36. Scarselli F and Tsoi AC (1998) Universal approximation using feed-forward neural networks: a survey of some existing methods, and some new results, *Neural Networks*, 11(1): 15-37.
37. Scofield RA and Oliver VJ (1977) A scheme for estimating convective rainfall from satellite images, *NOAA Technical Memorandum NESS 86*, US Department of Commerce, Washington, DC.
38. Scofield RA (1987) The NESDIS operational convective precipitation estimation technique, *Monthly Weather Review*, 115: 1773 - 1792.
39. Skapura DM (1996) *Building Neural Networks*, Addison-Wesley (ACM Press), Reading, MA.
40. Vecci LF, Piazza F, Uncini A (1998) Learning and approximation capabilities of adaptive spline activation function neural networks, *Neural Networks*, 11: 259-270.
41. Vemuri V and Rogers R (1994) *Artificial Neural Networks: Forecasting Time Series*, IEEE Computer Society Press, Piscataway, NJ.
42. Werbos P (1994) *The Roots of Backpropagation: from Ordered Derivatives to Neural Networks and Political Forecasting*, Wiley, New York.
43. White H (1989) Learning in artificial neural networks: a statistical perspective, *Neural Computation*, 1: 425-464.
44. Widrow B and Smith FW (1963) Pattern Recognition Control Systems, *Proc Computer & Information Science Symposium*, Spartan Books, Washington, DC.
45. Willcox C (1991) Understanding hierarchical neural network behavior: a renormalization group approach, *J Physics A*, 24: 2655-2644.
46. Xie J and Scofield RA (1989) Satellite-derived rainfall estimates and propagation characteristics associated with mesoscal convective systems (MCSs), *NOAA Technical Memorandum NESDIS,* 25: 0-49.

47. Zell A *et al* (1995) *Stuttgart Neural Network Simulator V4.1* University of Stuttgart, Institute for Parallel & Distributed High Performance Systems (ftp.informatik.uni-stuttgart.de).
48. Zhang M and Scofield RA (1994) Artificial Neural Network Techniques for Estimating Heavy Convective Rainfall and Recognition Cloud Mergers from Satellite Data, *Intl J Remote Sensing*, 15(16): 3241-3262.
49. Zhang M, Murugesan S and Sadeghi M (1995) Polynomial higher order neural network for economic data simulation, in *Proc Intl Conf Neural Information Processing*, Beijing, China, October 30 - November 3, 493-496.
50. Zhang M, Fulcher J and Scofield RA (1996) Neural network group models for estimating rainfall from satellite images, in *Proc World Congress Neural Networks*, San Diego, CA, September 15 -18, 897-900.
51. Zhang M and Fulcher J (1996) Face recognition using artificial neural network group-based adaptive tolerance (GAT) trees, *IEEE Trans Neural Networks*, 7(3): 555-567.
52. Zhang M, Fulcher J and Scofield R (1997) Rainfall estimation using artificial neural network group, *Neurocomputing*, 16(2): 97-115.
53. Zhang M, Zhang JC and Keen S (1999) Using THONN system for higher frequency non-linear data simulation & prediction, in *Proc IASTED Intl Conf Artificial Intelligence & Soft Computing*, Honolulu, Hawaii, USA, August 9-12: 320-323.
54. Zhang M, Zhang JC and Fulcher J (2000) Higher order neural network group models for data approximation, *Intl J Neural Systems*, 10(2): 123-142.
55. Zhang M, Xu S and Fulcher J (2002) Neuron-Adaptive Higher Order Neural Network Models for Automated Financial Data Modeling, *IEEE Trans Neural Networks*, 13(1): 188-204.

Appendix-A Second hidden layer (multiply) neurons

In Sect. 2.2.2 we derived weight update equations for sigmoid neurons; for the sake of completeness, we include derivations for multiplicative neurons below.

A.1   Second-Hidden Layer Neurons in THONN Model#1

The second hidden layer weights are updated according to:

$$w_{ji}^{h}(t+1) = w_{ji}^{h}(t) - \eta(\partial E_p / \partial w_{ji}^{h}) \tag{A.1}$$

where η = learning rate (positive & usually < 1)
j = input neuron index (1…L) (= ith neuron in 2nd hidden layer)
j = jth output neuron (1…M) (ref. Fig.2.3)
E = error
t = training time
h = hidden layer
$_h p$ = pth training vector
w_{ji}^{h} = weight connecting ith neuron in 1st hidden layer to jth neuron in 2nd hidden layer

The equations for the jth 2nd hidden layer node are:

$$net_{pj}^{h} = \prod_{i=1}^{N} w_{ji}^{h} x_{pi} \tag{A.2}$$

$$i_{pj} = f_{j}^{h}(net_{pj}^{h})$$

where i_{pj} = output from 2nd hidden layer (= input to the output neuron)
x_{pi} = input to 2nd hidden layer neuron
       (= output from the 1st hidden layer neuron)
f_{j}^{h} = hidden neuron activation function

The error of a single output unit will be:

$$\delta_{pk} = (y_{pk} - o_{pk}) \tag{A.3}$$

where y_{pk} = desired output value
o_{pk} = actual output from the kth unit

The total error is the sum of the squared errors across all output units, namely:

$$E_p = 1/2 \sum_{k=1}^{M} \delta_{pk}^2 = 1/2 \sum_{k=1}^{M} (y_{pk} - o_{pk})^2 \quad \text{(A.4)}$$

$$= 1/2 \sum_k (y_{pk} - f_k^o(net_{pk}^o))^2$$

$$= 1/2 \sum_k (y_{pk} - f_k^o(\sum_j w_{kj}^o i_{pj} + \theta_k^o))^2$$

The derivatives $f_j^{h}{}'(net_{pj}^h)$ are calculated as follows, for a linear function:

$$i_{pj} = f_j^h(net_{pj}^h) = net_{pj}^h \quad \text{(A.5)}$$

$$f_j^{h}{}'(net_{pj}^h) = 1$$

The gradient ($\partial E_p / \partial w_{ji}^h$) is given by:

$$\partial E_p / \partial w_{ji}^h = \partial(1/2 \sum_k (y_{pk} - o_{pk})^2) / \partial w_{ji}^h \quad \text{(A.6)}$$

$$= (\partial(\tfrac{1}{2} \sum_k (y_{pk} - o_{pk})^2) / \partial o_{pk})(\partial o_{pk} / \partial(net_{pk}^o))$$

$$(\partial(net_{pk}^o) / \partial i_{pj})(\partial i_{pj} / \partial(net_{pj}^h))(\partial(net_{pj}^h) / \partial w_{ji}^h)$$

$$\partial(\tfrac{1}{2} \sum_k (y_{pk} - o_{pk})^2) / \partial o_{pk} = -\sum_k (y_{pk} - o_{pk}) \quad \text{(A.7)}$$

$$\partial o_{pk} / \partial(net_{pk}^o) = (\partial f_{pk}^o / \partial(net_{pk}^o)) = f_k^{o}{}'(net_{pk}^o) \quad \text{(A.8)}$$

$$\partial(net_{pk}^o) / \partial i_{pj} = \partial(\sum_{j=1}^{L} (w_{kj}^o i_{pj} + \theta_k^o)) / \partial i_{pj} = w_{kj}^o \quad \text{(A.9)}$$

$$\partial i_{pj} / \partial(net_{pj}^h) = \partial(f_j^h(net_{pj}^h)) / \partial(net_{pj}^h) = f_j^{h}{}'(net_{pj}^h) \quad \text{(A.10)}$$

$$\partial(net_{pj}^h) / \partial w_{ji}^h = \partial(\prod_{i=1}^{N} (w_{ji}^h x_{pi})) / \partial w_{ji}^h = \delta_{ji}^{hm} x_{pi} \quad \text{(A.11)}$$

Combining Eqs. (A.6) through (A.11) we have, for the negative gradient:

$$-\partial E_p/\partial w_{ji}^h = \sum_k (y_{pk}-o_{pk})f_k^{o\prime}(net_{pk}^o)w_{kj}^o f_j^{h\prime}(net_{pj}^h)\delta_{ji}^{hm}x_{pi} \quad (A.12)$$

The weight update equations are formulated as follows:
- for sigmoid neurons, let:

$$\delta^{os}_{pk} = (y_{pk} - o_{pk})f^o{}_k{}'(net^o{}_{pk}) = (y_{pk} - o_{pk})o_{pk}(1 - o_{pk}) \quad (A.13)$$

Combining Eqs. (A.1), (A.5), (A.12) and (A.13):

$$w_{ji}^h(t+1) = w_{ji}^h(t) - \eta(\partial E_p/\partial w_{ji}^h) \quad (A.14)$$

$$= w_{ji}^h(t) + \eta(\sum_k (y_{pk}-o_{pk})f_k^{o\prime}(net_{pk}^o)w_{kj}^o)f_j^{h\prime}(net_{pj}^h)\delta_{ji}^{hm}x_{pi})$$

$$= w_{ji}^h(t) + \eta(\sum_k \delta_{pk}^{os} w_{kj}^o)f_j^{h\prime}(net_{pj}^h)\delta_{ji}^{hm}x_{pi})$$

$$= w_{ji}^h(t) + \eta(\sum_k \delta_{pk}^{os} w_{kj}^o)\delta_{ji}^{hm}x_{pi}$$

- for linear neurons, let:

$$\delta^{ol}_{pk} = (y_{pk} - o_{pk})f^o{}_k{}'(net^o{}_{pk}) = (y_{pk} - o_{pk}) \quad (A.15)$$

Combining Eqs. (A.1), (A.5), (A.12) and (A.15):

$$w_{ji}^h(t+1) = w_{ji}^h(t) - \eta(\partial E_p/\partial w_{ji}^h) \quad (A.16)$$

$$= w_{ji}^h(t) + \eta(\sum_k (y_{pk}-o_{pk})f_k^{o\prime}(net_{pk}^o)w_{kj}^o)f_j^{h\prime}(net_{pj}^h)\delta_{ji}^{hm}x_{pi})$$

$$= w_{ji}^h(t) + \eta(\sum_k \delta_{pk}^{ol} w_{kj}^o)f_j^{h\prime}(net_{pj}^h)\delta_{ji}^{hm}x_{pi})$$

$$= w_{ji}^h(t) + \eta\sum_k \delta_{pk}^{ol} w_{kj}^o)\delta_{ji}^{hm}x_{pi}$$

Appendix-B First hidden layer neurons

The 1st hidden layer weights are updated according to:

$$w_{im}^{h}(t+1) = w_{im}^{h}(t) - \eta(\partial E_p / \partial w_{im}^{h}) \quad (B.1)$$

where η = learning rate (positive & usually < 1)
  m = input neuron index (1…N) (ref. Fig.2.3)
  i = ith 2nd hidden layer neuron (1…L)
  E = error
  t = training time
  h = hidden layer
  p = pth training vector
  w^h_{im} = weight connecting mth input neuron to ith neuron in 1st hidden layer

The equations for the ith hidden node are:

$$net_{pi}^{h} = \sum_{m=1}^{R} w_{im}^{h} s_{pm}^{h} + \theta_i^{h} \quad (B.2)$$

$$x_{pi} = f_i^{h}(net_{pi}^{h})$$

where i_{pj} = output from 2nd hidden layer (= input to the output neuron)
  x_{pi} = output from the 1st hidden layer neuron
    (= input to 2nd hidden layer neuron)
  f_i^{h} = 1st hidden layer neuron activation function
  R = number of 1st hidden layer inputs
  s_{pm} = input to 1st hidden layer

The error of a single hidden unit will be:

$$\delta_{pk} = (y_{pk} - o_{pk}) \quad (B.3)$$

where y_{pk} = desired output value
  o_{pk} = actual output from the kth unit

The total error is the sum of the squared errors across all hidden units, namely:

2 Higher-Order Neural Networks for Satellite Weather Prediction 55

$$E_p = 1/2 \sum_{k=1}^{M} \delta_{pk}^{2} = 1/2 \sum_{k=1}^{M} (y_{pk} - o_{pk})^2 \quad \text{(B.4)}$$

$$= 1/2 \sum_{k} (y_{pk} - f_k^o(net_{pk}^o))^2$$

$$= 1/2 \sum_{k} (y_{pk} - f_k^o(\sum_{j} w_{kj}^o i_{pj} + \theta_k^o))^2$$

The derivatives $f_i^{h\prime}(net_{pi}^h)$ are calculated as follows, for a linear function:

$$x_{pi} = f_i^h(net_{pi}^h) = net_{pi}^h \quad \text{(B.5)}$$

$$f_i^{h\prime}(net_{pi}^h) = \partial x_{pi} / \partial(net_{pi}^h) = 1$$

Alternatively, for a power (square) function:

$$x_{pi} = f_i^h(net_{pi}^h) = (net_{pi}^h)^2 \quad \text{(B.6)}$$

$$f_i^{h\prime}(net_{pi}^h) = \partial x_{pi} / \partial(net_{pi}^h) =$$

$$\partial((net_{pi}^h)^2) / \partial(net_{pi}^h) = 2(net_{pi}^h)$$

For a sine function (and similarly for cosine):

$$x_{pi} = f_i^h(net_{pi}^h) = \sin(net_{pi}^h) \quad \text{(B.7)}$$

$$f_i^{h\prime}(net_{pi}^h) = \partial x_{pi} / \partial(net_{pi}^h) =$$

$$\partial \sin(net_{pi}^h)) / \partial(net_{pi}^h) = \cos(net_{pi}^h)$$

And for powers of sin(cos):

$$x_{pi} = f_i^h(net_{pi}^h) = \sin^n(net_{pi}^h) \quad \text{(B.8)}$$

$$f_i^{h\prime}(net_{pi}^h) = \partial x_{pi} / \partial(net_{pi}^h) =$$

$$\partial \sin^n(net_{pi}^h)) / \partial(net_{pi}^h) = n \sin^{n-1}(net_{pi}^h) \cos(net_{pi}^h)$$

The gradient ($\partial E_p / \partial w_{im}^h$) is given by:

$$\partial E_p / \partial w_{im}^h = \partial (1/2 \sum_k (y_{pk} - o_{pk})^2) / \partial w_{im}^h \qquad \text{(B.9)}$$

$$= (\partial (\sum_k (y_{pk} - o_{pk})^2) / \partial o_{pk})(\partial o_{pk} / \partial (net_{pk}^o))$$

$$(\partial (net_{pk}^o) / \partial i_{pj})(\partial i_{pj} / \partial (net_{pj}^h))(\partial (net_{pj}^h) / \partial x_{pi})$$

$$(\partial x_{pi} / \partial (net_{pi}^h))(\partial (net_{pi}^h) / \partial w_{im}^h)$$

$$\partial (1/2 \sum_k (y_{pk} - o_{pk})^2) / \partial o_{pk} = -\sum_k (y_{pk} - o_{pk}) \qquad \text{(B.10)}$$

$$\partial o_{pk} / \partial (net_{pk}^o) = \partial f_k^o / \partial (net_{pk}^o) = f_k^{o\prime}(net_{pk}^o) \qquad \text{(B.11)}$$

$$\partial (net_{pk}^o) / \partial i_{pj} = \partial (\sum_{j=1}^L (w_{kj}^o i_{pj} + \theta_k^o) / \partial i_{pj} = \sum_{j=1}^L (w_{kj}^o) \qquad \text{(B.12)}$$

$$\partial i_{pj} / \partial (net_{pj}^h) = \partial (f_j^h(net_{pj}^h)) \partial (net_{pj}^h) = f_j^{h\prime}(net_{pj}^h) \qquad \text{(B.13)}$$

$$\partial net_{pj}^h / \partial x_{pi} = \partial (\prod_{i=1}^N w_{ji}^h x_{pi}) / \partial x_{pi} = (\prod_{i=1}^N w_{ji}^h x_{pj}) / x_{pi} = \delta_{ji}^{hm} w_{ji}^h \qquad \text{(B.14)}$$

$$\partial x_{pi} / \partial (net_{pi}^h) = f_i^{h\prime}(net_{pi}^h) \qquad \text{(B.15)}$$

$$\partial (net_{pi}^h) / \partial w_{im}^h = \partial (\sum_{m=1}^R (w_{im}^h s_{pm} + \theta_i^h) / \partial w_{im}^h) = s_{pm} \qquad \text{(B.16)}$$

Combining Eqs. (B.9) through (B.16) we have, for the negative gradient:

$$-\partial E_p / \partial w_{im}^h = \sum_{i=1}^N (\sum_k^M (y_{pk} - o_{pk}) f_k^{o\prime}(net_{pk}^o) w_{kj}^o) \qquad \text{(B.17)}$$

$$f_j^{h\prime}(net_{pj}^h) \delta_{ji}^{hm} w_{ji}^h f_i^{h\prime}(net_{pi}^h) s_{pm}$$

The weight update equations are calculated as follows:
- for sigmoid output neurons:

$$\delta^{os}{}_{pk} = (y_{pk} - o_{pk}) f^o{}_k{}'(net^o{}_{pk}) = (y_{pk} - o_{pk}) o_{pk}(1 - o_{pk}) \quad (B.18)$$

- whereas for linear output neurons, this becomes:

$$\delta^{ol}{}_{pk} = (y_{pk} - o_{pk}) f^o{}_k{}'(net^o{}_{pk}) = (y_{pk} - o_{pk}) \quad (B.19)$$

In the case of sigmoid output neurons, and by combining Eqs. (B.1), (B.7), (B.17) and (B.18), we have, for a linear 1st hidden layer neuron:

$$w_{im}{}^h(t+1) = w_{im}{}^h(t) - \eta(\partial E_p / \partial w_{ji}{}^h) = \quad (B.20)$$

$$w_{im}{}^h(t) + \eta(\left[\sum_i (\sum_k \delta_{pk}{}^{os} w_{kj}{}^o) f_j{}^{h'}(net_{pj}{}^h) \delta_{ji}{}^{hm} w_{ji}{}^h f_i{}^{h'}(net_{pi}{}^h)\right] s_{pm})$$

$$= w_{im}{}^h(t) + \eta(\left[\sum_i (\sum_k \delta_{pk}{}^{os} w_{kj}{}^o) \delta_{ji}{}^{hm} w_{ji}{}^h\right] s_{pm})$$

Alternatively, for sigmoid output neurons and *power* 1st hidden layer neurons, we have, by combining Eqs. (B.1), (B.6), (B.17) and (B.18):

$$w_{im}{}^h(t+1) = w_{im}{}^h(t) + \eta(2\sum_i \sum_k \delta_{pk}{}^{os} w_{kj}{}^o)(net_{pi}{}^h) \delta_{ji}{}^{hm} w_{ji}{}^h] s_{pm}) \quad (B.21)$$

Similarly, for linear output neurons, and combining Eqs. (B.1), (B.6), (B.17) and (B.19), and for linear 1st hidden layer neurons, we have:

$$w_{im}{}^h(t+1) = w_{im}{}^h(t) + \eta(\left[\sum_i \sum_k \delta_{pk}{}^{ol} w_{kj}{}^o) \delta_{ji}{}^{hm} w_{ji}{}^h\right] s_{pm}) \quad (B.22)$$

For *power* 1st hidden layer neurons, this becomes:

$$w_{im}{}^h(t+1) = w_{im}{}^h(t) + \eta(2\sum_i \sum_k \delta_{pk}{}^{ol} w_{kj}{}^o) net_{pi}{}^h \delta_{ji}{}^{hm} w_{ji}{}^h] s_{pm}) \quad (B.23)$$

For *sinusoidal* 1st hidden layer neurons, this is:

$$w_{im}{}^h(t+1) = w_{im}{}^h(t) + \eta(\left[\sum_i \sum_k \delta_{pk}{}^{ol} w_{kj}{}^o) \cos(net_{pi}{}^h) \delta_{ji}{}^{hm} w_{ji}{}^h\right] s_{pm}) \quad (B.24)$$

with a similar formulation for cosine (ie. involving a sin term rather than cos). Finally, for powers of sine (& similarly for powers of cosine), we have:

$$w_{im}{}^h(t+1) = \quad (B.25)$$

$$w_{im}{}^h(t) + \eta(\left[\sum_i \sum_k \delta_{pk}{}^{ol} w_{kj}{}^o) n \sin^{n-1}(net_{pi}{}^h) \cos(net_{pi}{}^h) \delta_{ji}{}^{hm} w_{ji}{}^h\right] s_{pm})$$

3 Independent Component Analysis

Andrew D. Back[1]

[1] Windale Technologies, Brisbane, QLD 4075, Australia.
andrew@andrewback.com

3.1 Introduction

In this chapter we present a relatively new, but very powerful technique that has caused immense interest due to the significance it presents for intelligent systems. Consider a classical, discrete-time model

$$\mathbf{y}(t) = G(\mathbf{u}(t)) \qquad (3.1)$$

where G is some multi-input multi-output function of the multivariate input $\mathbf{u}(t)$. Typically, $\mathbf{y}(t) = [y_1 \cdots y_n]^T$ represents some multivariate measurement of physical quantities (for example, temperature, pressure, dollars, speed, position etc).

A common assumption is that $\mathbf{y}(t)$ can be observed, possibly with some additive noise. However real world modeling problems such as biomedical, financial, or environmental systems may have the follow difficulties:

- The measured data may contain large amounts of noise or extraneous information.
- It may not be possible to measure the output $\mathbf{y}(t)$ from the system directly.
- Complex systems may give rise to multiple outputs, mixed together in some way.
- It may not be possible to obtain the system output $\mathbf{y}(t)$ with conventional filtering techniques.

A clear goal for intelligent systems therefore, is to function in the presence of uncertainty such as with very noisy or missing data. Humans are adept at consciously and subconsciously processing enormous amounts of complex data to make meaningful decisions. For intelligent systems, it is necessary that they be able to handle situations when the incoming data is obscured in some manner.

In this Chapter, we consider a relatively new mathematical technique that is likely to be critical to future intelligent systems by enabling the isolation and extraction of desired information from a plethora of otherwise noisy data. This technique, known as *Independent Component Analysis* (ICA), enables signals to be extracted from what was previously thought to be an exceptionally difficult problem. In this Chapter we will provide an introduction and review of several algorithms that have been proposed for ICA.

3.2 Independent Component Analysis Methods

3.2.1 Basic Principles and Background

Let us assume that our objective is to measure and observe only some signals contained within $y(t)$, while the issue of estimating G and $u(t)$ is not required. For the purpose of this chapter, we will generally not consider Eq. 3.1 further. We assume that $y(t)$ exists, however it cannot be measured directly. Instead, it may be heavily masked or corrupted by other data so that we can only access some mixture of the signals $x(t) = [x_1...x_n]^T$ such that:

$$x(t) = As(t) \qquad (3.2)$$

where $s(t) = [s_1...s_m]^T$ is a vector of input data, comprised of column signal vectors representing time series, A is an $n \times m$ real-valued mixing matrix. Note that each measured signal $x_i(t)$ is the mixture of any of the true 'generative' source signals $s_j(t)$, such that each row of A gives a different mixture of $s(t)$. The goal is to determine the true input signal vector $s(t)$, and thereby access the desired signals $y(t)$, using only the measured signal vector $x(t)$.

If the mixing process is a linear transformation as in Eq. (3.2), it is possible to unmix them by applying an inverse linear such that:

$$z(t) = Wx(t) \qquad (3.3)$$
$$= WAs(t)$$

where W is an $m \times n$ demixing matrix. Given the correct choice of W, it is then possible to extract some linear combination of the inputs in $s(t)$.

This process is referred to as *blind source separation* (BSS). The term blind comes from the fact that we cannot observe the input signal vector $s(t)$, but we assume it exists and try to separate into $z(t)$ separate signals.

The blind source separation problem can be summarized as follows: We have a mixing system given by Eq. (3.2), and a demixing system given by Eq. (3.3). We seek to find a demixing matrix **W** to provide an output **Y** which is an estimate of **s**(t). If **W** = **A**$^{-1}$, then **z**(t) = **s**(t), and the input signals will be obtained without error (see Fig.3.1). The main approach to do this is based on the premise that the source signals will be statistically independent in some sense. For this reason, the problem is known as *Independent Component Analysis* (ICA).

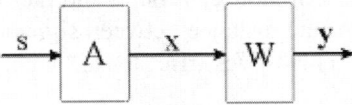

Fig. 3.1. General form of Independent Component Analysis: Input **s** is mixed by matrix **A** resulting in observed output **x**. The output **y** is obtained by finding the demixing matrix **W** and supplying it with the input **x**

Some of the basic assumptions included when deriving ICA algorithms are:
1. The measured signals consist of signals which are linearly combined in an instantaneous fashion.
2. The input signals, measured signals and mixture matrix are all stationary. In fact, this can be relaxed to quasi-stationarity, ie stationary for the duration of the measurement. Some recent ICA algorithms have been derived to utilize non-stationarity [29, 61, 88, 96-100].
3. For some ICA algorithms at most one of the signals is Gaussian. All other input signals are non-Gaussian. Algorithms which utilize only second order statistics to achieve source separation have also been derived [15, 81].
4. Usually, it is assumed that no prior knowledge of the signals, such as their probability distribution function – pdf – is available. If this is available, then one may consider deriving algorithms that can make use of it. A maximum likelihood approach may be more suitable in this case [17, 26, 97-100].
5. A sufficient amount of data is available so that the statistical characteristics of the signals can be estimated and hence the demixing matrix can be estimated.
6. The mixing process is linear and invertible. There should exist an inverse mixing operation that would enable the original signals to be

obtained[1]. Although there is interest in nonlinear mixtures, the invertibility issue is one of the main reasons why it is very difficult to demix signals passed through a nonlinear mixing process.
7. There should be little or no sensor noise. It is known that the presence of sensor noise will significantly hamper the separation performance of ICA algorithms.

The starting point for adaptive ICA algorithms is to specify a cost function-*f* which gives a measure of the signals independence. This is commonly referred to as the *contrast* function[2]. The terminology reflects the fact that the cost function should give a measure of the distance between signals, as opposed to the usual error criterion. This gives rise to an adaptive algorithm of the form:

$$\Delta \mathbf{W} = \frac{\partial \phi}{\partial \mathbf{W}} \qquad (3.4)$$

This means that we are adjusting the demixing matrix in the direction of the positive gradient of the change in independence with respect to the change in mixture weights. This will tend to maximize the independence of the signals, however it is based on the assumption of a smooth surface in terms of *f* with respect to **W**. This type of algorithm will tend to converge to a local minima in the parameter space. Unfortunately however, in ICA problems, this space may not be unimodal [40]. In the sections below, we consider some specific ICA algorithms. Other reviews of the field of ICA have been published previously, see for example [21, 28, 33, 62, 66, 85, 107].

3.2.2 Mutual Information Methods

Statistically independent signals can be defined in terms of their probability density functions – pdfs. Given a signal vector $\mathbf{y}(t) = [y_1...y_n]^T$, where the individual components of **y** are independent, the joint pdf of **y**, denoted as $p(\mathbf{y})$, can be factorized into the product of the marginal pdfs of the components:

---

[1] Clearly, it is possible to relax this assumption in terms of allow some approximate inverse only to exist, or to permit the extraction of less than the original number of inputs etc.

[2] Comon proposed that by minimizing the average mutual information, statistically independent signals could be extracted [39, 40]. This criteria was termed a contrast function, however the use is exactly the same as the usual cost or objective function employed within adaptive learning algorithms [30].

$$p(\mathbf{y}) = \prod_{j=1}^{n} p(y_j) \qquad (3.5)$$

One way to determine whether the components of **y** are independent or not, is to measure the difference between the joint pdf of **y** and the product of the marginal pdfs. Any discrepancy between the two will indicate the departure from independence.

A convenient and well known method for estimating the divergence between two distributions, is the Kullback-Leibler (K-L) divergence [71]. The Kullback-Leibler divergence measures the difference between the distributions of two random vectors and is defined as:

$$K(p_a | p_b) = \int p_a(x) \log\left(\frac{p_a(x)}{p_b(x)}\right) dx \qquad (3.6)$$

for the pdfs p_a, p_b. So, in order to compare the joint and marginal pdfs above, the following K-L divergence measure can be used:

$$K = \int p(\mathbf{y}) \log\left(\frac{p(\mathbf{y})}{\prod_{i=1}^{N} p_i(y_i)}\right) dy \qquad (3.7)$$

From this we can obtain [1]:

$$K = \int p(\mathbf{y}) \log p(\mathbf{y}) dy - \sum_{i=1}^{n} p_i(y_i) \log p_i(y_i) dy \qquad (3.8)$$

$$= \sum_{i=1}^{n} H(y_i) - H(\mathbf{y})$$

where $H(\mathbf{y})$ is the differential entropy, defined as:

$$H(\mathbf{y}) = -\int_{\mathbf{y}} p(\mathbf{y}) \log p(\mathbf{y}) dy \qquad (3.9)$$

and $H(y_i)$ is the marginal entropy, defined as:

$$H(y_i) = -\int p_i(y_i) \log p_i(y_i) dy_i \qquad (3.10)$$

Hence, we can relate the Kullback-Leibler divergence to the mutual information between the pdf of a random vector and the closest distribution which has independent entries. The mutual information of **y** is defined as:

$$I(y_1 \cdots y_n) = \sum_{i=1}^{n} H(y_i) - H(\mathbf{y}) \qquad (3.11)$$

Hence, to determine whether or not two signals are independent, the mutual information between the signals can be applied.

This formulation can be related to the entropies of the signals – that is, $H(\mathbf{y}) = -E[\log p(\mathbf{y})]$ and $H(y_i) = -E[\log p_i(y_i)]$ are the joint and marginal entropies respectively. The mutual information between two signals y_i and y_j is therefore given by is also known as the cross-entropy between the joint pdf of (y_i, y_j) given by $p(y_i, y_j)$ and the product of the marginal pdfs $p(y_i)$, $p(y_j)$.

The importance of ICA techniques in intelligent systems is clearly evident if we assume that the nature of intelligent systems is to broadly reduce entropy to achieve a given task. Below we review some specific ICA algorithms which use mutual information.

3.2.3 InfoMax ICA Algorithm

An ICA algorithm known as 'InfoMax' based on information maximization was proposed by Bell and Sejnowski [14]. They considered the problem of maximizing information flow through a neural network [102] with sigmoidal nonlinearities. Starting from this viewpoint, they suggested that the weights **W** should be adjusted by matching them to the mean and variance of the Gaussian input data **x**, so that the maximum amount of information is passed through the nonlinear functions of the network.

The information is expressed as the mutual information between the output **y** and the input **x**:

$$I(\mathbf{y},\mathbf{x}) = H(\mathbf{y}) - H(\mathbf{y}|\mathbf{x}) \qquad (3.12)$$

where $H(\mathbf{y}|\mathbf{x})$ is the entropy at the output not arising from the input data. In order to pass maximum information through to the outputs, the weights are adjusted so that the mutual information is maximized with respect to the **W**. Noting that

$$\frac{\partial H(\mathbf{y}|\mathbf{x})}{\partial \mathbf{W}} = 0 \qquad (3.13)$$

for a single layer neural network, this gives rise to an objective (or contrast) function of the form:

$$\phi = H(f(\mathbf{WX})) \qquad (3.14)$$

where $f: u \to v, u \in \Re$ and $v \subseteq (0,1)$ is a sigmoid function. Deriving a stochastic online learning algorithm for a single input shows that the weights are adjusted to align the steepest part of the sigmoid curve with the peak of the input density function.

This approach also overcomes the problem of slow learning or saturation that can occur in sigmoidal networks when the activation function[3] becomes too small or too large respectively.

Hence the final anti-Hebbian style algorithm can be derived as:

$$\Delta \mathbf{W} = \mathbf{W}^{-T} + (\mathbf{1} - 2\mathbf{y})\mathbf{x}^T \qquad (3.15)$$
$$\Delta \mathbf{W}_o = \mathbf{1} - 2\mathbf{y}$$

where $\mathbf{W}_0$ is the bias weight vector and $\mathbf{1}$ is a vector of ones.

The problem with the above form of the InfoMax algorithm is that it requires inversion of the weight matrix $\mathbf{W}$. This limits the speed and scalability of the algorithm for larger problems. A method of overcoming this limitation has been devised independently by Amari [1] and Cardoso [27]. Their approaches, known as Natural Gradient and Relative Gradient respectively, are described briefly in the following Section.

3.2.4 Natural/Relative Gradient Methods

Consider the standard form for many ICA algorithms:

$$\Delta \mathbf{W} = \eta(t) \frac{\partial \phi}{\partial \mathbf{W}} \qquad (3.16)$$

This is usually found to result in an algorithm which requires the inverse of the weight matrix $\mathbf{W}$, giving an algorithm such as:

$$\Delta \mathbf{W} = \eta(t) g(\mathbf{y}) \mathbf{W}^{-T} \qquad (3.17)$$

Based on principles from information geometry, the natural gradient and relative gradient methods suggest that Eq. (3.16) should be multiplied by $\mathbf{W}^T \mathbf{W}$ [2-4, 27]. Hence, instead of an ICA algorithm such as Eq. (3.17), this results in an algorithm of the form:

---

[3] The activation function v is defined as the total input supplied to the sigmoid function.

$$\Delta \mathbf{W} = \eta(t)g(\mathbf{y})\mathbf{W} \quad (3.18)$$

A consequence of using this approach is that there is no requirement to invert W at each iteration, resulting in significant savings in computational and memory requirements. For practical implementations, this is very important.

3.2.5 Extended InfoMax

The original InfoMax algorithm has been extended in various ways since its introduction. The algorithm is limited to sources that are super-Gaussian, and an extended version proposed in [74] overcomes this problem. Lee, Girolami and Sejnowski have proposed an algorithm that detects the pdf of the signals so that it can be used for general non-Gaussian signals.

The Extended InfoMax algorithm is defined as:

$$\Delta \mathbf{W} = \begin{cases} \eta(t)\left[I - f(y)y - \alpha yy^T\right]\mathbf{W} \\ \eta(t)\left[I + f(y)y - \alpha yy^T\right]\mathbf{W} \end{cases} \quad (3.19)$$

for super-Gaussian and sub-Gaussian pdfs respectively, and where $a > 0$ is a constant, and $f(\mathbf{y})$ is an odd, nonlinear, monotonically increasing function. It is defined to have the form of the derivative of the log density of an arbitrary super-Gaussian pdf of the sources. An example is $f(y) = \tanh(y)$.

3.2.6 Adaptive Mutual Information [1]

Amari, Cichocki & Yang derived an adaptive ICA algorithm based on mutual information as follows [1]. The aim in this case is to derive an algorithm that adjusts the demixing weights in **W** in the direction of the negative gradient of a cost function with respect to the weights. Hence, the starting point is to use Eq. (3.16), where the cost function is the mutual information.

To estimate the mutual information in Eq. (3.11), they estimated the marginal entropies by approximating the marginal pdfs using the truncated Gram-Charlier expansion. The Gram-Charlier expansion approximates the pdfs using cross cumulants of the signals of the signals [1]. For zero-mean independent components, the cross cumulant for n zero mean stationary processes $y_i(t)$ $i = 1,...,n$ is:

$$c_{y1\cdots yn}(\tau_1,\cdots,\tau_n) = E[y_1(t)\cdots y_n(t+n)] \quad (3.20)$$

Note that the second order cross cumulant is exactly the same as the second order cross correlation function, in other words:

$$c_{xy}(k) = E[x(t)y(t+k)] \tag{3.21}$$

Thus, the pdf of y can be estimated as:

$$p(y) = N(y)\left[1 + \frac{\kappa_3(y)}{3!}C_3(y) + \frac{\kappa_4(y)}{4!}C_4(y)\right] \tag{3.22}$$

where

$$N(y) = \frac{1}{\sqrt{2p}} e^{-\frac{y^2}{2}} \tag{3.23}$$

and $k_0(y)$ are the p-th order cumulants of y; $m_0(y)$ are the p-th order moments of y, where:

$$m_p(y) = E[y^p] \tag{3.24}$$

$$m_2(y) = 1$$

$C_k(y)$ are the Chebyshev-Hermite polynomials defined by:

$$(-1)\frac{d^k N(y)}{dy^k} = C_k(y)N(y) \tag{3.25}$$

Hence $k_3(y) = m_3(y)$ and $k_4(y) = m_4(y) - 3$. Thus, we can derive the marginal entropy as:

$$H(y) = \frac{1}{2}\log(2\pi e) - \frac{\kappa_3^2(y)}{12} - \frac{\kappa_4^2(y)}{48} + \frac{5\kappa_3^2(y)\kappa_4(y)}{8} + \frac{\kappa}{-} \tag{3.26}$$

The joint pdf can be estimated in a similar manner. Hence we obtain an expression for the mutual information between y_i, y_j obtained from the derived expressions for the Chebychev-Hermite polynomials. Hence, the Amari-Cichocki-Yang algorithm can be finally derived as:

$$\Delta \mathbf{W} = \eta(t)[\mathbf{I} - f(\mathbf{y})\mathbf{y}]\mathbf{W} \tag{3.27}$$

where:

$$f(y) = \frac{29}{4}y^3 - \frac{47}{5}y^5 - \frac{14}{3}y^7 + \frac{25}{4}y^9 + \frac{3}{4}y^{11} \tag{3.28}$$

Note that this algorithm uses the natural gradient method to improve convergence performance.

3.2.7 Fixed Point ICA Algorithms

The FastICA algorithm is based on the idea of using a fixed-point iterative scheme to optimize a contrast function [56-62]. The contrast function is based on a distance measure between the estimated pdf of the sources and a Gaussian distribution.

This algorithm, which is known as FastICA, performs ICA using a two step procedure. In the first stage, the signals are decorrelated in a whitening step:

$$\mathbf{u} = \mathbf{W}_a \mathbf{x} \qquad (3.29)$$
$$\mathbf{y} = \mathbf{W}_b \mathbf{u}$$

where $\mathbf{u}$ are the whitened outputs. Each output can be extracted by one row of the weight matrix $\mathbf{W}$, such that:

$$y_i = \mathbf{w}_i^T \mathbf{x} \qquad (3.30)$$

where $W = [w_1 ... w_m]^T$. Each of the m rows, corresponds to the demixing of one output. The FastICA algorithm uses a fixed point method to determine the weight vector such that the non-Gaussianity of the outputs is maximized.

A measure of Gaussianity is entropy. It can be shown that a Gaussian variable has maximal entropy when compared with all other random variables of equal variance. In FastICA, the contrast function used is based on the concept of negentropy. Negentropy or relative entropy, is derived by normalizing the differential entropy so that it is invariant under linear transformations [40] and is defined as

$$J(\mathbf{y}) = H(\mathbf{y}_g) - H(\mathbf{y}) \qquad (3.31)$$

where y_g corresponds to a Gaussian random vector having the same covariance matrix as $\mathbf{y}$. Negentropy can be estimated using higher order statistics, as:

$$J(y) = \frac{1}{12} E[y_i^3]^2 + \frac{1}{48} \kappa(y_i)^2 \qquad (3.32)$$

This approach for estimating negentropy suffers from the problem of not being particularly robust [62]. For example, outliers may give rise to significant errors, and hence a more accurate method is required to usefully estimate negentropy for ICA algorithms. A contrast function based on negentropy was proposed, and in general, is of the form:

$$J(y) = [E[G(y)] - E[G(v)]]^2 \qquad (3.33)$$

where G is some non-quadratic function, such as

$$G(y) = \frac{1}{a}\log(\cosh(ay)), \quad \text{or} \qquad (3.34)$$

$$G(y) = -\exp\left(-\frac{y^2}{2}\right)$$

where $1 \le a \le 2$. It has been found that this method of estimating entropy is reasonably robust and forms the basis for the FastICA algorithm.

The simplest way to understand FastICA is to consider initially just one output. Hence, from Eq. (3.30), we have the following iterative algorithm:

$$\mathbf{w}^+ = E[\mathbf{x}G(\mathbf{w}^T\mathbf{x})] - E[G(\mathbf{w}^T\mathbf{x})]^T\mathbf{w} \qquad (3.35)$$

$$\mathbf{w} = \frac{\mathbf{w}^+}{\|\mathbf{w}^+\|}$$

This algorithm is applied until the dot product of the current and previous $\mathbf{w}$ converges to almost unity.

The FastICA algorithm is derived by optimizing $E[\mathbf{x}G(\mathbf{w}^T\mathbf{x})]$ subject to the constraint $E[(\mathbf{w}^T\mathbf{x})^2] = \|\mathbf{w}\| = 1$. The interested reader may wish to consult [56, 59] for details of the derivation of this algorithm which we do not repeat here. The algorithm can be extended to multiple outputs.

Hyvärinen notes that finding $\mathbf{W}$ which minimizes mutual information is closely related to finding the directions in which negentropy is minimized [59]. The FastICA algorithm has been derived for complex signals [18]. Many ICA algorithms can be interpreted in a maximum likelihood framework [24, 97-100], and it is possible to view FastICA in this manner also [57].

3.2.8 Decorrelation and Rotation Methods

Another approach that can be used to separate mixed signals, is to note that for signals to be independent, requires that they must be decorrelated, and their higher order cross moments must also be zero. Hence it is possible to break the process of separation down into two distinct steps:

1. *Decorrelation of the measured signals.* This is achieved by diagonalizing the covariance matrix and gives '2nd order' independence.
2. *Orthogonalization of decorrelated signals.* In this stage, a unitary matrix rotation is used to minimize the higher order cross-moments. This provides higher order independence by ensuring that the outputs are as Gaussian as possible. The gaussianity is evaluated by means of a divergence from Gaussianity. The divergence functions are approximated using approaches such as the Edgeworth or Gram-Charlier expansion. Statistical moments only up to 4th order are required.

The aim of the whitening stage is to obtain the output $\mathbf{u}(t) = \mathbf{\Omega}\mathbf{x}(t)$, such that the covariance matrix $E[\mathbf{u}(t)\mathbf{u}^T(t)]$ equals the identity matrix. Various approaches can be used to perform the decorrelation. For example, it is possible to use Gram-Schmidt Orthogonalization, Singular Value Decomposition (SVD), or adaptive methods [72, 118]. Note that the decorrelation stage is often not used in the on-line learning approaches because it is assumed that the learning algorithm will learn to perform the whole required transformation.

The two stage approach of whitening followed by rotation can have some difficulties if the whitening stage has errors in it and if the decorrelation is imperfect, then the separation will be poor. This in turn means that the apparently optimal unitary rotation matrix might not give the best results. Consider the following:

$$\mathbf{W} = \mathbf{A}^+ \tag{3.36}$$
$$= \mathbf{U}\mathbf{\Omega}$$
$$\tilde{\mathbf{A}}_1^+ = \mathbf{U}\tilde{\mathbf{\Omega}}$$
$$\tilde{\mathbf{A}}_2^+ = \tilde{\mathbf{U}}\tilde{\mathbf{\Omega}}$$

Now, while $\tilde{\mathbf{A}}^+$ may be the perfect matrix inverse (or pseudo-inverse of $\mathbf{W}$), suppose that the decorrelation stage is less than perfect. In this case, even if $\mathbf{U}$ is the best unitary matrix possible, it is still possible that $\tilde{\mathbf{A}}_2^+$ may be a better approximation to $\tilde{\mathbf{A}}^+$ than $\tilde{\mathbf{A}}_1^+$ since $\tilde{\mathbf{U}}$ is in fact the best matrix possible for the overall approximation. For this reason, rotation is not the best approach for independent component analysis. However, if it is to be used, then a better approach than attempting to find the 'perfect' $\mathbf{U}$ matrix, is to decorrelate and then find the 'best' rotation matrix.

The decorrelation and rotation approach has been considered by various authors using various approaches, includng: a cumulant method [73], orthonormalization and quadratic weighting of covariances used to obtain 4th order moments [31, 32], higher order cumulants [40], and heuristic methods for the rotations [21]. We describe the Comon algorithm [40] below.

3.2.9 Comon Decorrelation and Rotation Algorithm

Comon proposed minimizing the contrast function of average mutual information [40]. The mutual information can be written down in terms of the second order moments and the higher order moments. This means that it is possible to apply a decorrelation to the data, followed by a stage which cancels the higher order moments.

The second stage is carried out by means of an orthogonal transformation that is carried out by a rotation. The rotation stage is more complex to derive and implement. There have been various approaches to this proposed in the literature. One approach is to obtain the rotation angle by optimization of a function of cumulants.

For many signals, the aim is to consider pairwise independence. For every pair of all output signals, there is a corresponding contrast function, which can be expressed as:

$$\Psi_{ij}(\mathbf{Q}) = K_{iiii}^2 + K_{jjjj}^2 \qquad (3.37)$$

where K are the cumulant functions and Q is the Givens rotation matrix, defined by

$$\mathbf{Q} = \frac{1}{\sqrt{1+\theta^2}} \begin{bmatrix} 1 & \theta \\ -\theta & 1 \end{bmatrix} \qquad (3.38)$$

Hence, the angle θ is adjusted to find a closed form solution.

The pdfs in the contrast function are evaluated using cumulants, and hence it is possible to show that:

$$I(y) = f(\text{higher order cumulants}) + g(\text{2nd order moments}) \qquad (3.39)$$

The first part represents the rotation, while the second part is the decorrelation. Another way to view this is that the cancellation of the 2nd order moments is part of the overall process of canceling higher order moments. Only 3rd or 4th order cumulants are generally required.

3.2.10 Temporal Decorrelation Methods

In the source separation methods based on instantaneous statistical independence, the time-dependence of the signals is not considered.

Another approach is to use the temporal correlation structure of the signals. In this approach, the idea is to use the covariance matrix at different time instants to determine the sources.

Temporal decorrelation ICA algorithms generally use an optimization function based on the time-delayed second order correlations, for example:

$$J = \sum_{i \neq j} E[y_i(t)y_j(t)]^2 + \sum_k \sum_{i \neq j} E[y_i(t)y_j(t-k)]^2 \qquad (3.40)$$

The key advantage of this approach is that for Gaussian signals, it is capable of providing much better source separation than other ICA algorithms which require that all source signals except for one, must be non-Gaussian. Algorithms in this class have been devised by a number of authors [5, 15, 45, 46, 81, 86-89, 124].

The assumptions made when using temporal correlation for ICA, are that each source is uncorrelated with every other source, the cross correlations of independent signals are close to zero, and the autocorrelation level is not constant for each signal, so the normalized correlation coefficients $r_i(k)$ are distinct for all sources [89]. In other words, the n source signals should have different autocorrelation functions, or spectra, leading to unique n eigenvalues. Note that the statistical independence and non Gaussianity conditions required by previous methods are not necessary in this case.

3.2.11 Molgedey and Schuster Temporal Correlation Algorithm

The basic principle of the temporal correlation ICA algorithm proposed by Molgedey and Schuster [89], is to determine the demixing matrix by solving an eigenvalue problem which involves the joint diagonalization of two symmetric matrices, corresponding to the time delayed autocorrelation matrices, where $k = 0$ and $k \neq 0$. We show this in further detail below.

Consider the time delayed autocorrelation matrices $\mathbf{R}_{xx}(k)$ defined as:

$$\mathbf{R}_{xx}(k) = E[\mathbf{x}(t)\mathbf{x}(t-k)^T] \qquad (3.41)$$

for some time delay k. Using the mixing relationship, we have that:

$$\mathbf{R}_{xx}(k) = E[\mathbf{A}\mathbf{s}(t)\mathbf{A}\mathbf{s}(t-k)^T] \qquad (3.42)$$
$$= \mathbf{A}\mathbf{R}_{ss}(k)\mathbf{A}^T$$

Assuming source independency, we have:

$$\frac{1}{m}\lim_{m\to\infty} \mathbf{R}_{ss}(k) = \Gamma(k) \tag{3.43}$$

where $\Gamma(k)$ is the diagonal cross correlation matrix (or autocorrelation matrix for $k = 0$). Hence, given sufficient data so that Eq. (3.43) can be applied, the eigenvalue problem can be solved as follows[4] [43, 49]:

$$\begin{aligned}
\mathbf{R}_{xx}(k)[\mathbf{R}_{xx}(0)]^{-1} &= \mathbf{A}\Gamma(k)\mathbf{A}^T[\mathbf{A}\Gamma(0)\mathbf{A}^T]^{-1} \\
&= \mathbf{A}\Gamma(k)\mathbf{A}^T[\mathbf{A}^T\Gamma(0)\mathbf{A}]^{-1} \\
&= \mathbf{A}\Gamma(k)\mathbf{A}^T\mathbf{A}^{-T}\Gamma(0)^{-1}\mathbf{A}^{-1} \\
&= \mathbf{A}\Gamma(k)\Gamma(0)^{-1}\mathbf{A}^{-1} \\
&= \mathbf{A}\Sigma\mathbf{A}^{-1}
\end{aligned} \tag{3.44}$$

where $\Sigma = \Gamma(k)\Gamma(0)^{-1}$. It is easy to relate $\mathbf{A}\Sigma\mathbf{A}^{-1}$ to an eigenvalue decomposition. Hence, we may apply an eigenvalue decomposition method to the product $\mathbf{R}_{xx}(k)[\mathbf{R}_{xx}(0)]^{-1}$ to directly determine the mixing matrix $\mathbf{A}$. The demixing matrix $\mathbf{W}$ is then simply found as the inverse of $\mathbf{A}$.

Note that in the above derivation, we only used a single time delay value k, to obtain the time delayed correlation matrix, even though there is a multivariate time series $\mathbf{x}$ available. It is clear that the performance of this type of algorithm depends on the correlation matrices being non singular. Under some conditions, such as fast sampling [7], it is known that the correlation matrix can approach singularity. Hence there is a need to consider alternative methods to overcome this potential difficulty. One method suggested is to use a number of cross correlation matrices over different time delays [15]. Other variations of this type of algorithm have been proposed in [91, 125]. An ICA algorithm based on time-frequency methods was derived in [16].

3.2.12 Spatio-Temporal ICA Methods

Since ICA can be performed using either spatial independence, or temporal independence, it is natural to consider methods which utilize both spatial and temporal properties. While there has been less work in this area, one of the more well known methods is that proposed by Pearlmutter and Parra [96]. They proposed the contextual ICA (cICA) algorithm which is suitable for sources having colored Gaussian distributions or low kurtosis.

---

[4] Since this method is not as widely used in the literature as other ICA algorithms, we give a few extra steps to clarify how the method is derived.

The cICA algorithm generalizes the Infomax algorithm [14], where it is assumed that the sources are generalized autoregressive processes which include additive non Gaussian noise. The algorithm is derived by considering the pdf of the souce model as:

$$p(\mathbf{y}(t)|\mathbf{y}(t-1),\cdots,\mathbf{y}(t-h)) = \prod_i f_i(y_i(t)|y_i(t-1),\cdots,y_i(t-h)) \tag{3.45}$$

where f_i are pdfs parameterized by a vector $\mathbf{w}_1$. This gives rise to a maximum likelihood ICA learning algorithm of the form:

$$\Delta \mathbf{w}_j = -\frac{f_j'(y_j;\mathbf{w}_j)}{f_j(y_j;\mathbf{w}_j)} \tag{3.46}$$

where the derivative of the pdf is

$$\frac{df_j(y_j(t)|y_j(t-1),\cdots,\mathbf{w}_j)}{d\mathbf{W}} = \tag{3.47}$$

$$\frac{\partial f_j}{\partial y_j(t)} \frac{dy_j(t)}{d\mathbf{W}} + \sum_k \frac{\partial f_j}{\partial y_j(t-k)} \frac{dy_j(t-k)}{d\mathbf{W}}$$

The algorithm is implemented with the natural gradient method which improves the convergence. In practice, the cICA algorithm has been shown to be able to separate signals which cannot be separated by other ICA algorithms which do not take the temporal information into account.

Other ICA algorithms have been proposed for separating convolved signals, see for example [64, 68, 92, 101, 113].

3.2.13 Cumulant Tensor Methods

The JADE (Joint Approximate Diagonalization of Eigenmatrices) algorithm uses the data in a two-stage batch form [32]. In the first stage of the algorithm, the data is whitened by finding the sample covariance matrix of the input using an eigendecomposition. In the second stage, a rotation matrix is found which performs the joint diagonalization of the matrices obtained obtained from parallel slices of the fourth order cumulant tensor. As with other ICA algorithms relying on spatial independence, the JADE algorithm assumes that the signals are non-Gaussian and uses their joint pdfs.

3.2.14 Nonlinear Decorrelation Methods

Higher order statistics have been demonstrated as a fundamental characteristic of many ICA algorithms. Another class of ICA algorithm is the nonlinear Principal Component Analysis (PCA) or nonlinear decorrelation methods. These algorithms use nonlinear functions which indirectly result in the use of higher order statistics to achieve source separation. In fact, the algorithms are very similar to those derived using mutual information, however the nonlinear functions in that case were the result of approximating the densities.

The early algorithm proposed by Herault and Jutten [51-53, 66, 87] can be written as:

$$\frac{d\mathbf{W}}{dt} = \eta \mathbf{W}(t) G(\mathbf{y}) \quad (3.48)$$

where $G(\mathbf{y})$ is a nonlinear function of y. The particular form of the Herault-Jutten algorithm is:

$$\frac{dw_{ij}}{dt} = \eta f(y_i) g(y_j) \quad (3.49)$$

$$f(y_i) = y^3$$

$$g(y_j) = \tanh(y)$$

Note that in this methodology, the use of nonlinear terms implicitly introduces higher order moments into the equation. Thus, the adjustment of **W** makes the signals more independent. This work is also related to the earlier application of neural networks to PCA [9, 22, 72, 105, 118].

3.3 Applications of ICA

3.3.1 Guidelines for Application of ICA

To apply ICA algorithms in practice requires more than simply using the first algorithm at hand. It is important that the characteristics of the source signals are used to determine the most suitable algorithm. The key points to consider are:

1. *Gaussianity*. This will influence the ICA algorithm that is suitable for the sources. If the source signals are all close to Gaussian, then it may be necessary to consider alternative algorithms, for example temporal correlation algorithms.

2. *Probability Density Function.* Many of the algorithms rely on the sources having a particular type of pdf, for example super-Gaussian or sub-Gaussian. Examples of super-Gaussian pdfs include speech, music. The longer tails of financial time series may also fall into the category of super-Gaussian pdfs.
3. *Temporal Correlation.* If the source signals have strong temporal correlations and have unique spectra, then it may be possible to use ICA algorithms employing this characteristic. In some cases, for example, shorter data lengths, these types of algorithms may give superior results than those based on higher order statistics, which require larger amounts of data [43].
4. *Stationarity.* If the sources with non stationary pdfs, particular care needs to be taken with algorithm selection. For example, if a block based algorithm is chosen, the demixing matrix **W** is likely to be completely different from one update to the next. The ICA algorithm itself should take into account the fact that the source pdfs can vary.

3.3.2 Biomedical Signal Processing

Biomedical signal processing involves the interpretation of highly noisy multivariate data taken from a number of sensors which are usually non-invasive. In most cases, it is desired to observe some particular signals while ignoring others. Techniques in this area have been developed over a long period of time. For example, early methods of listening to the fetal heartbeat required adaptive signal processing techniques in order to remove the mother's heartbeat from electrocardiographic (ECG) measurements [117]. A recent application of ICA to fetal heart rate monitoring permits the resolution of time-varying heart rate frequency [12].

However, in contrast to the relatively simple task of ECG noise removal where there are only two sources, each of which can be readily isolated by adaptive filtering and sensor placement, other biomedical procedures can involve a large number of signals for which there is no convenient method of separating the desired source signals from unwanted ones.

For example, electroencephalography (EEG) measures signals from the brain using a large number of sensors. Here, the aim is usually to diagnose patient condition based on the analysis of electrode recordings from the scalp of a patient. It is assumed that the signals recorded are generated from specific nervous centers within the head. Hence the task for ICA is to isolate signals which come from different regions within the brain. These different signals may be due to different functions being performed and hence we may expect there to be a degree of statistical independence.

ICA was first applied to EEG signal analysis by [83, 84]. In this case, the Infomax algorithm was applied. They were able to successfully separate artifacts due to eye movements and muscle noise. More importantly, the use of ICA was able to isolate spatially separable event-related potential (ERP) components and other overlapping phenomena.

A related task to EEG, is magnetoencephalography (MEG) which has the same issues with noise removal as EEG [32, 117]. In [126] the authors applied ICA to the problem of reducing 150 Hz noise artifacts in MEG. Conventional techniques such as notch filtering were able to also provide effective performance when the noise signal is known, but when this is not the case, ICA will provide better performance.

ICA has also been applied to ECG, MRI, fMRI, PET and other biomedical signal analysis tasks [24, 50].

3.3.3 Extracting Speech from Noise

ICA has been applied to the task of speech enhancement by numerous authors [23, 37, 104, 114, 116]. An overview of the field and examination of the problems is given by Torkkola [114]. He makes the following key points:
- Some ICA algorithms assume stationarity, however, speech is nonstationary in a longer time scale, but quasi-stationary in shorter time scales. This may lead to local minima in the optimization process.
- In practical applications such as separating car noise in mobile communications, the number of noise sources may be very large, hence simple systems such as a 2x2 noise-source separation system will not work well.
- The sources may have a large dynamic range. For example, due to speaker movement, the optimum model to separate the sources will vary from one time instant to the next. Hence, improved ICA algorithms are required which take into account the dynamic nature of the system.
- Convergence in a short time frame is required, but this may be difficult for speech applications where filter lengths of many thousands of samples are frequently used.
- Using *a priori* knowledge of the speech signals is an important aspect that should be considered in the derivation of appropriate ICA algorithms [23].

Frequency domain ICA algorithms appears to be one method by which the problem of high order filters can be overcome. There have been several frequency domain ICA algorithms proposed. One of the main problems to avoid in this type of algorithm is transferring the permutation and scaling that normally occurs into the frequency domain. There have been some novel algorithms proposed which apply the nonlinear functions in the time domain, but still use the frequency domain to speed up the calculations [6, 54, 78]. This hybrid time-frequency domain approach therefore overcomes the permutation and scaling problems.

3.3.4 Unsupervised Classification Using ICA

Clustering or unsupervised classification is important in a range of fields where there is a need to classify data into some number of classes when there is no training set. Recently there has been some interest in determining how ICA can be applied to this area. We examine a few approaches to unsupervised classification and improvements using ICA below.

A common method of unsupervised classification is to use the Gaussian mixture model (GMM). In this approach, it is assumed that the data consists of n mutually exclusive classes. It is further assumed that each class can be approximated by a multivariate Gaussian density given by:

$$p(\mathbf{x}|\Theta) = \sum_{i=1}^{n} p(\mathbf{x}|C_i, \theta_i) p(C_i) \qquad (3.50)$$

where $p(\mathbf{x}|C_i, \theta_i)$ are the component multivariate Gaussian densities having parameters θ_i and prior probabilities $p(C_i)$ for n classes. The Gaussian mixture model provides a method of obtaining a smooth approximation to an arbitrary density. A standard method of determining the parameters in a GMM model is to use a maximum likelihood algorithm such as EM (expectation maximization).

A generalization of the Gaussian mixture model was proposed by [77], such that the component densities are non-Gaussian and the mixture can be described by an ICA model. In their approach, each class is comprised of independent components; however it is possible for there to be dependencies between classes. This results in an ICA application where the independence assumption is weakened, yet the overall experimental results are as good or better than other approaches such as AutoClass [111] on benchmark problems. An EM (expectation maximization) learning algorithm can be derived for this task in a similar manner to the Gaussian mixture model.

A maximum likelihood method of clustering where the output classes can be further dichotomized using a structured form of ICA has been proposed by Girolami [45]. This method is based on the extended InfoMax algorithm and is intended for interactive data analysis to effectively partition multi-class data. The novelty of this approach is to structure the demixing matrix **W** so that it dichotomizes the data. That is, the ICA algorithm is used to arrive at the original sources, since it is assumed that there is some transformation taking place between the 'true' sources of the data and the measured data. The ICA algorithm permits these true sources to be approximated. But, furthermore, since the goal is to cluster the data, this can be done by hypothesizing a model of the pdf of those sources.

Consider for a moment the ICA algorithm in [45]. This algorithm is derived as:

$$\Delta \mathbf{W}(t) = \eta(t)\left(\mathbf{I} + \mathbf{f}(\mathbf{y}(t))\mathbf{y}(t)^T\right)\mathbf{W}(t) \quad (3.51)$$

where:

$$f(y) = \frac{p'(y)}{p(y)} \quad (3.52)$$

and $\mathbf{f}(\mathbf{y}) = [f(y_1)... f(y_m)]^T$ for m sources. Now, in this case, it is assumed that the particular form of the non Gaussian pdf is known. In particular, [45] assume that it is a simple mixture of Gaussians, otherwise known as a Pearson model, where:

$$p(y_i) = a_i p_G(y_{ij}) + (1-a_i) p_G(y_{ij}) \quad (3.53)$$

and $p_c(y)$ is a normal pdf. The case where $a_i = 0.5$ results in a symmetric pdf, with negative kurtosis. This case divides the data into two clusters. Each source therefore has two clusters, so that for n sources, there are $2n$ possible clusters that can be obtained using this approach.

The ICA algorithm attempts to find the sources which maximize the likelihood of the output probabilities being of the hypothesized structure, in other words:

$$p(\mathbf{y}_j) = \prod_{i=1}^{n} p_i(y_{ij}) \quad (3.54)$$

$$= \prod_{i=1}^{n} a_i p_i(y_{ij}) + (1-a_i) p_i(y_{ij})$$

where y_i represents the j-th observation in the **y** data set.

This approach has been tested on some interesting applications:
- *Noninvasive monitoring of oil flow.* In this application, considered previously in [19], the aim is to solve the inverse problem of estimating the quantity of oil flowing in a pipeline carrying a multiphase mixture of oil, water and gas. There are twelve measurement variables, and three types of flows that can occur within the pipeline: laminar, annular and

homogeneous. The task is to cluster each of the data points into one of three classes corresponding to the flow type in a two dimensional space, corresponding to the fraction of oil and water. This clustering is then able to provide the required information on the amount of oil present in the pipeline. In comparison with other techniques such as PCA and GTM (Generative Topographic Mapping) [20], the proposed ICA method performs well.
- *Swiss Banknote Fraud detection.* In this application [44], the data has two classes, corresponding to the two possible outcomes: legal and forged. It is known that PCA can provide reasonably effective classification, however it is found that the ICA based method provides further insight into the data set. In particular, by using the hierarchical approach, it is found that there are two clusters within the forgeries group. Hence, this indicates that there are two different sets of forgers. The value of improved clustering is clearly evident.

The mathematically intuitive and appealing nature of this method means that it is a leading candidate for future work on unsupervised classification.

The contextual ICA algorithm has been applied to the classification of hyperspectral image data [13]. In this application, images are taken of geographic regions which result in images containing a large number of spectral channels for each image pixel. It is assumed that there are a number of mineral types having different spectral characteristics. Hence, by demixing an assumed number of input sources, it becomes possible to obtain images of the mineral content in different geographic locations. Previous approaches generally make a number of a priori assumptions about the physical characteristics of the data set which may lead to errors of interpretation. Without precise a priori knowledge of the scenes, the task is difficult to solve and makes an ideal candidate for the unsupervised approach of ICA.

In a set of experiments, cICA was applied firstly to synthetic data in [13] and the correlation between unmixed spectral data and the known spectral data of minerals to determine the classification at the output. Due to similarities in correlation and the presence of negative correlations, this process requires careful attention, but yielded acceptably good performance in extracting the required original spectral image data.

The authors then applied cICA to actual AVIRIS image data which was compared against ground truth maps taken from the Cuprite, Nevada region. In this case, although the algorithm was able to unmix the sources and provide some results, there are some drawbacks to the method. In particular, it was found in these experiments that using correlation to classify the outputs did have problems. The authors only used correlation and not dependence to classify the outputs. It may be possible that improved results could be obtained by using dependence (which will then utilize higher order statistics).

An interesting outcome of this work is that there is a demand for ICA methods which can be used to more accurately classify the outputs against known data. In other words, in some applications, it is not sufficient just to be able to separate the data, but to be able to classify the data. The criteria to assess the ICA algorithm is slightly different – instead of the maximum amount of independence in the outputs, we are more interested in maximizing the classification accuracy – possibly as measured by the dependence of the outputs on some known inputs. Hence, future work in this application area evidently requires ICA methods suitable for accurately classifying the outputs against known inputs.

3.3.5 Computational Finance

ICA has significant appeal when it comes to understanding forces which may be driving the financial markets. In the financial markets, we may be only able to observe for example, stock price movements and transaction volumes, yet many observers argue for the existence of market forces that may be responsible for moving the market. Is it possible to uncover evidence of this using only the observed stock price and volume movements?

Small improvements in trading models can make a big difference in the overall profit or loss, and so it is of interest to determine whether or not ICA can be used to advantage when applied to financial data. A number of groups have published promising results obtained using ICA in this field.

One of the earliest applications was by Baram and Roth who applied ICA to the problem of estimating the pdf of a random data set [10, 103]. The method used was to maximize the output entropy of a neural network with sigmoidal units. In [103] the authors applied this methodology to diamond classification.

Back and Weigend applied the JADE algorithm to a set of daily closing prices from 28 of the largest stocks on the Tokyo Stock Exchange, from 1986 until 1989 [8]. The results showed that the extracted components can be classed as either large infrequent shocks which cause most of the major changes in the stock prices, or smaller, more frequent fluctuations which have little effect on the overall market price. In this case, due to the spatial independence assumption, ICA proved useful in being able to give insight into the physical nature of the data in terms of the actual shape and level of the price data. Other techniques such as PCA which are based on the variance of the data only are unsuited to this type of approach.

ICA has been applied to the task of removing noise from the observation process using a state space model for interbank foreign exchange rates by [90]. Financial bid-ask price data obtained from market makers is non-binding and so in a sense, is not the 'true' underlying price. In other words, if we wish to develop a trading model or prediction system, it is more appropriate to determine the actual price rather than a price which may have some built-in noise due to the market makers. Hence, the aim in this task is to find an underlying 'true' price from the noisy bid-ask price quotes (or observations).

Solving this type of estimation problem is often performed using a state space model where the state is estimated with a Kalman filter [108]. That is:

$$q(t) = p(t) + e(t) \qquad (3.55)$$

where $p(t)$ is the true price and $e(t)$ is assumed to be zero mean random noise. Moody and Wu proposed a series of models governing $p(t)$, including a random walk model, random trends model, and a fractional Brownian motion (FBM) model. The first two models were estimated using a Kalman filter, while the FBM model was estimated using an EM algorithm and wavelet decomposition.

However due to the non Gaussian signals (both noise and price series), the Kalman filter and EM algorithm are not able to perform effectively. Hence in each case, ICA was used to further isolate the noise from the estimated true price. Significantly, the estimated true price is much smoother than the quoted prices, and shows signs of behaviour that appear to relate to expected events in the financial markets, for example, slow periods, interrupted by waves of activity which may be due to news events.

ICA methods developed using the Bayesian Ying Yang (BYY) system has been applied to the analysis of financial data [120-123].

It is well known that the outputs from an ICA algorithm may be re-ordered with respect to the true input signals. Therefore, an important consideration can be how to decide the most appropriate ordering of the outputs. A criteria based on the L_i norm was used in [8], however this only considered each output separately. A different approach for output ordering in financial time series analysis was proposed in [33,74] where the outputs are ordered based on their joint contribution to the overall data reconstruction.

Other financial applications of ICA have been proposed: for example, Chan and Cha proposed an independent factor model [35].

ICA has been applied to a diverse range of other applications, including Nuclear Magnetic Resonance (NMR) spectroscopy [93], magnetoencephalography [106], rotating machine vibration analysis [124], statistical process control [65], separation of signals from seismic sources [112], document indexing using topic extraction [69], gene classification using microarray data [54, 80], multimedia signals [48], blind deconvolution and equalization [2, 42, 46, 47, 75, 92], image coding [11].

A brief list of resources has been included at the end of this chapter. We recommend that the interested reader should follow up on the available conference proceedings for details of more applications and further theory.

3.4 Open Problems in ICA Research

While there have been many advances made in the derivation of new ICA learning algorithms, there are still numerous open problems. Many of these are particularly relevant to the use of ICA in practical applications. We present a brief overview of some of these problems here which are relevant for applying ICA to practical problems.

ICA algorithms tend to be either on-line 'neural' learning type algorithms or batch type algorithms which use all the data at once. The problem with on-line algorithms is that they can be slow to learn and may also become trapped in local minima. In addition, if the statistical characteristics change over time, that is, the generative model for the signals is nonstationary, then such algorithms may have some trouble converging also.

The usual problem of choosing a suitable learning rate in on-line learning algorithms is also present in on-line ICA algorithms. There have been numerous methods proposed in the literature for linear models and various neural networks such as the multilayer perceptron, however further work is required to establish the validity of such approaches when used in ICA algorithms. It is not known if these algorithms will introduce special problems of their own.

The computational complexity of ICA algorithms remains an important issue. Some newer algorithms such as FastICA provide efficient implementation of ICA, however further work is required in this area.

Many of the assumptions made in deriving ICA algorithms, ie stationarity [95], non-Gaussian sources, linear additive mixtures, no knowledge of the input pdfs etc, are done so on the basis of academic convenience. However for practical problems, these assumptions may not be true. For example, it may be that there is some a priori knowledge of the source pdf. In such cases, it makes sense to be able to use this knowledge.

One would expect intelligent systems to be able to make full use of the information at hand by utilizing all available techniques. Further research is required for hybrid methods, combining ICA algorithms with other techniques to be able to yield better source separation results.

Nearly all of the work on ICA algorithms assumes that the number of sensors is greater than or equal to the number of sources. In practice, this may not be the case and so it is necessary to consider approaches which overcome this limitation.

There has been some work done on the problem of nonlinear mixing. In general, this is a very difficult area to work with. The range of nonlinear mixtures is infinite, and an inverse may not exist. Yet there are likely to be practical problems for which some forms of nonlinearity in the mixing process are present. For a review on this area, see [67].

Domain specific algorithms should be derived. Some work in this area has been done already. The extended InfoMax algorithm is known to be very effective at removing artifacts from biomedical source separation problems such as EEG recordings. In this case, the artifacts may be sub-Gaussian, while the sources are super-Gaussian. With domain knowledge such as this, it may be possible to introduce other algorithms which offer specific advantages to real world problems.

Similarly, with the wide range of ICA algorithms available, it is important to determine the suitability of the various algorithms for different tasks. Although there have been some comparative studies done, there is a need for wider ranging studies to be done so that the effectiveness of different ICA algorithms can be assessed on some benchmark problem sets. These problems should reflect real world situations, though they need not necessarily consist of only measured real world data. Note that we are not simply referring to the best performance as measured in terms of some demixing criterion. Most ICA algorithms perform similarly given sufficient time. The real issue to consider is their performance under similar conditions and for a similar amount of computational effort. Benchmarking can also be performed relative to other real world difficulties such as added noise, memory limitations, number of independent sources vs number of sensors and so on.

Noise, including noise from sensors can present special difficulties to ICA. In order to obtain better performance, there is a necessity to derive ICA algorithms which are not subject to the problem of poor convergence when sensor noise is present. If the noise is present in an additive form, then it may be possible to treat it as simply an additional input. However noise can also disrupt the learning process. For example, consider the on-line learning algorithm in Eq. (3.24). If there is noise v in the measurement process of y such that:

$$\frac{d\mathbf{W}}{dt} = \eta \mathbf{W}(t) G(\mathbf{y} + v) \tag{3.56}$$

it is clear that there will be error introduced into the estimation of W by noise of this type. Therefore it is important to consider at least the following aspects:
1. Minimization of the possible sources of noise in the ICA process - measurement, calculation and other sources.
2. Derivation of algorithms which are more robust in the possible presence of noise.

An open problem exists with mutual information algorithms in knowing how well they will perform for different types of non-Gaussian source distributions. The polynomial estimation procedures are reasonably accurate for pdfs close to Gaussian, however if they are far from Gaussian, it is considerably more difficult to estimate the densities with some degree of accuracy. A large amount of data may be required. Usually up to fourth order cumulants have been used in the published ICA algorithms. However to approximate a pdf far from Gaussian, a larger number of terms may be required [110]. There seems to be a need to derive ICA algorithms able to cope with a wider range of distributions, especially those far from Gaussian.

3.5 Summary

Independent component analysis is a powerful mathematical technique, embodied in the form of many algorithms that enables what was once considered an impossible problem, to be readily solved. Intelligent systems are aimed at dealing with the complexities of real world data and may be required to extract information from a jumble of sources. ICA is ideally suited for separating out data that is useful from data that is either not required or which can be considered as noise.

In this relatively recent field, much effort has been devoted to deriving new algorithms based on differing assumptions. As the field matures, the range of applications is increasing, as evidenced by recent conference publications. It is clear that ICA is certain to form a necessary part of the tools required to develop intelligent systems.

References

1. Amari S, Cichocki A and Yang H (1996) A new learning algorithm for blind signal separation, in *Advances in Neural Information Processing Systems* 8, Tesauro G, Touretzky DS and Leen TK (eds), MIT Press, Cambridge, MA: 757-763.
2. Amari S, Douglas SC, Cichocki A, and Yang HH (1997) Multichannel Blind Deconvolution and Equalization Using the Natural Gradient, *Proc IEEE Workshop on Signal Processing Advances in Wireless Communications*, Paris, France, April: 101-104.
3. Amari S and Douglas SC (1998) Why Natural Gradient, *Proc IEEE Intl Conf Acoustics, Speech & Signal Processing*, Seattle, WA, May, II: 1213-1216.
4. Amari S (1998) Natural gradient works efficiently in learning, *Neural Computation*, 10(4): 251-276.
5. Amari S (1999) ICA of temporally correlated signals - learning algorithm, *Proc 1st Intl Workshop Independent Component Analysis & Signal Separation (ICA99)*, Aussois, France: 13-18.
6. Back AD and Tsoi AC (1994) Blind deconvolution of signals using a complex recurrent network, *Neural Networks for Signal Processing 4*, IEEE Press, Piscataway, NJ: 565-574.
7. Back AD and Cichocki AC (1997) Blind source separation and deconvolution of fast sampled signals, *Proc Intl Conf Neural Information Processing (ICONIP-97)*, Dunedin, New Zealand, Kasabov N, Kozma R, Ko K, O'Shea R, Coghill G and Gedeon T (eds), Springer Verlag, Singapore, I: 637-641.
8. Back AD and Weigend AS (1998) A first application of independent component analysis to extracting structure from stock returns, *Intl J Neural Systems*, 8(4): 473-484.
9. Baldi P and Hornik K (1989) Neural Networks and Principal Component Analysis: Learning from Examples Without Local Minima, *Neural Networks*, 2: 53-58.
10. Baram Y and Roth Z (1995) Forecasting by density shaping using neural networks, *Proc Conf Computational Intelligence for Financial Engineering (CIFEr)*, IEEE Press, Piscataway, NJ: 57-71.
11. Barret M and Narozny M (2003) Application of ICA to Lossless Image Coding, *Proc 4th Intl Symp Independent Component Analysis & Blind Signal Separation (ICA2003)*, Nara, Japan, April 1-4: 855-859.
12. Barros AK (2002) Extracting the fetal heart rate variability using a frequency tracking algorithm, *Neurocomputing*, 49(1-4): 279-288.
13. Bayliss JD, Gualtieri JA and Cromp RF (1997) Analyzing hyperspectral data with independent component analysis, *Proc SPIE Applied Image & Pattern Recognition Workshop*, Selander JM (ed), Keszthely, Hungary, October 9-11.
14. Bell AJ and Sejnowski TJ (1995) An information-maximization approach to blind separation and blind deconvolution, *Neural Computation*, 7: 1129-1159.

15. Belouchrani A, Abed-Meraim K, Cardoso J-F and Moulines E (1997) A blind source separation technique using second-order statistics, *IEEE Trans Signal Processing*, 45: 434-444.
16. Belouchrani A, Amin MG (1998) Blind Source Separation Based on Time-Frequency Signal Representations, *IEEE Trans Signal Processing*, 46: 2888-2897.
17. Belouchrani A and Cardoso JF, (1994) A maximum likelihood source separation for discrete sources, *Proc EUSIPCO*, 2: 768-771.
18. Bingham E and Hyvärinen A (2000) A fast fixed-point algorithm for independent component analysis of complex-valued signals, *Intl J Neural Systems*, 10(1): 1-8.
19. Bishop C and Tipping M (1997) A Hierarchical Latent Variable Model for Data Visualisation, Aston University Technical Report NCRG/96/028.
20. Bishop CM, Svensén M and Williams CKI (1998) GTM: the generative topographic mapping, *Neural Computation*, 10(1): 215-234.
21. Bogner RE (1992) Blind Separation of Sources, Defence Research Agency, Malvern, England, Technical Report 4559, May.
22. Bourlard H and Kamp Y (1988) Auto-Association by Multilayer Perceptrons and Singular Value Decomposition, *Biological Cybernetics*, 59: 291-294.
23. Brandstein MS (1998) On the use of explicit speech modeling in microphone array applications, *Proc Intl Conf Acoustics Speech & Signal Processing*, Seattle, WA; 3613-3616.
24. Calhoun VD, Adali T, Pearlson GD and Pekar JJ (2001) Spatial and temporal independent component analysis of functional MRI data containing a pair of task-related waveforms, *Human Brain Mapping*, 13(1): 43-53.
25. Capdevielle V, Serviere Ch and Lacoume J (1995) Blind Separation of Wide-Band Sources in the Frequency Domain, *Proc IEEE Intl Conf Acoustics Speech & Signal Processing*, Detroit, Michigan, May 9-12, USA, 3: 2080-2083.
26. Cardoso J-F (1997) Infomax and maximum likelihood for source separation, *IEEE Letters on Signal Processing*, 4: 112-114.
27. Cardoso J-F and Laheld BH (1996) Equivariant adaptive source separation, *IEEE Trans Signal Processing*, 44(12): 3017-3030.
28. Cardoso J-F (1998) Blind signal separation: statistical principles, *Proc IEEE (Special issue on blind identification and estimation)*, Liu R-W and Tong L (eds), 86(10): 2009-2025.
29. Cardoso J-F (2001) The Three Easy Routes To Independent Component Analysis, Contrasts And Geometry, *Proc 3rd Intl Conf Independent Component Analysis and Blind Signal Separation - ICA 2001*, San Diego, CA.
30. Cardoso J-F, (1999) Higher-order contrasts for ICA, *Neural Computation*, 11(1): 157-192.
31. Cardoso J-F (1989) Source Separation using Higher Order Moments, *Proc Intl Conf Acoustics Speech & Signal Processing,* Glasgow, Scotland, May: 2109-2112.

32. Cardoso J-F and Souloumiac A (1993) Blind beamforming for non-Gaussian signals, *IEE Proc F*, 140(6): 771-774.
33. Cheung YM and Xu L (2001) Independent Component Ordering in ICA Time Series Analysis, *Neurocomputing*, 41: 145-152.
34. Cichocki A and Amari S (2003) *Adaptive Blind Signal and Image Processing: Learning Algorithms and Applications*, Wiley, New York.
35. Chan L-W and Cha S-W (2001) Selection of Independent Factor Model in Finance, *Proc 3rd Intl Conf Independent Component Analysis & Blind Signal Separation*, San Diego, CA, December 9-12.
36. Choi S and Cichocki A (1997) Blind signal deconvolution by spatio-temporal decorrelation and demixing, *Neural Networks for Signal Processing VII (Proc IEEE NNSP'97 Workshop*, Amelia Island Plantation, FA, September 24-26, Principe J, Giles L, Morgan N, Wilson E (eds): 426-435.
37. Choi C, Choi S and Kim S-R (2001) Speech Enhancement Using Sparse Code Shrinkage and Global Soft Decision, *Proc Intl Conf Independent Component Analysis & Blind Signal Separation*, 2001, San Diego, CA, December 9-12.
38. Cichocki A, Bogner RE, Moszczynski L and Pope K (1997) Modified Herault-Jutten algorithms for blind separation of sources, *Digital Signal Processing*, 7: 80-93.
39. Comon P, Jutten C and Herault J (1991) Blind separation of sources, Part II: Problem statement, *Signal Processing*, 24: 11-20.
40. Comon P (1994) Independent component analysis - a new concept?, *Signal Processing*, 36: 287-314.
41. Delfosse N and Loubaton P (1995) Adaptive blind separation of independent sources: a deflation approach, *Signal Processing*, 45: 59-83.
42. Douglas SC and Cichocki A (1997) Neural Networks for Blind Decorrelation of Signals, *IEEE Trans Signal Processing*, 45(11): 2829-2842.
43. Douglas SC (2002) Blind signal separation and blind deconvolution, *Handbook of Neural Network Signal Processing*, Hu YH and Hwang J-N (eds), CRC Press, Boca Raton FL, (Chapter 7).
44. Girolami M (1999) Hierarchic dichotomizing of polychotomous data - an ICA based data mining tool, *Proc Intl Workshop Independent Component Analysis & Signal Separation (ICA'99)*, Aussois, France: 197-202.
45. Girolami M, Cichocki A and Amari S (1998) A common neural network model for exploratory data analysis and independent component analysis, *IEEE Trans Neural Networks*, 9(6): 1495-1501.
46. Gorokhov A and Loubaton P (1996) Second order blind identification of convolutive mixtures with temporally correlated sources: A subspace based approach, *Proc Signal Processing VIII, Theories and Applications*, Triest, Italy, Elsevier, North Holland: 2093-2096.
47. Gorokhov A and Loubaton P (1997) Subspace based techniques for second order blind separation of convolutive mixtures with temporally correlated sources, *IEEE Trans Circuits & Systems*, 44: 813-820.
48. Hansen L-K, Larsen J and Kolenda T (2000) On Independent Component Analysis for Multimedia Signals, in Guan L, Kung SY and Larsen J (eds.) *Multimedia Image and Video Processing*, CRC Press, Boca Raton, FL.

49. Hansen L-K and Kolenda T (2002) Independent Component Analysis, Technical University of Denmark, Lecture Notes, Advanced DSP (PhD Course).
50. He T, Clifford G and Tarassenko L (2002) Application of ICA in removing artifacts from the ECG, *Neural Computing & Applications*, 12: (in press).
51. Herault J, Jutten C and Ans B (1985) Détection de grandeurs primitives dans un message composite par une architecture de calcul neuromimétique en apprentissage non supervise, *Proc GRETSI'85*, Nice, France: 1017-1022 *(in French)*.
52. Herault J and Jutten C (1986) Space or time adaptive signal processing by neural network models, Neural Networks for Computing, Denker JS (ed), *Proc AIP Conf*, Snowbird, UT, American Institute of Physics, New York, 151: 206-211.
53. Herault J and Jutten C (1991) Blind separation of sources, Part I: An adaptive algorithm based on neuromimetic architecture, *Signal Processing*, 24: 1-10.
54. Hori G, Inoue M, Nishimura S and Nakahara H (2001) Blind Gene Classification Based On ICA Of Microarray Data, *Proc Intl Conf Independent Component Analysis & Blind Signal Separation*, San Diego, CA, December 9-12.
55. Hotelling H (1933) Analysis of a complex of statistical variables into principal components, *J Educational Psychology*, 24: 417-441, 498-520.
56. Hyvärinen A and Oja E (1997) A fast fixed-point algorithm for independent component analysis, *Neural Computation*, 9(7): 1483-1492.
57. Hyvärinen A (1999) The fixed-point algorithm and maximum likelihood estimation for independent component analysis, *Neural Processing Letters*, 10(1): 1-5.
58. Hyvärinen A (1998) New approximations of differential entropy for independent component analysis and projection pursuit, *Advances in Neural Information Processing Systems*, MIT Press, Cambridge, MA, 10: 273-279.
59. Hyvärinen A (1999) Fast and robust fixed-point algorithms for independent component analysis, *IEEE Trans Neural Networks*, 10(3): 626-634.
60. Hyvärinen A (1996) Simple one-unit algorithms for blind source separation and blind deconvolution, *Proc 1996 Intl Conf Neural Information Processing (ICONIP-96)*, Hong Kong, September 24-17, 1996, Springer Verlag, Berlin, 2; 1201-1206.
61. Hyvärinen A (2001) Blind Source Separation by Nonstationarity of Variance: A Cumulant-Based Approach, *IEEE Trans Neural Networks*, 12(6): 1471-1474.
62. Hyvärinen A and Oja E (2000) Independent Component Analysis: Algorithms and Applications, *Neural Networks*, 13(4-5): 411-430.
63. Jutten Ch and Taleb A (2000) Source separation: From dusk till dawn, *Proc ICA 2000*, Helsinki, Finland, June: 15-26.
64. Jutten C, Nguyen THL, Dijkstra E, Vittoz E and Caelen J (1991) Blind separation of sources: an algorithm for separation of convolutive mixtures, *Proc Intl Workshop Higher Order Statistics*, Chamrousse, France, July: 275-278.

65. Kano M, Tanaka S, Ohno H, Hasebe S and Hashimoto I (2002) The Use of Independent Component Analysis for Multivariate Statistical Process Control, *Proc Intl Symp Advanced Control of Industrial Processes (AdCONIP'02)*, Kumamoto, Japan, June 10-11: 423-428.
66. Karhunen J (1996) Neural approaches to independent component analysis and source separation, *Proc 4th European Symp Artificial Neural Networks - ESANN'96*, Bruges, Belgium, April 24-26: 249-266.
67. Karhunen, J (2001) Nonlinear Independent Component Analysis in R. Everson and S. Roberts (eds.), *ICA: Principles and Practice*, Cambridge University Press, Cambridge, UK:113-134.
68. Kawamoto M, Barros AK, Mansour A, Matsuoka K and Ohnishi N (1998) Blind separation for convolutive mixtures of non-stationary signals, *Proc 5th Intl Conf Neural Information Processing (ICONIP'98)*, Kitakyushu, Japan, October 21-23: 743-746.
69. Kim Y-H and Zhang B-T (2001) Document Indexing Using Independent Topic Extraction, *Proc Intl Conf Independent Component Analysis & Blind Signal Separation*, San Diego, CA, December 9-12: 557-562
70. Kiviluoto K and Oja E (1998) Independent Component Analysis for Parallel Financial Time Series, *Proc 5th Intl Conf Neural Information Processing (ICONIP'98)*, Kitakyushu, Japan, October 21-23: 895-898.
71. Kullback S and Leibler RA (19510 On information and sufficiency, *Annals Mathematical Statistics*, 22; 76-86, 1951.
72. Kung SY and Diamantaras KI (1990) A neural network learning algorithm for adaptive principal component extraction (APEX), *Proc IEEE Intl Conf Acoustics Speech & Signal Processing*, Albuquerque, NM, April 3-6: 861-864.
73. Lacoume JL and Ruiz P (1988) Sources identification: a solution based on cumulants, *Proc 4th Acoustics Speech & Signal Processing Workshop on Spectral Estimation & Modeling*, Minneapolis, August: 199-203.
74. Lai ZB, Cheung YM and Xu L (1999) *Independent Component Ordering in ICA Analysis of Financial Data, Computational Finance*, Abu-Mostafa YS, LeBaron B, Lo AW and Weigend AS (eds), MIT Press, Cambridge, MA, Chapter 14: 201-212.
75. Lambert R (1996), Multichannel blind deconvolution: FIR matrix algebra and separation of multipath mixtures. PhD Thesis, University of Southern California, Department of Electrical Engineering.
76. Lee T-W, Girolami M and Sejnowski TJ (1999) Independent component analysis using an extended infomax algorithm for mixed sub-gaussian and super-gaussian sources, *Neural Computation*, 11(2): 417-441.
77. Lee T-W, Lewicki MS and Sejnowski TJ (1998) Unsupervised Classification with Non-Gaussian Mixture Models using ICA, *Advances in Neural Information Processing Systems 11*, MIT Press, Cambridge, MA: 508-514.
78. Lee T-W, Bell A, Orglmeister R and Sejnowski TJ (1997) Blind source separation of real world signals, *Proc Intl Conf Neural Networks*, Houston, TX, June 9-12.

79. Li S and Sejnowski TJ (1995) Adaptive Separation of Mixed Broad-Band Sound Sources with Delays by a Beamforming Hérault-Jutten Network, *IEEE J Oceanic Engineering*, 20(1): 73-79.
80. Liao X and Carin L (2002) Constrained Independent Component Analysis of DNA Microarray *Signals, Proc IEEE Workshop on Genomic Signal Processing & Statistics (GENSIPS)*, Raleigh, NC, October 12-13.
81. Lindgren U and Broman H (1998) Source separation using a criterion based on second-order statistics, *IEEE Trans Signal Processing*, 46(7): 1837-1850.
82. Loéve M (1963) *Probability Theory*, Van Nostrand, New York.
83. Makeig S, Bell AJ, Jung T-P and Sejnowski TJ (1996) Independent component analysis of electroencephalographic data, *Advances in Neural Information Processing Systems 8*, MIT Press, Cambridge, MA: 145-151.
84. Makeig, S, Jung TP, Bell AJ, Ghahremani D, Sejnowski TJ (1997) Blind separation of auditory event-related brain responses into independent components, Proceedings of the National Academy of Sciences, 94: 10979-10984.
85. Mansour A, Barros AK and Ohnishi N (2000) Blind separation of sources: Methods, assumptions and applications, *IEICE Trans Fundamentals of Electronics, Communications & Computer Sciences*, E83-A(8): 1498-1512.
86. Mansour A (1999) The blind separation of non stationary signals by only using the second order statistics, *Proc 5th Intl Symp Signal Processing & Its Applications (ISSPA'99)*, Brisbane, Australia, August 22-25: 235-238.
87. Marsman H (1995) A Neural Net Approach to the Source Separation Problem, University of Twente, Netherlands, Masters Thesis.
88. Matsuoka K, Ohya M, and Kawamoto M (1995) A neural net for blind separation of nonstationary signals, *Neural Networks*, 8(3): 411-419.
89. Molgedey L and Schuster G (1994) Separation of a mixture of independent signals using time delayed correlations, *Physical Review Letters*, 72(23): 3634-3637.
90. Moody J and Wu L (1997) What is the True Price? - State Space Models for High Frequency FX Rates, in *Decision Technologies for Financial Engineering*, Abu-Mostafa Y, Refenes AN and Weigend AS (eds), World Scientific, London.
91. Muller PPKRR and Ziehe A (1999) JADE TD Combining higher order statistics and temporal information for blind source separation (with noise), *Proc 1st Intl Workshop Independent Component Analysis & Signal Separation (ICA99)*, Aussois, France, January 11-15: 87- 92.
92. Nguyen TH-L and Jutten C (1995) Blind source separation for convolutive mixtures, *Signal Processing*, 45(2): 209-229.
93. Nuzillard D and Nuzillard J-M (1999) Blind Source Separation Applied to Non-Orthogonal Signals, *Proc 1st Intl Workshop Independent Component Analysis & Signal Separation (ICA99)*, Aussois, France, January 11-15: 25-30.
94. Oja E (1989) Neural networks, principal components and subspaces, *Intl J Neural Systems*, 1: 61-68.

95. Parra LC, Spence C and Sajda P (2000) Higher-Order Statistical Properties Arising from the Non-Stationarity of Natural Signals, *Advances in Neural Information Processing Systems 13 (Papers from Neural Information Processing Systems - NIPS'2000*, Denver, CO, Leen TK, Dietterich TG and Tresp V (eds), MIT Press, Cambridge, MA, 2001: 786-792.
96. Pearlmutter BA and Parra LC (1997) Maximum likelihood blind source separation: A context-sensitive generalization of ICA, in *Advances in Neural Information Processing Systems 9 (NIPS*96)*, Mozer MC, Jordan MI and Petsche T (eds), MIT Press, Cambridge, MA: 613-619.
97. Pham DT, Garat P and Jutten C (1992) Separation of a Mixture of Independent Sources Through a Maximum Likelihood Approach, in *Signal Processing VI: Theories and Applications*, Vandevalle J, Boite R, Moonen M and Oosterlink A (eds), Elsevier, North Holland: 771-774.
98. Pham DT and Cardoso J-F (2000) Blind separation of instantaneous mixtures of non stationary sources, *Proc ICA 2000 Conference*, Helsinki, Finland, June 19-22, Pajunen P and Karhunen J (eds): 187-192.
99. Pham DT, Garat P and. Jutten C (1992) Separation of a mixture of independent sources through a maximum likelihood approach, *Signal Processing VI: Theories and Applications*, Vandevalle J, Boite R, Moonen M and Oosterlink A (eds), Elsevier, North Holland: 771-774.
100. Pham DT and Garat P (1997) Blind separation of mixture of independent sources through a quasi-maximum likelihood approach, *IEEE Trans Signal Processing*, 45(7): 1712-1725.
101. Pope KJ and Bogner RE (1996) Blind Signal Separation. II: Linear, Convolutive Combinations, *Digital Signal Processing*, 6: 17-28.
102. Ripley B (1996) *Pattern Recognition and Neural Networks*, Cambridge University Press, UK.
103. Roth Z and Baram Y (1996) Multi-dimensional density shaping by Sigmoids, *IEEE Trans Neural Networks*, 7(5): 1291-1298.
104. Rutkowski T, Cichocki A and Barros AK (2001) Speech Enhancement From Interfering Sounds Using CASA Techniques And Blind Source Separation, *Proc Intl Conf Independent Component Analysis & Blind Signal Separation (ICA 2001)*, San Diego, CA, December 9-12: 728-733.
105. Sanger TD (1990) Optimal unsupervised learning in a single layer linear feed forward neural network, *Neural Networks*, 2: 459-473.
106. Särelä J, Valpola H, Vigário R and Oja E (2001) Dynamical Factor Analysis Of Rhythmic Magnetoencephalographic Activity, *Proc Intl Conf Independent Component Analysis & Blind Signal Separation*, San Diego, CA, December 9-12.
107. Schobben DWE, Torkkola K, Smaragdis P, (1999), Evaluation of Blind Signal Separation Methods, *Proc Intl Workshop Independent Component Analysis & Blind Signal Separation*, Aussois, France, January 11-15: 261-266.
108. Söderström T and Stoica P (1989) *System Identification*, Prentice-Hall, London.

109. Stewart GW (1973) *Introduction to matrix computations*, Academic Press, New York.
110. Stuart A and Ord K (1994) *Kendall's Advanced Theory of Statistics: Volume 1, Distribution Theory (6th ed)*, Edward Arnold, New York.
111. Stutz, J and Cheeseman, P (1994), AutoClass - a Bayesian Approach to classification, in *Maximum Entropy and Bayesian Methods*, Skilling J and Sibisi S (eds), Kluwer Academic Publishers, Dordrecht, The Netherlands.
112. Thirion N, Mars J and Boelle JL (1996) Separation of seismic signals: A new concept based on a blind algorithm, *Proc Signal Processing VIII: Theories and Applications*, Triest, Italy, September, Elsevier, North Holland: 85-88.
113. Torkkola K (1996) Blind Separation of Convolved Sources Based on Information Maximization, *IEEE Workshop Neural Networks for Signal Processing*, Kyoto, Japan, September 4-6: 423-432.
114. Torkkola K (1999) Blind separation of audio signals: Are we there yet?, *Proc 1st Intl Workshop Independent Component Analysis & Signal Separation (ICA99)*, Aussois, France, January 11-15: 13-18.
115. Vigário R, Jousmäki V, Hämäläinen M, Hari R and Oja E (1998) Independent component analysis for identification of artifacts in magnetoencephalographic recordings, *Advances in Neural Information Processing Systems 10*, Jordan MI, Kearns MJ and Solla SA (eds), MIT Press, Cambridge, MA: 229-235.
116. Visser E, Lee T-W and Otsuka M (2001) Speech Enhancement In A Noisy Car Environment, *Proc Intl Conf Independent Component Analysis & Blind Signal Separation*, San Diego, CA, December 9-12: 272-277.
117. Welling M and Weber M (2001) A Constrained EM Algorithm for Independent Component Analysis, *Neural Computation*, 13(3): 677-689.
118. Weingessel A and Hornik K (1997) SVD algorithms: APEX-like versus Subspace Methods, *Neural Processing Letters*, 5:177-184.
119. Widrow B and Stearns SD (1985) *Adaptive Signal Processing*, Prentice-Hall, Englewood Cliffs, NJ.
120. Xu L (2002) Ying-Yang Learning, in *The Handbook of Brain Theory and Neural Networks, 2nd edition*, Arbib MA (ed), MIT Press, Cambridge, MA: 1231-1237.
121. Xu L (2001) BYY Harmony Learning, Independent State Space and Generalized APT Financial Analyses, *IEEE Trans Neural Networks*, 12(4): 822-849.
122. Xu L (2000) Temporal BYY Learning for State Space Approach, Hidden Markov Model and Blind Source Separation, *IEEE Trans Signal Processing*, 48(7): 2132-2144.
123. Xu L (1998) Bayesian Ying-Yang Dimension Reduction and Determination, *J Computational Intelligence in Finance*, Special issue on Complexity and Dimensionality Reduction in Finance, 6(5):6-18.
124. Ypma A and Pajunen P (1999) Rotating machine vibration analysis with second-order independent component analysis, *Proc 1st Intl Workshop*

Independent Component Analysis & Signal Separation (ICA99), Aussois, France, January 11-15: 37-42.
125. Ziehe A and Müller K-R (1998) TDSEP - an efficient algorithm for blind separation using time structure, *Proc Intl Conf Artificial Neural Networks (ICANN'98)*, Skövde, Sweden, September 2-4: 675-680.
126. Ziehe A, Nolte G, Sander T, Müller K-R, and Curio G (2001) A comparison of ICA-based artifact reduction methods for MEG. in *Recent Advances in Biomagnetism: Proc 12th Intl Conf Biomagnetism*, Espoo, Finland, Jukka Nenonen (ed), Helsinki University of Technology: 895-898.

Appendix - Selected ICA Resources

Academic Web Sites

Mathematical Neuroscience, RIKEN – www.mns.brain.riken.go.jp
Brain Signal Processing, RIKEN – www.bsp.brain.riken.go.jp
CIS, HUT – www.cis.hut.fi
Lee Lab – inc2.ucsd.edu/~tewon
ICA at CNL – www.cnl.salk.edu/~tewon/ica_cnl.html
Girolami Research Group – www.cis.paisley.ac.uk/giro-ci0/group.html
ICA Central – www.tsi.enst.fr/icacentral
ICA-99 Benchmark Data and Code – www2.ele.tue.nl/ica99

Conferences

ICA 2003 – ica2003.jp
ICA 2001 – www.ica2001.org
ICA 2000 – www.cis.hut.fi/ica2000/cfpfinal.html
NIPS99 Workshop on ICA – www.cnl.salk.edu/~tewon/ICA/ *(nips_workshop.html)*
NIPS96 Workshop on Blind Signal Processing – andrewback.com/nips96ws

Source Code and Toolkits

ICA:DTU Toolbox (Matlab) – mole.imm.dtu.dk/toolbox
ICALAB Toolboxes (Matlab) – www.bsp.brain.riken.go.jp/ICALAB
The FastICA Package for Matlab – cis.hut.fi/projects/ica/fastica
EEGLAB Toolbox for Matlab – sccn.ucsd.edu/~scott/ica.html
Infomax Algorithm (Matlab) – ftp.cnl.salk.edu/pub/tony/sep96.public
Extended Infomax Algorithm (Matlab) – inc2.ucsd.edu/~tewon/ica_cnl.html

Commercial Software

Windale Technologies ICA/X (ActiveX/COM) – windale.com

4 Regulatory Applications of Artificial Intelligence

Howard Copland[1]

1   Program Review Division, Health Insurance Commission, 134 Reid St, Tuggeranong ACT 2900, Australia, howard.copland@hic.gov.au

4.1 Introduction

Simplicity and effectiveness guide commercial and government applications of Artificial Intelligence, with a strong emphasis on the simplicity. Basically, if a system is too complicated or demands too high a level of maintenance, then it will probably fail in most commercial or government environments. Furthermore, if a moderate level of understanding of the solution 'engine' is required in order to interpret the results, then it will also probably fail in most commercial or government environments.

The primary reason for all of this is that *how* a problem is solved is of little concern to any beyond the typically few people responsible for generating a solution. The ideal solution is implemented in such a way that it meets the business requirements, is achieved at minimum cost, is achieved in minimum time and requires negligible maintenance. It's also important that the solution can be supported in an environment that can in practice be one where much of the IT support has been outsourced, with attendant constraints on available hardware and software.

The Australian Health Insurance Commission – HIC – is responsible for, among other things, the annual payment of hundreds of millions of transactions to the Australian health care sector and the Australian public. The two largest components of this activity are the programs for the Medicare (the national health cover scheme) and the Pharmaceutical Benefits Scheme – PBS (through which selected prescription drugs are subsidized by the Federal Government). Associated with the payment activity is a regulatory activity, undertaken by the Program Review Division of the HIC. This Division employs a number of AI-based techniques for compliance purposes, primarily for anomaly detection and techniques for Expert System formulation.

4.2 Solution Spaces, Data and Mining

Data Mining (DM) has become an accepted part of business analysis with the advent of high volume electronic transactions, particularly by the finance and marketing sectors. Other high volume data sources where Data Mining has been used include areas like genomics and health informatics. Given the decade of commercial effort devoted to the area, it is surprising how inadequate much of it is, frequently leading to fairly spectacular failures of initiatives like advanced functionality for data warehouses and online transactional analysis systems.

The actual stages involved in DM are quite straightforward:
- data validation;
- statistical analysis;
- feature generation;
- feature selection;
- technique selection;
- other algorithms; and,
- interpretation.

With real-time systems, and depending on the data source, there may be an additional filtering or pre-processing step during acquisition as well. Despite being presented in a seemingly logical order above, the more usual place to start is with a proposed solution. Solution spaces are basic maps of where to go to get a solution to any given problem. They can be a little domain dependent, so the solution map for health informatics (which is the basis of the work described here) isn't likely to be entirely the same as the one required for, say, deep-sea submersibles engineering. Given the time constraints on most systems development, avoiding trial and error approaches to generating solutions is a necessity.

There is, alas, no rule-of-thumb as to which kind of solutions map to which kind of problem, and with some problems it doesn't make too much difference. Knowing exactly where to start comes only with experience in a domain of problems. However, some things are immediately obvious. If precisely how a solution works is part of the business requirements, then Artificial Neural Networks (ANNs) are out of the running, since while they're not the 'black boxes' they're often claimed to be, they're not that amenable to having their internal processes formally specified. Formal statistical models and Expert Systems will be the order of the day. If the data quality is poor, then expert systems will be much harder to achieve and, at the very least, statistical methods for interpolation will likely prove useful no matter what the final solution is. If the problem is highly dynamic or non-linear, then statistical techniques can rapidly escalate in complexity and ANNs are often the best approach. Also, it is not uncommon to combine different techniques in achieving a workable solution.

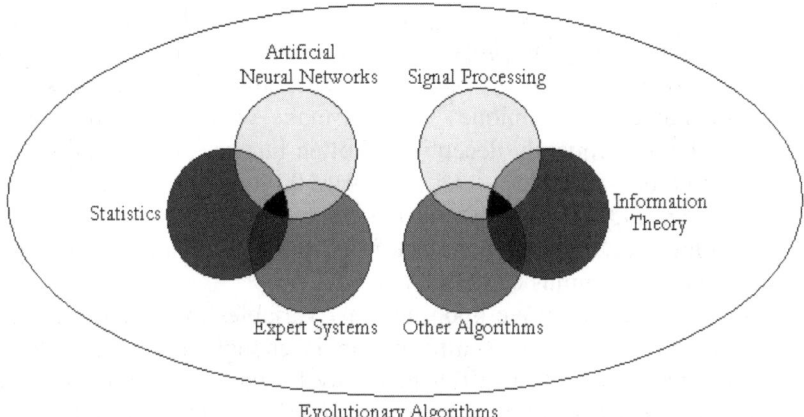

Fig. 4.1. A Solution Space for Health Informatics Problems

Note that 'Artificial Intelligence' (AI) is a much-used umbrella term for many different kinds of techniques, some genuinely from the field of AI research, such as Artificial Neural Networks, Evolutionary Algorithms and Expert Systems, and some from other fields of numerical analysis such as information theory and signal processing. Often, a lot of work is required to get the data to the point where AI techniques *can* be used.

In that regard, from the outset of any systems development, there are a few things to watch for – common presumptions that have the power to undermine an entire Data Mining or Artificial Intelligence project. First, there is a common presumption that there is actually useful data to be mined. Second, there is the common presumption that the data at hand is both accurate and complete. Third, there is the common presumption that all the data must be used when Data Mining. Typically, the first is true but the second and third are horribly wrong.

First and foremost, in deciding what techniques are most applicable, it is critical to understand not only the data in terms of quality and dynamism, but also its nature. Data quality issues aside, it is important to know what the limitations of the data are. Variables can possess a range of qualitative and quantitative properties and these properties determine the applicability and constraints on what (and how) techniques can be used. Many organizations have invested considerable resources in trying to create solutions where it simply isn't possible to devise one working with the available data.

Nominal variables are purely qualitative; they are classification variables and contain no quantitative information whatsoever. An example might be an industry classification scheme: forestry, tourism, automotive manufacturing, commercial fishing, and so on. It is possible to say that any individual entity belongs to *this* or *that*, but it isn't possible to say, on the basis of *this* or *that*, that *this* is bigger or smaller than *that*. This constrains the type of numerical techniques that can make sensible use of nominal variables. This is frequently deceptive as often numerical designations are given to nominal variables – 01 forestry, 02 tourism, 03 automotive manufacturing, and so on. It would not be sensible to sum or transform such variables – 'automotive manufacturing' minus 'tourism' doesn't equal 'forestry' even if 03 minus 02 does equal 01.

Ordinal variables improve upon nominal variables by containing some quantitative information. This information is enough to say that *this* is bigger than *that*, but it is insufficient to say by how much *this* is bigger than *that*. An example of this kind of variable is celebrity fame. It is possible to say that Arnold Schwarzenegger is *more* famous than Johnny Depp, but it isn't possible to say *how much* more.

Interval variables contain still more quantitative information. It is possible to ascertain that *this* is bigger than *that*, and the difference between *this* and *that* is *this big*. The classic example of this kind of variable is the Celsius temperature scale. It's possible to say that 10°C is less than 30°C and that the difference between 10°C and 30°C is the same as the difference between 70°C and 90°C. However, it isn't possible to say that -10°C is twice as cold as 10°C.

Ratio variables make up for the lack of an absolute reference point in interval variables by incorporating an identifiable zero point. Similar to the above, the Kelvin temperature scale is a ratio variable-based scale. Here, it is possible to say that 30 K is higher than 10 K and that it is 3 times higher.

Most statistical and AI techniques, being essentially numerical in nature, tend to assume they're working with interval and ratio variables. It is possible to use other variables, but it has to be done very carefully and may require restructuring of the variable into a form more amenable to numeric manipulation. Again, many organizations have invested considerable time and resources implementing poor quality systems by not discriminating between the properties of different variables.

With the question of what data is available resolved, the next question is how good is it? Poor data quality can undermine even the best techniques. A survey of each field in the data being examined to establish the range (and lack) of entries is critical. Often, databases have been around for over a decade and the meaning of the fields has drifted over time, or fields intended for one use have ended up being used for something rather different. Relying on dated, formal documentation describing the content of a database is likely to result in problems.

If some analysis or system development is being undertaken based on a sample of data, then ensure the sample has the same distribution properties as the whole of the data set. Don't simply sample randomly or select the first x observations in the data set.

Typically, raw data is summarized in some way or another before presentation to an analysis system. Statistics are the usual means of generating feature sets for further analysis, but techniques from data fusion, signal processing and information theory can often prove useful. One key consideration is the number of transforms applied to the raw data to generate any given feature. Ideally, the number of transforms should be minimized; too many transforms and the feature risks becoming rather 'ornate' – firstly, difficult to understand and, secondly, much more susceptible to noise in the data. Where possible, try to avoid opinion-based selection of features; rather, let the data speak for itself.

With many data sets, interpolation of missing fields is often problematical, particularly if they're a classification variable of some kind, such as a pharmaceutical, industry code or similar. Deducing, for example, from incomplete data whether an antibiotic or an anti-malarial was dispensed is well beyond the capability of any interpolation algorithm. However, throwing away poor quality data isn't always possible when a sizeable component of the data is doubtful or ambiguous to some degree. This means employing techniques able to work with a high degree of ambiguity and, if possible, reducing reliance on the poorer quality fields.

One means of rapidly identifying an optimal feature set that takes into account many of these issues is through selection via an Evolutionary Algorithm (EA). The evolutionary task is, from a large set of input features, to evolve a 'system' best able to perform the task at hand (be it prediction, classification, filtering or some other), as described by one or more members from a large set of desired output features. Next, weight the evolutionary process to select as few input features as possible consistent with maintaining good performance of the 'system', then select those output features that give best relative performance. Other constraints can also be readily incorporated. Often, for assorted business reasons, the inclusion of particular features is required whether or not they contribute to the solution. Use of an Evolutionary Algorithm allows such inclusion in the least disruptive way.

The advantages of using an Evolutionary Algorithm are:
- no reliance on the opinion of experts but rather on the available data;

- exclusion of poor quality fields not contributing successfully to a solution;
- optimal or near-optimal selection of features based on multiple criteria;
- rapid reduction of large sets of input parameters to those most relevant; and,
- useful in conjunction with nearly any other DM technique.

Simply stated, successful Data Mining requires that the data be known intimately. Successful DM is also about throwing redundant or irrelevant data away so it doesn't have to be analyzed – a key and often forgotten part of Data Mining is data winnowing. Evolutionary Algorithms are an enormous boon here. Let the approach to problems be data-driven and not model-driven. Either way, there is no guarantee of a successful outcome, but insisting on a particular model at the outset will reduce the probability of success.

Also, there is no gain in using DM techniques for small data sets. At the very least, data sets should contain at least a dozen fields and be well into the GigaByte range before it makes sense to invest the time and resources required for Data Mining. Data Mining is done for a reason, and it's best if the reasons are good ones if organizational support for the activity is required.

4.3 Artificial Intelligence in Context

Intelligent tools working with data do not exist in a vacuum. They exist to support a better business outcome, and are usually the most minor component of that outcome. Having the most sophisticated AI system for identifying fraud and abuse is immaterial if the system is not used, nor is it efficient if it isn't integrated into the surrounding payments infrastructure.

Fig.4.2 outlines the basic structure of one possible approach to an organization-wide automated risk analysis system for an organization dealing with high volumes of transactions.

Transaction data arrives and is evaluated by the *Validation Module*. This simply checks field content against a series of templates as to what is valid for any given field of the transaction. Erroneous transactions can be rejected or diverted for further attention at this stage. By insisting on data quality in the first step, subsequent analyses and reporting are vastly simplified.

The *Synthesis Module* serves to create causal chains with respect to the transactions, interpolate as possible or where necessary, as well as incorporating any other data from other available databases, such as registration details or names and addresses. More general transaction rules can be applied at this point, such as examining whether there are duplicate transactions or whether the regulatory framework is violated by any transaction. In effect, this module 'value adds' to the transaction data.

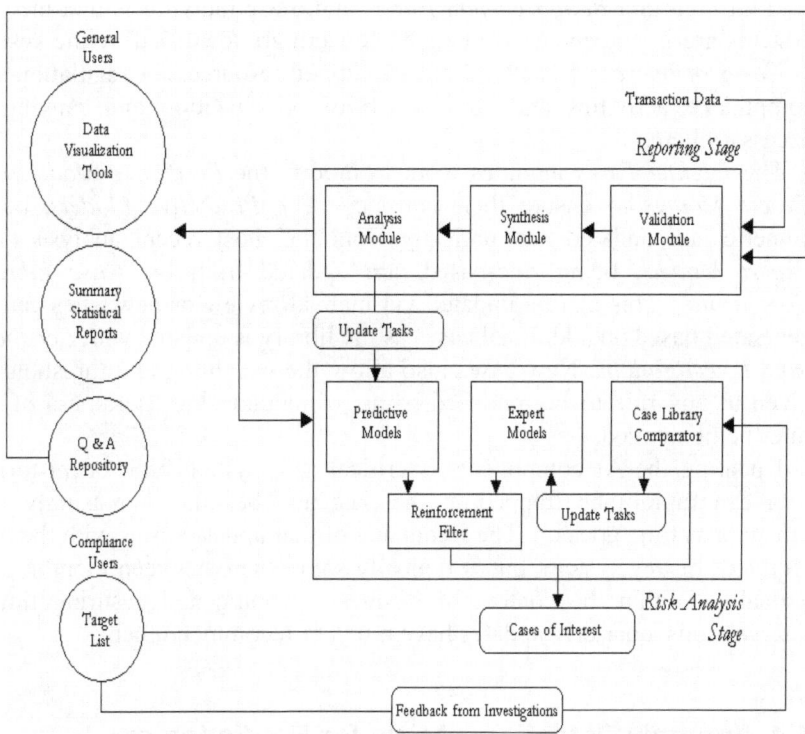

Fig. 4.2. A Global System for Data Analysis and Compliance Purposes

The *Analysis Module* is where basic statistics are performed. At this point summary reports can be produced and the analysis viewed, typically using a variety of *Data Visualization Tools* and *Summary Statistical Reports* - essentially a suite of standardized reports tailored to different business areas. There is a *Question & Answer Repository* that allows rapid interrogation of the system with respect to specific fields, analysis or individuals, and is cumulative over time.

The analysis also flows down to the 'risk control' modules, which is the domain of the AI components of the system. *Predictive Models* identify individuals unlike their peers and the reasons why they are unlike them. *Expert Models* incorporate rules-based measures for identifying individuals of interest and tend to examine the transactional data stream at a more detailed level than the *Predictive Models*. The *Case Library Comparator* maintains a history of individuals with known regulatory problems and examines whether or not others are behaving similarly.

For purposes of prioritization, *Predictive Models* and *Expert Models* feed through to a *Reinforcement Filter*, and those individuals that the two systems are in agreement as being of concern are identified by the system as *Cases of Interest*. Usually, there are limited resources for regulation and compliance activities and some form of prioritization and ranking is necessary here.

The *Update Tasks* modules work to modify the *Predictive Models* and *Expert Models* to ensure their currency. The *Predictive Models*, being numeric, are updated automatically from the most recent analysis. The *Expert Models*, being rule-based, are updated from the *Case Library Comparator*. This can be updated via manual review or new rules can be generated based on additional cases as the library is updated with *Feedback from Investigations*. New cases also allow the weighting and thresholding given to any rule to be modified so as to maintain the usefulness of the rules being applied.

Updating the AI components is critical as a system built on historical data can rapidly be trapped in the past and become increasingly less effective as time goes on. The frequency of that update varies with the data required for any system, but is typically somewhere between monthly and annually. Within the context of business planning and ensuring timely interventions, quarterly updates have much to recommend them.

4.4 Anomaly Detection: ANNs for Prediction or Classification?

One means of identifying anomalies is by using BackPropagation ANNs as heuristic models for feature prediction. A suite of ANNs, each effectively an expert on some particular feature of the system being modelled, can rapidly identify anomalous individuals. In general, no matter the problem, a minimum of at least five ANNs is required for adequate problem resolution. The underlying logic is that there is a good overlap between individuals who are found to be anomalous and those who have inappropriate behaviour of interest. Another approach, also using BackProp' ANNs, is to classify profiles of individuals and determine an 'inappropriateness' rating of some kind. For reasons that will be explored, prediction is to be preferred over classification.

4.4.1 Training to *Classify* on *Sparse* Data Sets: the Problem is the Solution

All BackProp' Artificial Neural Networks start life in some random state described by the small, initial values assigned to their weights. Presentation of the relationship to be learned drives the ANN to change its weights until a solution is generated. The solution, however, can vary widely depending on how dense or sparse the data is; the more sparsely distributed the data, the wider the variety of possible solutions. The problem illustrated below is to classify the input points as positive (+1), neutral (0) or negative (-1), based on only a handful of points. The data set is:

Table 4.1. Example Classification Data

		\multicolumn{5}{c}{Output}		
	2	-1	1	1	1	1
Input 2	1	1	-1	0	0	undefined
	0	1	0	-1	0	undefined
	-2	1	0	0	-1	undefined
	-2	1	undefined	undefined	undefined	-1
		-2	-1	0	1	2
		\multicolumn{5}{c}{Input 1}	

Each solution in Fig.4.3 is 'correct' but *which* overall solution – meaning classification of the space between the points – is most representative of the broader data set? Short of exhaustive analysis of the data in its own right, there is no way to tell. This is the essential difficulty with the classification approach.

With such a system, a sample of cases or profiles must first be identified and rated to varying degrees of inappropriateness or anomaly so that an ANN has something to learn from. However, given that the reason for devising such a system in the first place is that the volume of data to be evaluated is enormous, this sample is itself very large. In a system intended to evaluate 20,000 individuals, for example, a sample size of at least 1,000 would be required. This initial rating must be undertaken by domain experts – that is, people. However, people vary and a number of different perspectives – say, three to five at minimum – would be required. As any given person also varies from time to time, it is wise to duplicate 20% of the data profiles being evaluated to ensure consistency of evaluation. In addition, the domain experts must be provided with only the data that the system will see and which is de-identified to prevent them using external knowledge of individuals they may have. Performed with any rigour, this process is time consuming in the extreme. As a consequence, it is rarely done with any rigour. Furthermore, updating the ANN requires the entire process be undertaken again as the ANN is unable to update itself directly from the input data.

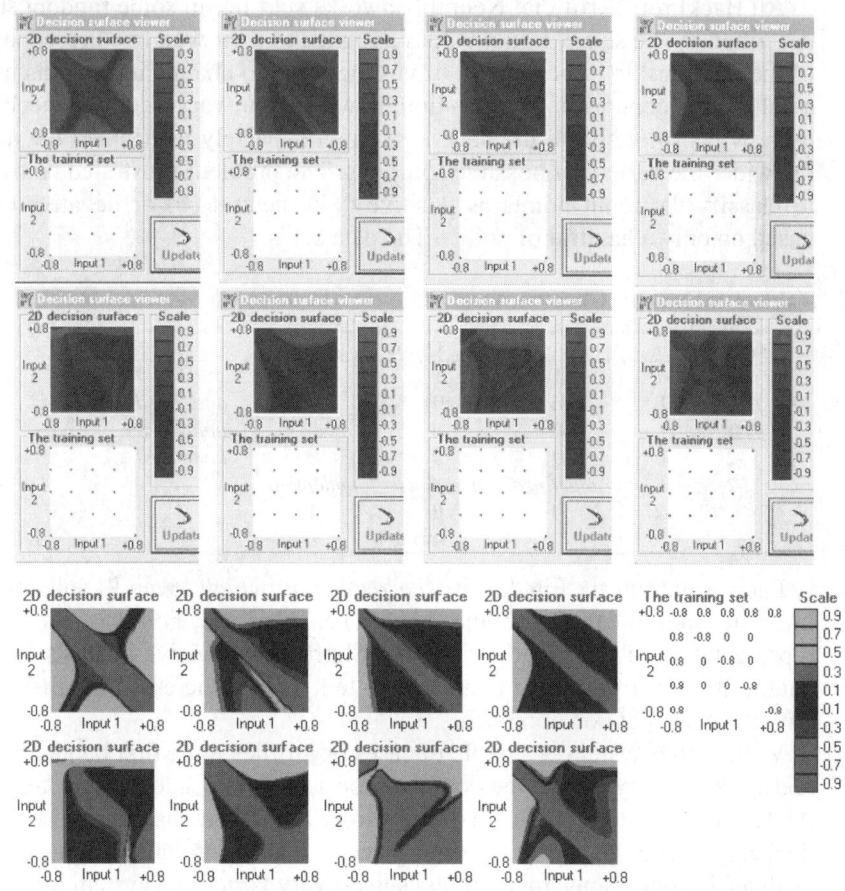

Fig. 4.3. Sparse Classification Solutions

More generally, the evaluation of feature sets by domain experts during the design stage of a system is an approach more applicable to Expert Systems and not Artificial Neural Networks. If ANNs are used, then Radial Basis Function networks [4, 5] are likely to perform better than BackProp'-style ones. Also, while domain experts have their place in every analysis, it is generally at the end of the analysis and not at the beginning.

4.4.2 Training to *Predict* on *Dense* Data Sets: the Problem is the Data

In the example given in Fig.4.4, where 5,000 training examples are presented to an ANN the task is not to classify them, but rather to predict the value of another variable. The generated solutions are either 'partitioned', as demonstrated in the first three (from top left), or 'banded', as in the next four in sequence; the last solution (bottom right) is something of an interesting compromise between these two.

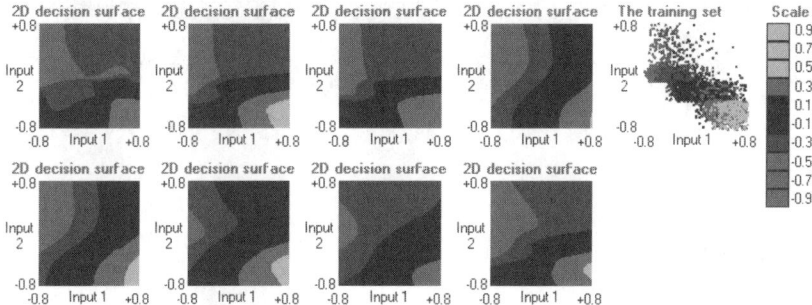

Fig. 4.4. Sparse Classification Solutions

The major differences between the heuristic models lie in the unrealizable areas of the feature space. What the ANN makes of these regions is not too important since the data is highly unlikely ever to go there. For example, except under exceptional circumstances, a doctor cannot have more patients than services. What is important is that the ANN has a good representation of the feature space that real data is likely to inhabit, and that this representation is consistent. There are no widely differing solutions here.

However, if the same ANN is trained to predict on sparse data, the same problems as for the classification example occur once more. Again, working from a small sample of the data, the ANN is unable to deduce the best representation of the broader data set. Even, as in Fig.4.5 where the data are genuinely representative of the actual distribution, its sparseness allows for a variety of possible solutions, which would give odd answers in realizable parts of the feature space.

For purposes of anomaly detection, training on sparse data often doesn't allow for a valid generalization of the feature space for either classification or prediction. There is dogma that data for ANNs should be divided into training and testing sets. However, there are any number of problems for which this is plainly nonsense since generalization capabilities (which is what the division of the data set is intended to test) is not required for every application. A simple problem where it would not be sensible to divide the data set would be XOR[1]. Similarly, multidimensional problems such as anomaly detection require the totality of the useable data set.

However, as noted in the title of this section ('Training to *Predict* on *Dense* Data Sets: the Problem is the Data'), training with a maximum of data brings a trouble of its own. Quite simply, while the feature space is well described and good heuristic models generated, *individual* examples may be sufficiently ill-defined to reliably fit *any* model.

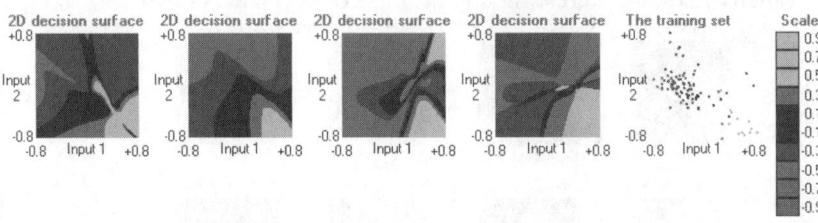

Fig. 4.3. Sparse Prediction Solutions

ANNs employed as heuristic models for feature prediction systems work with summarized data rather than directly with the transactional data. A typical spread of features might include features about locality, demographics, benefits and services. As the Artificial Neural Network can only remember a limited number of sub-models, the sub-models being generated simply cannot describe the whole of the feature space if that particular feature space is overly complex. The whole of summarized Medicare data, concerning well over 200 million transactions annually, falls into the category of overly complex. Contributing much to this complexity is the distribution associated with practice sizes.

Consider, say, a set of heuristic models which successfully predicted assorted practice statistics for 500 doctors, each doctor rendering 5,000 to 6,000 services per annum. If into this set of models is provided the summary statistics about a doctor who rendered a single service, the ANNs are simply not able to make a realistic appraisal of whether or not that single service was anomalous in terms of the practice models they have learned. What this means is that the more sparse the transaction data for a given individual, the less confidence one can have in the predictions about the individual being made by the ANNs.

As such examples influence the ANNs' models during update, they essentially equate to noise. The solution is to threshold at a minimum level

---

[1] In the predicate sense, XOR requires second order logic – in other words, a means of handling exception or contradiction.

of activity required by the examples that are contributing to the models. For any given anomaly detection problem, there will be a minimum of necessary data required for any individual to be evaluated usefully. This is not to suggest the individuals undertaking less than this minimum level of activity be ignored. They can be evaluated using other statistical means.

One such BackProp' ANN suite used by the Health Insurance Commission to evaluate a particular medical specialty, in the prototype stage, consisted of twelve Artificial Neural Networks, each modelling some particular feature of medical practice and capable of making predictions about what a particular doctor should be like. Systems like this can identify peer differences in particular features of the individuals being modelled. The primary advantage, in this case, of using ANNs is they learn from new data which means that as the specialty changes, the Artificial Neural Networks automatically modify themselves to keep up to date and maintain their performance. This substantially reduces the need for ongoing manual intervention in the update process.

Where the doctor's actual practice feature and the prediction of it diverge markedly – either above or below the prediction – they are judged to be anomalous. To be judged to be anomalous is not necessarily a problem in itself – many anomalies have legitimate explanations. However, it must be said, that many do not.

During the field test of the prototype, carried out in 2003, the number and degree of anomalies equated to the overall ranking of concerns for review. The prototype's success at identifying inappropriate or possibly inappropriate practice requiring intervention was a little over 50%.

The final suite for the specialty in question – the prototype being refined by the field test – consists of around 26 ANNs, each a fully connected 3:7:5:3:1 feedforward network. The architecture for all ANNs is common for simplicity, though the prototypes, being the product of an Evolutionary Algorithm, varied in size and had asymmetrically connected nodes. Over time, as the ongoing performance of each ANN and the redundancy between ANNs becomes more apparent, this may reduce. The time frame of this evaluation is estimated to be around 5 years.

4.4.3 Feature Selection for and Performance of Anomaly Detection Suites

In order to successfully build models that allow for good predictions of the feature variables in question, each Artificial Neural Network needs to take into account a number of factors about each doctor's practice. In the case of the suite described above, for that particular specialty, the number of inputs required ranged from 18 inputs for one feature to a maximum of 34 inputs for another (as this is an active system, the features cannot be identified here).

It is important not to provide a means of allowing the ANN to *simply* calculate the desired output. If a ratio is being sought, then the set of inputs cannot include both variables which form the ratio. The idea is to allow the ANN to only indirectly model the output feature, so gaining a better understanding of the range of parameters affecting it via a *complex* calculation. In part, it is in this way that anomalies are detected.

Table 4.2. Correlation Co-efficients for Prediction *versus* Actual Values

Quarter	Feature 1	Feature 2	Feature 3	Feature 4	Feature 5	Feature 6
Y00Q1	0.9709	0.9979	0.9163	0.8151	0.9174	0.7575
Y00Q2	0.9736	0.9982	0.9195	0.8253	0.9126	0.7549
Y00Q3	0.9741	0.9973	0.9150	0.8198	0.9100	0.7533
Y00Q4	0.9701	0.9962	0.9226	0.8385	0.9042	0.7640
Y01Q1	0.9691	0.9980	0.9244	0.8385	0.8986	0.7614
Y01Q2	0.9669	0.9959	0.9161	0.8357	0.8980	0.7807
Y01Q3	0.9702	0.9985	0.9176	0.8193	0.8942	0.7880
Y01Q4	0.9626	0.9983	0.9166	0.8222	0.8989	0.7913
Y02Q1	0.9591	0.9984	0.9222	0.8215	0.9016	0.7857
Y02Q2	0.9610	0.9973	0.9160	0.8129	0.8970	0.7952
Quarter	Feature 7	Feature 8	Feature 9	Feature 10	Feature 11	Feature 12
Y00Q1	0.9140	0.9722	0.9636	0.8158	0.9897	0.8924
Y00Q2	0.9073	0.9727	0.9634	0.8349	0.9885	0.9029
Y00Q3	0.9085	0.9724	0.9644	0.8339	0.9868	0.8993
Y00Q4	0.8731	0.9705	0.9620	0.8275	0.9870	0.8948
Y01Q1	0.8696	0.9701	0.9609	0.8274	0.9888	0.8987
Y01Q2	0.8691	0.9700	0.9620	0.8340	0.9844	0.8961
Y01Q3	0.8632	0.9693	0.9632	0.8362	0.9842	0.9017
Y01Q4	0.8559	0.9699	0.9614	0.8332	0.9862	0.8925
Y02Q1	0.8470	0.9685	0.9620	0.8262	0.9881	0.8975
Y02Q2	0.8493	0.9693	0.9622	0.8409	0.9869	0.9057

Variables that are too tightly coupled to the output variable are forbidden from the input variables. This is relatively easy to achieve, requiring few additional constraints on ANN creation. As noted previously, employing an Evolutionary Algorithm to optimize the selection of input features is extremely effective, particularly where up to several thousand potential features are available. Also useful is to allow the Evolutionary Algorithm to select the features being modelled, again, based on performance. This limits the resources being devoted to features that cannot be effectively modelled. One additional advantage of the approach is that multivariate statistical analysis, such as discriminant factor analysis, of the variables, while of academic interest, is not required.

It is apparent from Table 4.2 that the performance of the different ANNs varies. Artificial Neural Networks with coefficients greater than 0.60 are typically considered useable. All of the models are quite good – most, in fact, are excellent. The performance relates directly to the threshold at which deviation from the prediction will be deemed anomalous. It is possible, however, for a certain element of deception to occur when many thousand results are wrapped up into a single number, such as a correlation co-efficient. This is a phenomenon known as 'deceptive global performance'.

As already shown, there is more than one way for an ANN to generate its heuristic model, and even on dense data sets, some of them are just not useful for purposes of anomaly detection no matter how good they are globally. Ideally, whatever variable is being modelled by the Artificial Neural Network should be modelled exceedingly well so the majority of examples presented have minimal error and only a handful have high error – this is what allows the identification of the anomalies.

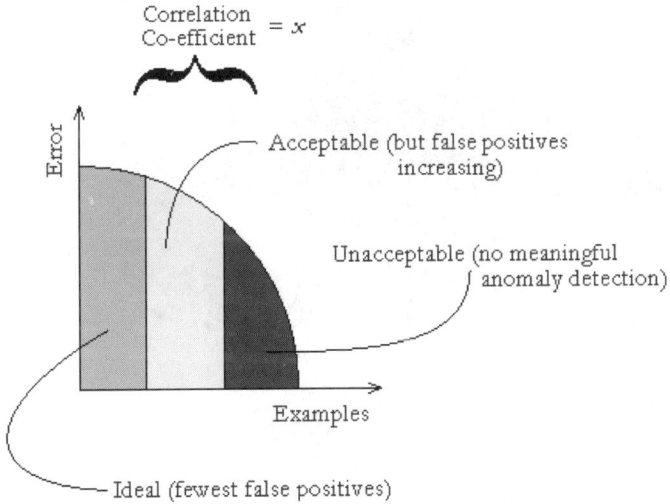

Fig. 4.4. Deceptive Global Performance – Right and Wrong Ways of Achieving the Same Performance

In Fig.4.6, such an ideal ANN would be operating in the region to the left. Sometimes though, the edges of the heuristic model – which are actually collections of sub-models – are 'blurry' and significant error gets attributed to a broader range of examples than in the ideal case. Usually, the ANN is still useable – this is the middle region of Fig.4.6.

In the worst case for effective anomaly detection, the ANN can simply not make the edges of the heuristic model 'blurry' but rather the entire model – in effect, it makes everything just a little bit wrong. This is the right-most region Fig.4.6. In this case, the ANN is not useable for anomaly detection, no matter how good its global performance, and the problem must be remedied by corrective re-training to drive it into either of the other two regions.

A final consideration is that of prediction resolution.

The difference between network precision and network accuracy is quite important. Whereas accuracy may be determined by comparing network output and desired output, precision is more about the resolution to which the network is operating; these are quite different concepts. To illustrate this, imagine two terrain mapping satellites measuring the position of a mountain and reporting its co-ordinates. Each satellite orbits five times, reporting the co-ordinates of the mountain each time.

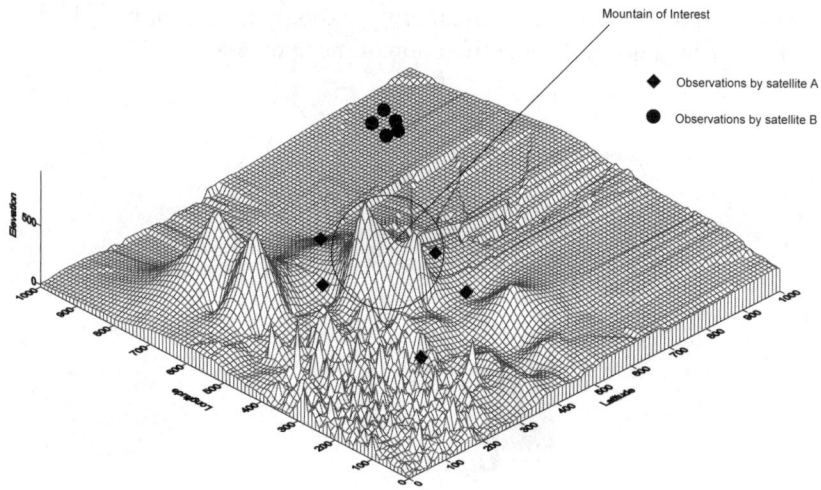

Fig. 4.5. Precision versus Accuracy

As can be seen, the coordinates reported are quite different for each satellite. Basically, precision has to do with the repeatability of a measurement, while accuracy has to do with how close the reported measurement is to the truth. In Fig.4.7, satellite-A, while less precise than satellite-B, is considerably more accurate. Satellite-B, with its much tighter clustering of observations, is much more precise, but considerably less accurate. While it reports its co-ordinates with a greater consistency, they contain greater error.

Networks can perform similarly. For some problems, they can be very precise and very accurate, or accuracy may dominate over precision or vice versa. Imagine three doctors, each actually performing 5,113, 5,118, and 5,121 services, respectively. The surrounding practice features, which describe the means by which they achieved their levels of service, will be different. If the network were to predict 5,187.17, 5,283.84 and 5,190.05 services, respectively, then while very precise, it would not be terribly accurate. On the other hand, if for each doctor the network were to predict 5,110, 5,110 and 5,110 services, then it would be much *more* accurate if somewhat *less* precise. While there is not exactly a trade-off between accuracy and precision, one or the other will set the limit to the resolution of the prediction. If, say, the network can only 'see' to within ±20 services with respect to accuracy, having precision down at hundredths of a service is meaningless – not that fractions of a service have much meaning, anyway – and that level of resolution certainly should not be used to discriminate between answers which are effectively the same. In most implementations an ANN will be much more precise than accurate.

4.4.4 Interpreting Anomalies

The approach outlined so far is intended to create ANN systems for the detection and quantification of anomalies.

An anomaly is, simply, a substantial difference between what one entity – be they an individual or a company – is doing, as described by a summary of their associated data, and what almost all other entities are doing. An anomaly, in the context of medical practice, is not necessarily inappropriate practice, fraud or abuse. It may be that a perfectly reasonable and legitimate clinical practice causes an anomaly.

An anomaly is not the same as an outlier.

There may well be good and many reasons, all reflected in the servicing and patient demographic, for a doctor to be an outlier in some particular feature or even several features. In many cases, such outlying providers will fit the heuristic model of the Artificial Neural Network and so not be flagged as anomalous; they are numerically explicable. ANNs can identify only those doctors for whom the numbers, in effect, don't quite add up.

It is important to keep in mind that ANNs are purely numerical engines, concerning themselves only with numbers in and numbers out. They have no 'general knowledge' of what it is they are doing. They don't understand nor have knowledge of what an appropriate transaction is. If, for example, one of the ANNs considers a doctor anomalous, it is likely a matter of small significance. But if many of the ANNs all agree that, in their respective areas of expertise, a provider is anomalous, there is something very different going on in that doctor's practice.

An important point to note is that suites of ANNs for anomaly detection won't detect inappropriate transactions. It is not evaluating the data at a resolution that would allow for such identification (other techniques allow for this, and these will be explored later in the Chapter). The anomalies identified are at the entity level, not the transactional level. Further, were a system designed to look at the transactional level, widespread or systematic inappropriate transactions would not be detected, since, by definition, anything widespread and systematic is not anomalous.

It is also important to limit the scope of result interpretation to that which the data can support. At first instance, for example, it would seem reasonable to suppose that having near global capture of transaction level data for medical and pharmaceutical servicing would provide insight into the effectiveness of clinical treatments of disease.

In health informatics, there is a world of difference between transaction data and clinical data. This difference means that it is usually not possible to derive detailed clinical information from purely transactional databases such as the Medicare database. Quite simply, knowing that a pathology test, for example, was performed, in no way tells you *what* the result of the pathology test was. Even if subsequent items for medical servicing seem to be indicative of a particular condition *if* the pathology test was positive for some criterion or other, is the subsequent servicing so unique that it could not be treatment for an alternative condition for which no pathology was required? Typically, there remains a high degree of ambiguity. Essentially, transaction data is to an event as a shadow is to an object. Depending on perspective and lighting, it can be very hard to determine what an object is from its shadow alone. Likewise, it can be very hard to determine the complete details of a real-world event.

The difficulty is that many different clinical conditions may be represented by the same transaction data. No system – not an Expert System, not a suite of ANNs, not a Bayesian approach, no statistical technique – is capable of integrating that transactional data to the point of deducing what the clinical condition of the patient was: the relative loss of information is simply too great. Put another way, in going from a higher dimensional mapping to a lower dimensional mapping, detail is lost in such a way that the higher dimensional mapping cannot be reconstructed.

Be it for anomaly detection or any other application, ANNs cannot provide more actual information than what was present in the input feature set.

4.4.5 Other Approaches to Anomaly Detection

Outliers are the simplest way of identifying anomalies, but are often only useful if the distribution is simple and a single dimension meaningful. When it comes to real data, genuine outliers are often difficult to identify using single variables only. This difficulty can be compounded by the use of percentile rankings for some particular variable or other, as such rankings provide no information on the real level of variation in the data set.

In Fig.4.8, the kind of variable distribution where simple outlier detection works is illustrated by variable *V4*, in the bottom plot. Here, a single outlier is readily identified.

Usually, it is of more interest to identify outliers in multiple dimensions and for reasonably linear data sets this is readily done, as illustrated in Fig.4.9. Here, the values for the *y*-axes are given by the rows and the values for the *x*-axes are given by the columns. The relationship between different pairs of variables is readily determined, as are any outliers in the multiple set of dimensions being examined.

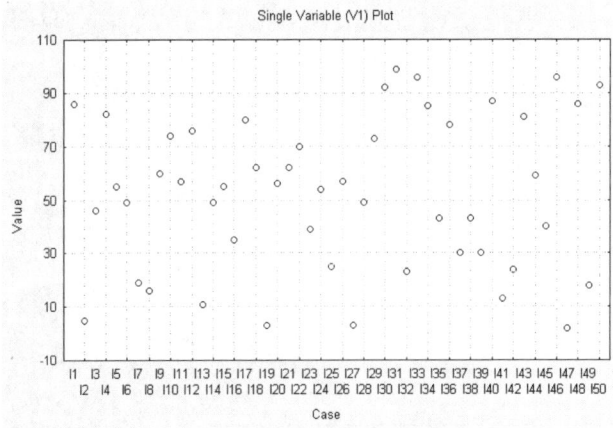

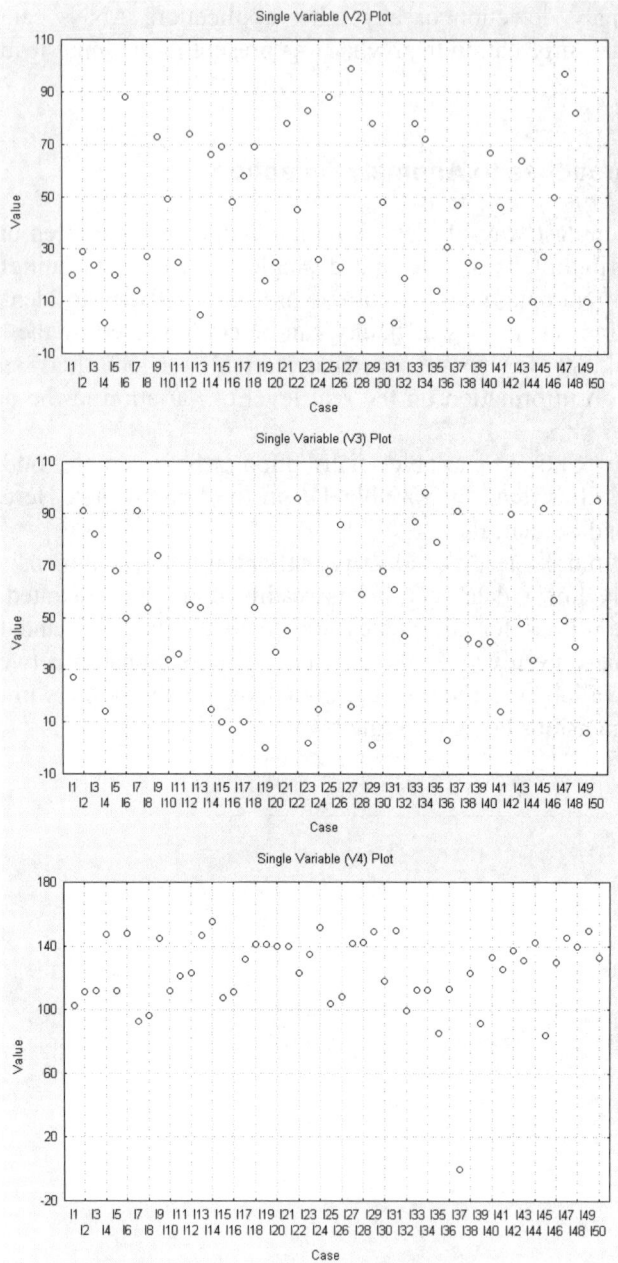

Fig. 4.6. Outliers in Different Distributions

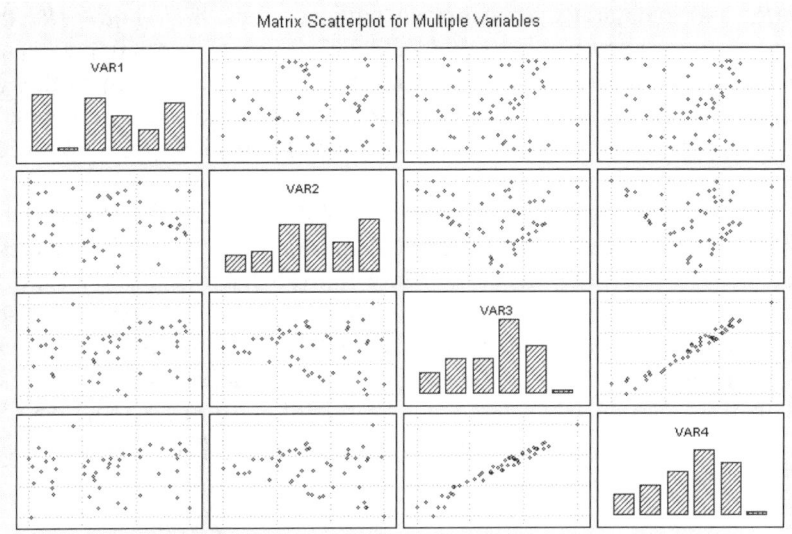

Fig. 4.7. Identifying Outliers in Multiple Dimensions – Distributions

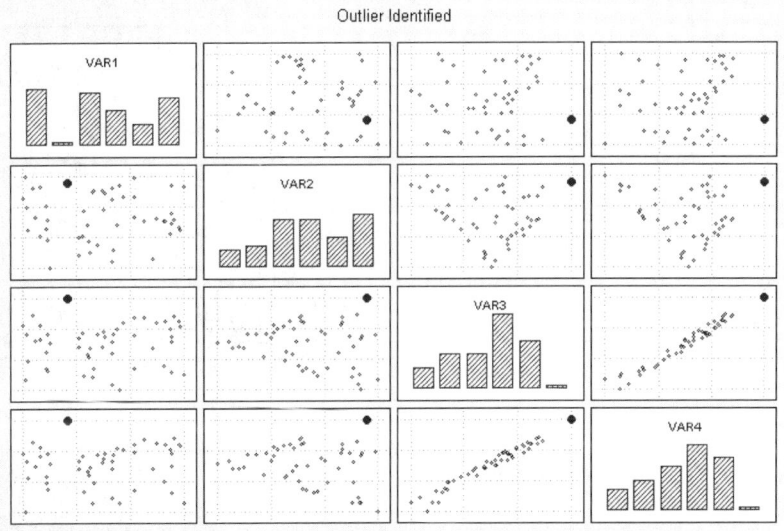

Fig. 4.8. Identifying Outliers in Multiple Dimensions – Highlights

In Fig.4.10 the same data is presented as in Fig.4.9, however, the same single case is identified in all the charts. This case is something of an outlier for variables *VAR4* and *VAR3*, well within the typical range for

VAR2 and anomalous – but not an outlier – for *VAR1*. For *VAR1* there appears to be a noticeable 'forbidden zone' of values that may not be taken and in which is situated the anomalous case. A feature such as this may indicate a data error, that the data is not descriptive of a single population but descriptive of (at least) three distinct populations, or that there is genuinely something odd about that case.

Returning to Fig.4.8, what is evident is that the distributions of all the presented variables (the first 4 of 112) are fairly symmetrical (skewness is less than 0.3 at most) and reasonably flat (all have a negative kurtosis). This data, therefore, is not normally distributed, and hence statistical techniques assuming normally distributed data should be used with caution.

Visual inspection of data sets is often an invaluable first step in any kind of analysis, particularly where identification of clustering is required. Other means of identifying clusters are k-means clustering [7] and tree clustering [9].

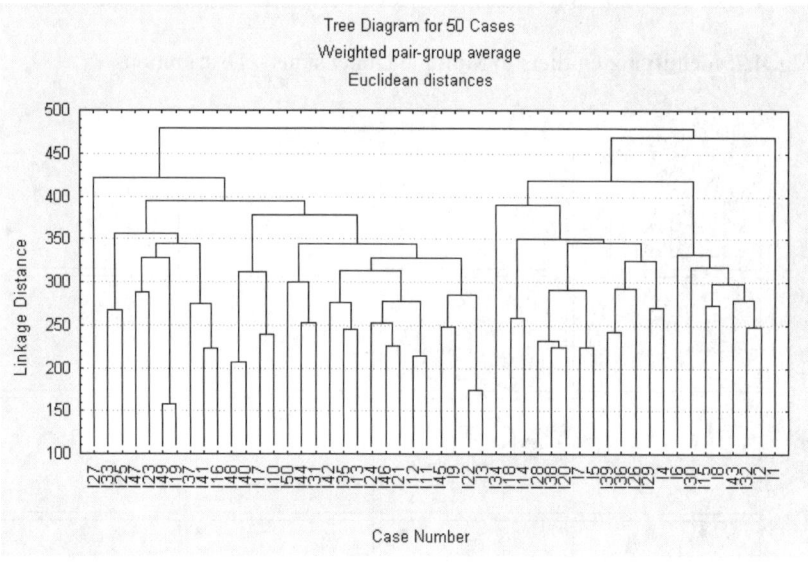

Fig. 4.9. Tree Clustering of 50 Cases

The principal difficulty with k-means clustering is that the number of clusters to be generated has to be specified *a priori* – the data doesn't get the chance to 'speak for itself', as it were. Once completed, the algorithm seeks to maximize and minimize inter-cluster and intra-cluster variability, respectively. What can be of interest, however, are the outlying cases that fail to belong to any cluster.

Often, a more interesting means of clustering data is by tree clustering, illustrated in Fig.4.11. Here, each case is initially by itself and the set of constraints defining differentiating or distinctive elements are broadened until cases begin to build up into branches based on similarity. The distance until cases join is a measure of their similarity. In this example, the data breaks down into two broad classes. Within this hierarchical structure, the outliers are readily observed. Case *I1* is least similar to anyone else, followed by case *I27*, while *I49* and *I19* are the two most similar.

There are several advantages to this form of representation. First, cases described by a large number of variables can be readily evaluated with respect to their overall similarity. In the example of Fig.4.12, cases are clustered by similarity across some 112 variables; using scatterplots or other visual methods to examine such a large data set in order to identify outliers is often impractical. Second, where there are known cases of interest, this kind of representation can identify the 'branch' where other, similar cases may be. Third, where there is a good sample of known cases, by clustering on different subsets of data until the known cases are on the same 'branch', the set of variables best able to identify the sought for issue can be rapidly identified. Again, an Evolutionary Algorithm can be very effective in undertaking such a task. Fourth, it can reveal previously unknown and often interesting associations.

All the techniques discussed so far work well with aggregated or summarized raw data accumulated over a month (or quarter or year). This will allow identification of some classes of anomalies, and some kinds of outlier, but not all. These techniques are less successful at identifying *temporal* anomalies. While some particular statistic may look reasonable, it is *how* it came to be that can be where the problem lies. A statistic suggesting that a doctor serviced some 4,000 patients in the course of a year looks quite reasonable. Such a statistic would look far less reasonable if there were only four dates of service, thereby averaging some 1,000 patients per working day.

Temporal analysis is useful. To give a few examples, in health informatics for identifying spatiotemporal properties of epidemics, in neuroscience for examining neural activation, in meteorology for examining events such as cloud formation, and in geology for examining events such as earthquakes. It can also be quite useful for identifying emerging trends.

Identifying temporal patterns in order to characterize clusters, anomalies and outliers is a significantly more difficult task than working with data where the temporal dimension can be ignored [2, 14]. Generally, apart from techniques drawn from signal processing, temporal analysis techniques tend very much to be in the developmental stage and are often problem-specific. Other approaches, such as the Box-Jenkins method [6] are highly constrained as to the kind of time series data they are appropriate for – but it can be an area worth persisting with if the business requirement has sufficient priority.

Pattern recognition and matching to established templates are common approaches [3, 8]. The difficulty with these techniques arises where little or no knowledge of the desired templates is available. A method of identifying key elements of time series events of interest without the use of templates and employing an Evolutionary Algorithm is outlined in [15].

While an area of ongoing research, the Health Insurance Commission currently fields no AI systems based on temporal analysis.

4.4.6 Variations of BackProp' ANNs for Use with Complex Data Sets

The Artificial Neural Networks in the anomaly detection suites are all BackProp' ANNs – very probably the most common kind of ANN implementation of all. There are, however, many different kinds of ANN each with particular – and, at times, peculiar – strengths and weaknesses [16]. A readily accessible overview and introduction to the many different kinds of ANN available is to be found in the regular FAQ postings to the `comp.ai.neural-nets` Usenet newsgroup (also available at ftp://ftp.sas.com/pub/neural/FAQ.html). Alternatively, for a detailed examination of BackProp' ANNs themselves, `Backpropagator's Review` at http://www.dontveter.com/bpr/bpr.html is an excellent resource.

Two variations to the standard BackProp' implementation relevant to Data Mining applications are worth mentioning, particularly for working with large and complex data sets.

Often a data set contains relationships that no single ANN can effectively learn. A partial solution is to break down the problem and spread it across several ANNs, each required to learn a very specific relation. The complexity of the data is, in part, one reason this approach is undertaken for the suites (the other reason that a single, multiple output ANN is not used is that the variables being modelled by some of the ANNs are required as inputs by other ANNs – and it is pointless to provide the ANN directly with the answer it is attempting to generate).

However, trouble can arise where – even for a very specific relation – the relation to be learned is too complex for a single ANN, or the ANN must be 'just right' in order to learn the relation [17]. The use of ANNs has often been referred to as a 'black art' given the uncertainties when it comes to choosing the type, size and associated parameters of the ANN, so stumbling across the correct ANN for a complex problem is unlikely [1].

A different approach is to do away with the uncertainties altogether by using a 'standard unit' BackProp' ANN, and employ a divide-and-conquer technique to generate a solution [11]. Each network learns a different part of the problem space, and in combination they learn a relationship that matches to the data exactly.

To generate such a solution, the first network is presented with the whole of the data set and training continues until the network is stable – which doesn't necessarily mean it has learned a great deal. The network is tested to establish which points that the ANN has learned. These learned points can then be removed from the training set. Those that remain unlearned serve as a reduced input data set for the next ANN and so forth until a 'modular ensemble' of ANNs is formed that has learned the whole of the data. Re-presentation of some learned points is often advantageous for creating good boundary definition within the problem space.

While something of a brute force approach, the success of this technique has been proven for both digital [12] and analogue [13] problems.

The second variation to BackProp' implementation with good application to Data Mining involves an enhanced node design called a 'micro-net' [10]. Micro-net architecture effectively blurs the distinction between ensemble-based and modular ANNs [17]. Essentially, these ANNs allow for significantly more complex connections between nodes in a network that, in turn, allow micro-net ANNs to solve far more complex tasks than equivalently resourced conventional ANNs. Networks of this type are also capable of learning relationships that defeat even modular ensembles.

Where the complexity of the problem is such that it seems likely to defeat – or has defeated – other approaches, micro-net ANNs are worth trying. However, conventional approaches should be exhausted first. Without exception, the simplest possible solution that meets the business requirement should be used.

4.5 Formulating Expert Systems to Identify Common Events of Interest

Anomaly detection, as mentioned, doesn't identify events that are relatively common. Where there are widespread or endemic problems to be identified, different techniques are called for. One particular problem is that knowing about some particular event is often not at all revealing. Rather, the problem is not the single event itself but a sequence of events in combination.

Within the context of compliance, it is not possible to bring to bear domain knowledge – medical and pharmaceutical expertise – to evaluate the transactions recorded against 20 million individuals. Likewise, examining overall usage of the items in the Medicare or Pharmaceutical Benefits Schedule fails to address the actual episode of care received by the patient.

The task, then, is to identify common problems buried in a broad and diverse range of servicing, as described by transaction data, and use this as the basis for an Expert System. Identification calls for domain knowledge, so the question is, then, how to automatically evaluate the millions of sequences to draw out those that are likely to contain problems?

The properties of the sequences are important determinants:
- Are the transactional elements of the sequence legitimately repeatable in a sequence?
- Is the order they actually happened in known or is only the order they were processed in known?
- Is a sequence of transactions of a fixed or variable length?
- Does a sequence of transactions cover a fixed or variable time?
- Do the properties associated with any given transaction (e.g. cost, location) vary in combination with others?
- Can multiple sequences be associated with a single identifier (e.g. doctor, company, bank account)?
- Is there a consistent and specific interpretation as to what each transaction is actually for?

The most complicated data sets to work with tend to be those where:
- the transactions within any sequence may be legitimately repeated;
- only the order of the processing is known;
- the sequence can be of any length;
- the sequence can occur over any time;
- the properties of any given transaction may depend on what else it is in combination with;
- multiple sequences can be associated with a single identifier; and,
- there is no consistent use or interpretation of any particular transaction type.

With any data set, the first step is to limit it to those transactions from within a particular class of transactions that is of interest. In principle, no such constraint is required, but most regulatory activities tend to be fairly specific in nature, seeking to address particular issues rather than everything at once. These sequences of transactions may relate to servicing by a medical specialty, a particular grouping of financial investments, classes of insurance claims – whatever is best descriptive of the area under review in the context of transactions.

All sequences to be evaluated, then, will be known to contain at least one transaction of interest. All combinations derived thereof, also, will contain at least one transaction of interest.

Grouped by identifier (typically, the identifier which will draw the intervention), these sequences are broken down into all possible combinations or 'modes' together with as much 'structural' information about the data as possible – for example, demographic, geographic, medical, financial or historical. Appending this information is crucial as it is the basis of reducing the modal data set – with potentially tens of millions of combinations – to the few hundred to a thousand that a domain expert(s) can evaluate in a timely manner.

It is important to stress that the sequences will exist in a kind of superposition. The *first* mode contains information, by identifier, only on the designated set of transactions.

The *second* mode contains information, by identifier, on the designated set of transactions in combination with every other transaction that they exist in combination with (including other transactions from the designated set).

The *third* mode contains information, by identifier, on the designated set of transactions in combination with every other pair of transactions that they existed in combination with (including other transactions from the designated set).

This process of summarization continues until a mode length equal to the maximum sequence length of unique transactions is reached. Note that there is considerable overlap between the modes. With respect to target transactions, all the information in higher modes is contained within lower modes, albeit in a quite distributed form. The *second* mode is the most dispersed in this regard, as it contains all target and all associated non-target transactions – however, the *second* mode does exclude instances where the target transaction occurred in isolation.

Consider a Doctor (X) who undertakes the following servicing, where A, B, C, D, E and F represent different items from the Medicare Benefits Schedule.

Here, in Tables 4.3 and 4.4, there are five sequences of servicing for five different patients, containing either one or two of the target items, A and E. The sequence lengths range from 2 to 6, and there are duplicate items of servicing. Note that the benefit varies with the sequence of servicing. Typically, to make the data set more manageable, it is necessary to contract the data to remove duplicate transactions, but flag that duplicate transactions were present. Note the effect this has on the benefit, which is calculated as a sum of means in the case of duplicates; if duplicate items are of specific concern, then a separate analysis based around data matching should be undertaken to evaluate them directly.

Table 4.3. Example Sequence Data, Raw (servicing by Dr. X – target transaction items designated as **A** and **E**)

	Item 1 - $Ben	Item 2 - $Ben	Item 3 - $Ben	Item 4 - $Ben	Item 5 - $Ben	Item 6 - $Ben
Patient 1	A - **$83.70**	A – **$62.80**	B - $111.50	C - $38.10	D - $54.80	E - **$103.20**
Patient 2	A - **$83.70**	B - $111.50	B - $83.65			
Patient 3	B – $111.50	D - $54.80	C - $38.10	E - **$103.20**	F - $8.60	
Patient 4	A - **$83.70**	F - $8.60				
Patient 5	C - $38.10	C - $28.60	E - **$206.40**			

Table 4.4. Example Sequence Data, Contracted (servicing by Dr. X – duplicate transaction items denoted by *)

	Item 1	Item 2	Item 3	Item 4	Item 5
Patient 1	A* - **$73.25**	B - $111.50	C - $38.10	D - $54.80	E - **$103.20**
Patient 2	A - **$83.70**	B* - $97.56			
Patient 3	B - $111.50	C - $38.10	D - $54.80	E - **$103.20**	F - $8.60
Patient 4	A - **$83.70**	F - $8.60			
Patient 5	C* - $33.35	E - **$206.40**			

Table 4.5. Servicing by Dr. X – *First* Mode (target items only)

Comb	$	Tot	Pat	Sub	FSR	SSR	Len	Com
A	$80.22	3	3	1	0	0.33	3.67	171
E	$137.60	3	3	1	0	0	4.67	204

This data would break down into 5 modes. A basic set of variables of interest by mode is given below, most of which (or near equivalents) would likely be appropriate for many transactional data sets. For medical servicing data, for example, detailed demographic information would also be appended. The modal breakdown would appear as outlined in Tables 4.5 through 4.9.

The variables listed are:
- *State*, being the State in which the doctor's practice is located (NSW here – but omitted from the Tables).
- *Spec*, being the speciality of the doctor (in this instance, 'D' is simply for doctor – i.e. General Practitioner; again omitted from the Tables).
- *Comb*, being the combination of items.
- *SeqBen*, being the mean Medicare rebate for those items as rendered by Dr. X.
- *SeqTot*, being the total number of occurrences of the combination in sequences of any length.
- *SeqPat*, being the total number of patients to whom the combination was rendered (this may be different to *SeqTot* if, as occasionally happens, the same or similar procedures are carried out more than once).
- *SeqSub*, being the fraction of the occurrences of the combination that are part of a larger sequence of servicing.
- *SeqFSR*, being the fraction of the occurrences of the combination where that combination is the complete sequence of servicing, and one or more items of the combination were duplicated.
- *SeqSSR*, being the fraction of the occurrences of the combination where that combination is part of a larger sequence of servicing, and one or more items of the combination were duplicated.
- *SeqLen*, being the mean length of the sequences of servicing, including duplicates, of which the combination is part.
- *SeqCom*, being the total number of doctors that render the combination.

(Note: the prefix *Seq* has been omitted from the Tables).

Table 4.6. Servicing by Dr. X – *Second* Mode (target items in association with all others)

Comb	$	Tot	Pat	Sub	FSR	SSR	Len	Com
AB	$183.00	2	2	0.5	0.5	0.5	4.5	123
AC	$111.35	1	1	1	0	1	6	116
AD	$128.05	1	1	1	0	1	6	116
AE	$176.45	1	1	1	0	1	6	116
AF	$92.30	1	1	0	0	0	2	154
EB	$214.70	2	2	1	0	0	5.5	116
EC	$174.12	3	3	0.67	0.33	0	4.67	116
ED	$158.00	2	2	1	0	0	5.5	112
EF	$111.80	1	1	1	0	0	5	112

This kind of summarization is repeated for every doctor, since the purpose of the exercise is to identify the distribution of all combinations against every individual. Note that the summaries generated are significantly larger than the raw data. Within Medicare servicing data, mode length can extend to well over 20 unique items.

Table 4.7. Servicing by Dr. X – *Third* Mode (target items in association with all others)

Comb	$	Tot	Pat	Sub	FSR	SSR	Len	Com
ABC	$225.85	1	1	1	0	1	6	116
ABD	$239.55	1	1	1	0	1	6	116
ABE	$287.95	1	1	1	0	1	6	116
ACD	$166.15	1	1	1	0	1	6	116
ACE	$214.55	1	1	1	0	1	6	116
ADE	$231.25	1	1	1	0	1	6	112
EBC	$252.80	2	2	1	0	0	5.5	116
EBD	$269.50	2	2	1	0	0	5.5	112
ECD	$196.10	2	2	1	0	0	5.5	112
EFB	$223.30	1	1	1	0	0	5	112
EFC	$149.90	1	1	1	0	0	5	112
EFD	$166.60	1	1	1	0	0	5	58

Table 4.8. Servicing by Dr. X – *Fourth* Mode (target items in association with all others)

Comb	$	Tot	Pat	Sub	FSR	SSR	Len	Com
ABCD	$277.65	1	1	1	0	1	6	112
ABCE	$326.05	1	1	1	0	1	6	110
ABDE	$342.75	1	1	1	0	1	6	110
ACDE	$269.35	1	1	1	0	1	6	88
EBCD	$307.60	2	2	1	0	0	5.5	62
ECBF	$261.40	1	1	1	0	0	5	71
ECDF	$204.70	1	1	1	0	0	5	58
EDBF	$278.10	1	1	1	0	0	5	58

Table 4.9. Servicing by Dr. X – *Fifth* Mode (target items in association with all others)

Comb	$	Tot	Pat	Sub	FSR	SSR	Len	Com
ABCDE	$380.85	1	1	0	1	0	6	1
BCDEF	$316.20	1	1	0	0	0	5	2

Note that the summaries generated are significantly larger than the raw data. Within Medicare servicing data, mode length can extend to well over 20 unique items. From a designated range of 250 target items, for example, over the course of a year it might be the case that 1,800 doctors would be found to have rendered 1,200,000 target items in combination with 3,000,000 additional items. Typically, summary files of 12 million to 14 million lines would result.

Initial analysis of these files answers a numbers of questions with often regulatory import:

- Which individuals are doing high volumes of unique combinations that no other person is doing (i.e. what combinations of high volume have $SeqCom = 1$)?
- What combinations are common to many (too many?) sequences? Does there appear to be endemic 'add on' transactions?
- What combinations are frequently repeated (i.e. for what combinations is $SeqTot > SeqPat$)? Do particular individuals favour particular combinations on a routine basis?
- Which individuals have a large number of different, moderately unique combinations? Who, generally, is practicing very differently to their peers?
- What combinations would seem to be 'incomplete'? Where it is known that certain transactions should be antecedent or consequent to certain others, is this always the case (i.e. $SeqSub < 1$ for incomplete combinations)?
- Which individuals are undertaking combinations already known to be less than compliant?
- Which individuals have atypically long or short episodes of servicing (i.e. $SeqLen$ is regularly at the tail end of distributions for many combinations)?

A number of these simple checks may serve as rules for incorporation into an Expert System, even without further domain knowledge than evidenced by past cases.

The more sophisticated analysis revolves around the additional 'structural' information. Here, peculiarities at a scale greater than an individual but less than the whole population can be identified. The basis of that identification is clustering of the combinations about one or more variables in the 'structural' information:

- Are certain combinations clustered around particular practices, postcodes or States?
- Are combinations typical of a particular demographic in one State typical of different demographics in other States?

- Are certain combinations clustered around particular specialties? (And, conversely, for physicians outside the cluster, does it suggest that some physicians are undertaking procedures for which they have inadequate training?)
- Are certain combinations clustered towards older or younger physicians, or clustered by gender?

Table 4.10. An Example of Combinations Clustering by State

		Comb	AOC	AOR	PQR	APC	AOP
Total Number of	Occurrences of Combination	ACT	46			144	35
		NSW	8,761	712	1,984	2,376	949
		NT					
		QLD	1,726	2,666	306	126	235
		SA			7		
		TAS		7	8	10	
		VIC		48	228		35
		WA	224	113	293	65	88
		Grand Total	10,757	3,546	2,826	2,721	1,342
Total Number of	Patients	ACT	39			111	29
		NSW	7,087	651	1,936	2,091	893
		NT					
		QLD	1,438	2,456	299	102	191
		SA			7		
		TAS		7	8	10	
		VIC		36	228		32
		WA	190	106	290	46	72
		Grand Total	8,754	3,256	2,768	2,360	1,217
Total Number of	Doctors	ACT	1			1	1
		NSW	35	30	77	19	22
		NT					
		QLD	13	25	15	5	9
		SA			1		
		TAS		1	1	1	
		VIC		2	20		3
		WA	3	4	11	3	2
		Grand Total	52	62	125	29	37

For most combinations, it is likely that no clustering will be readily identifiable. Often, also, where clustering is identifiable, such as with

States, it is readily accounted for simply by the difference in State populations. It is instances where the clustering cannot be readily accounted for that are of interest. An example of State-based clustering where the differences are attributable to State differences in the practice of medicine rather than differences in population is given in Table 4.10

Combination *AOC*, for example, is predominantly a New South Wales event – some 81% of occurrences. Combination *APC*, by contrast, is predominantly a Queensland event – some 87% of occurrences. Is item-*O* to be preferred over item-*P* when in combination with *A* and *C*? And why is neither combination practiced in Victoria or South Australia? These questions are of the kind which only a domain expert can answer.

This approach has served as the basis for creating a number of Expert Systems currently being deployed by the Health Insurance Commission. Approximately 80% of the combinations identified in this way were judged by the appropriate domain experts to be worthy of some kind of compliance intervention. The recommended intervention varied with the scope of the issue, but ranged from general education for some endemic problems concerning item use to investigations for possible fraud and inappropriate practice.

Once reviewed by a domain expert, their comments about combinations can be directly mapped back to the sequence of concern. Identification of sequences is thereafter automatic, with changes only required when the set of target transactions itself changes. Even then, some time must pass to allow the new transaction set to establish stable patterns of use. As a means of facilitating identification of issues and individuals in complex data sets, the approach has much to recommend it.

If the time series of the actual events creating the transaction sequence is known, then the data can be treated as order sensitive. In the outlined example, the data was assumed to be order insensitive, such that *ABC* was seen to be equivalent to *BAC*, *CAB* or *CBA*. While useful information can often be gleaned from such an analysis, it does come at a price. Suppose there is a sequence of 10 transactions, each different and each within the designated target set. For both order sensitive and order insensitive calculations, the first mode will comprise 10 entries. The second mode will have 90 and 45 combinations, respectively. The *third* mode will have 720 and 120 combinations, respectively. The *fourth* mode will have 5,040 and 210 combinations, respectively, and so on.

Simply, order sensitive data sets will scale roughly as the cumulative sum of ^nP_r whereas order insensitive data sets will scale roughly as the cumulative sum of ^nC_r. If the sequence length is more than moderate (i.e. 4-5 transactions), the number of combinations can simply be too large to generate or process in reasonable time. Explicitly keeping track of duplicate transactions in sequence compounds this growth still further.

Note that the way the information is summarized will typically not be of direct use when carrying out the intervention. While required to identify the issues at hand, reasons must be mapped back to the raw sequence data as it is practically impossible to identify the singular sequence of concern from the summary data.

In combination, anomaly detection systems centred around individuals combined with Expert Systems based on modal analysis at closer to a transaction level provide readily implementable means of identifying issues and individuals of regulatory concern.

A Note on Software

The ANN software used in this work is the CISCP AI Library BackPropagation Module (www.swin.edu.au/bioscieleceng/ciscp). The decision surface maps in Figs.4.3 through 4.5 were created using the same program. The surface map illustrating the difference between precision and accuracy in Fig.4.7 was generated using Golden Software's Surfer program. Finally, the scatterplots and tree in Figs.4.9 through 4.12 were created using Statsoft's Statistica program.

Acknowledgements

In writing this Chapter, I would like to acknowledge the support of the Program Review Division of the Australian Health Insurance Commission for supporting the work described herein. I would also like to acknowledge the Centre for Intelligent Systems and Complex Processes at Swinburne University of Technology, particularly Tim Hendtlass and Gerard Murray.

References

1. Baum E and Haussler D (1989) What size net gives valid generalization? *Neural Computation*, 1(1): 151-160.
2. Berger G and Tuzhilin A (1998) Discovering unexpected patterns in temporal data using temporal logic, in Etzion O, Jajodia S and Sripada S (eds) *Temporal Databases - Research and Practice (Lecture Notes in Computer Science 1399)*, Springer-Verlag, Berlin: 281-309.
3. Berndt DJ and Clifford J (1996) Finding Patterns Dynamic Programming Approach, in Fayyad UM, Piatetsky-Shapiro G, Smyth P, and Uthursamy R (eds) *Advances in Knowledge Discovery and Data Mining*, AAAI Press, Menlo Park, CA: 229-248.
4. Bors AG (2001) Introduction to the Radial Basis Function (RBF) Networks, Online Symposium for Electronics Engineers, 1(1), DSP Algorithms: Multimedia, http://www.osee.net/ February 13: 1-7.
5. Bors AG and Pitas I (2001) Robust RBF Networks, in Howlett RJ and Jain LC (eds), *Radial basis function neural networks: design and applications*, Physica-Verlag, Heidelberg: 125-153.
6. Box GEP, Jenkins GM and Reinsel GC (1994) *Time Series Analysis, Forecasting and Control (3rd ed)*, Prentice Hall, Englewood Cliffs, NJ.
7. Hartigan JA and Wong MA (1978) Algorithm 136. A k-means clustering algorithm, *Applied Statistics*, 28: 100.
8. Keogh E and Smyth P (1997) A Probabilistic Approach to Fast Pattern Matching in Time Series Databases, *Proc 3rd Intl Conf Knowledge Discovery and Data Mining*, Newport Beach, CA, 14-17 August: 20-24.
9. Loh W-Y and Shih Y-S (1997) Split selection methods for classification trees, *Statistica Sinica*, 7: 815-840.
10. Murray G and Hendtlass T (2001) Enhanced Artificial Neurons for Network Applications, *Proc 14th Intl Conf Engineering & Industrial Applications AI & Expert Systems*, Budapest, 4-7 June, *(Lecture Notes in Artificial Intelligence 2070)*, Springer-Verlag, Berlin: 281-289.
11. Philpot D and Hendtlass T (1996) A Cascade of Neural Networks for Complex Classification, *Proc 9th Intl Conf Engineering & Industrial Applications AI*, Fukuoka, 4-7 June, Springer-Verlag, Berlin: 791.
12. Philpot D and Hendtlass T (1997) Ensembles of Networks for Digital Problems, *Proc 3rd Intl Conf Artificial Neural Networks and Genetic Algorithms (ICANNGA97)*, Norwich, 1-4 April, Springer-Verlag, Berlin: 31-34.
13. Philpot D and Hendtlass T (1998) Ensembles of Networks for Analogue Problems, *Proc 11th IEA/AIE Conf (Methodology and Tools in Knowledge-Based Systems)*, Benicassim, 1-4 June, Springer-Verlag, Berlin: 625-631.

14. Povinelli RJ (2000) Using Genetic Algorithms to Find Temporal Patterns Indicative of Time Series Events, *Proc GECCO 2000 Workshop: Data Mining with Evolutionary Algorithms*, Las Vegas, Nevada, 8-12 July: 80-84.
15. Povinelli RJ and Feng X (2003) A New Temporal Pattern Identification Method For Characterization And Prediction Of Complex Time Series Events, *IEEE Trans Knowledge and Data Engineering*, 15(2): 339-352.
16. Sarle WS (1994) Neural Networks and Statistical Models, *Proc 19th Annual SAS Users Group Intl Conf*, Cary, NC, 10-13 April: 1538-1550.
17. Sharkey AJC (1996) On Combining Artificial Neural Networks, *Connection Science*, 8: 299-314.

5 An Introduction to Collective Intelligence

Tim Hendtlass [1]

1   Centre for Intelligent Systems and Complex Processes, Swinburne University of Technology, Hawthorn VIC 3125, Australia, thendtlass@swin.edu.au

5.1 Collective Intelligence

Collective intelligence is a term applied to any situation in which individuals in some community indirectly influence the behaviour of their fellows in that community. This differs from an Evolutionary Algorithm in which the individuals (via their performance-based breeding) influence the following generations.

The influence used in collective intelligence is indirect as individuals send a general 'message' rather than a 'message' directed specifically at one or more of their fellows. The nature of the 'message' can be anything:
- A chemical trail left on the ground that influences others who come across it (an example of how this is used by ants is described below),
- An audio signal conveying information to all within audible range,
- A dance or other stylised movement that conveys information, or even
- Just the way one individual alters the environment, which will in turn modify the behaviour of others (for example, if an insect removes all food from a region, other insects arriving later at this region will move on through it. Had the first insect not removed all food the later insects would have stayed. The earlier insect presence modified the subsequent exploration paths of the later insects).

The term *stigmergy* is used to describe any indirect communication that allows the activities of social insects to be directed to a common goal.

We use nature for our inspiration and build our own versions of a collective intelligence algorithm so there is no need to exactly copy any one natural example. As a result we may make the nature of the stigmergy anything that we find suitable for a particular problem.

5.1.1 A Simple Example of Stigmergy at Work

Before considering any full algorithm, we first look at a very simple example in which stigmergy influences behaviour. In this case we will model our process on ants that leave a pheromone trail behind them that slowly evaporates with time. Imagine a group of ants, say 100, released from a nest, constrained to a single exit path that divides into two paths, which later recombine at a food source. This is shown in the Fig. 5.1. The ants leave the nest, collect food and then return to the nest.

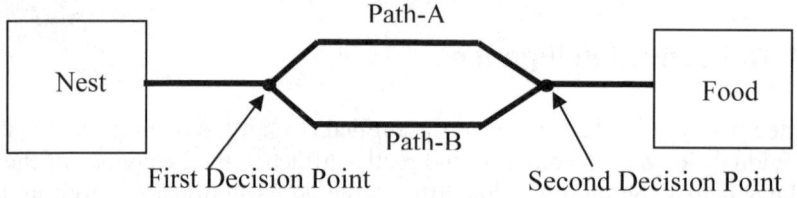

Fig. 5.1. The simple two-path route from a nest to a food source

For a moment, assume that the ants leave no pheromone. As the 100 ants arrive at the first decision point, on average, 50 will take path A and 50 will take path B. If the ants do an about turn when they have collected a load of food, equal numbers on average will return along each path. Of course, because this is a statistical process, it is highly improbable in any individual event that exactly 50 ants will take each path initially. Even if they did, the probability that they will continue to divide themselves between the paths is vanishingly small. But, over a number of events, about half will use each path as there will be nothing to recommend one path over the other.

However, now consider the effect of leaving a pheromone trail. As each ant reaches a decision point, the probability that they will follow a path is influenced by the amount of pheromone left on that path by other ants. The more pheromone on the path the more likely it is that an ant will take that path. Since pheromone evaporates in time, an ant reaching a decision point will be more likely to follow the path taken *recently* by the largest number of other ants.

If the first ant arrived at the first decision point and took path-A, it is slightly more probable that the second ant will also take path-A and, if both of those take path-A, the probability that the third ant will also take path-A is definitely more than 50%. Of course, as we are dealing with probabilities that are less than one, not all ants will immediately take path-A. Maybe the first two ants arrived at the decision point together and took different paths. However, sooner or later the number of ants travelling on the two paths will become unbalanced and thus the pheromone trail on the two paths will start to differ.

Now a positive feedback mechanism kicks in. Later ants are more likely to take the more travelled path, which further reinforces the imbalance in the amount of pheromone on the two paths. This will lead to an even higher percentage of ants taking the more travelled path in future with a concomitant decrease in the number of ants on the less travelled path. The pheromone will steadily evaporate from each path. It will be replenished (if not augmented) on the highly travelled path but will fade on the unpopular path as the evaporation rate exceeds the replenishment rate. The net effect is that the more travelled path becomes even more travelled while the less travelled path gets travelled less and less by ants until the pheromone imbalance between the two paths becomes so large that almost all the ants take the more travelled path.

There was nothing inherently better about what turned out to be the highly travelled path as the paths are of equal length. It is just that some random initial ant distribution favoured one path and the pheromone allowed it to reinforce its most travelled status. If there is a real difference between the paths, the process becomes quicker - consider the modified arrangement in Fig. 5.2.

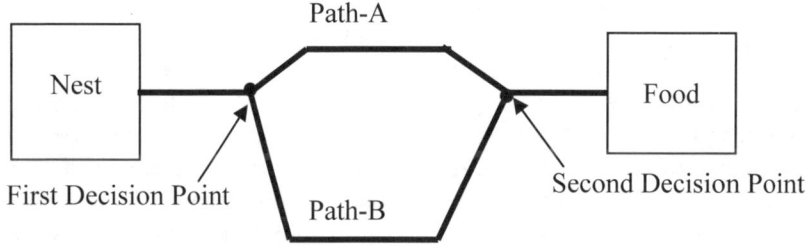

Fig. 5.2. Two dissimilar length paths from nest to food

Even if slightly more than half the ants take the longer path-B at the first decision point, by the time the first ants arrive at the food (via the shorter path-A) and decide at the second decision point which path to return by there will only be pheromone on path-A. As a result most of the ants will also return by this path, further reinforcing the pheromone there. By the time the ants on path-B are ready to return to the nest they see the pheromone they just laid on path-B, while path-A has the pheromone left by the ants that took that path both coming and going (less some evaporation).

Some of the ants that reached the food via path-B will return via path-A, augmenting the pheromone on that path. Those that do return via path-B will take longer to come back and by the time they arrive back the pheromone they laid down at the start of path-B will have faded more than that on path-A, which will have been reinforced by the ants that went to the food via path-A and have already returned the same way, plus the ants who deserted path-B at the second decision point and have already arrived back via path-A.

As a result, more ants setting out will now take path-A, and the more frequent reinforcing by the larger number of ants will soon lead to (perhaps) almost all ants following the shorter path. This, from the point of view of food gathering for the colony, is the most efficient path to take.

5.2 The Power of Collective Action

The word 'intelligence' has a wealth of meanings that often causes confusion when discussing intelligent systems. Even the word 'collective' may mean different things to different people. As a result, it should come as no surprise that the term 'collective intelligence' requires careful elucidation.

In this chapter, the term 'collective intelligence' will be used to describe the way that a number of relatively unintelligent individuals are able to achieve feats that would be far beyond any one of them individually as a direct result of the way that they interact together. We are not talking about a number joining together to simultaneously perform the same action and thus augmenting a particular capability of any one of them by a factor equal to the number assisting: for example two individuals joining together to move an object too heavy for either to move on their own. We are talking about a number of individuals performing different, although often similar, tasks – the net effect of which is to collectively achieve a result that is different to (and hopefully better than) the simple sum of these individual actions.

The inspiration for this work comes from biology, especially the actions of ants and birds, and does not require any directed communication between specific individuals. While direct negotiation between individuals could be allowed the objective is to see what can be achieved without it. As we are in no way bound by biology, when it suits our purpose we permit our individuals to do very un-biological things. The fact that ants and birds are able to achieve such amazing results augers well for our artificial systems provided that we do not depart fundamentally from what appears to give these real systems their power.

In particular we impose the following constraints on our artificial systems[1]:

1. *Intra Generation Learning Only* – In the algorithms described in this chapter all the 'learning' takes place within the current and only generation of individuals, even if these individuals may from time to time be erased and restarted at a random position. Importantly, without generations there are no performance selected parents and breeding and so obviously there cannot be any transfer of information between parent and child. As a result we can be sure that the effects observed are a result of the organizational aspects of the algorithm and cannot be the result of a traditional evolutionary process,

2. *Inter Changeability And Simplicity Of Individuals* – Ideally all the individuals that we will use will be identical in basic capability, although of course each individual will have its own particular internal parameter values (position, velocity and so on) that will change with time[2]. The power of coevolution in producing very effective sets of symbiotic 'species' is well known and there is little doubt that the use of multiple teams, each team consisting of task specific individuals, may offer similar advances. However, at the present stage of development it is not generally clear how to apportion ability between multiple teams. A few specific instances (e.g. ACS and Max-Min as described below) can be considered as using two teams, one being a small team of specialist individuals (often one) that perform one very specific purpose only (in the cases mentioned to reinforce the best solution found so far). This may be a first step towards symbiotic teams in the future.

3. *A Reliance On Indirect Communication Only* – While all individuals will broadcast information they do this either to the whole 'colony' or to

---

[1] It is tempting to speculate what might happen if we endow our individuals with greater individual ability. It is this author's contention that until we understand how collectives of simple individuals work we will not be in the position to know just what kind of extra ability to add to usefully augment the performance of this simple collective behaviour. It is inevitable that as this field develops we will be in a position to add individual capacity to augment the simple collective behaviour, and indeed some examples are discussed in this chapter, but the fundamentals must be right first.

[2] Naturally it is tempting to speculate how a combined approach would fare - that is allowing successive generations of individuals to inherit from their forebears and also cooperatively learn within each generation. Again, in this chapter such speculation is firmly ignored in favour of first getting the basics right.

all those within some specific distance from themselves. Direct peer-to-peer communication is not allowed. While this may seem very restrictive from a human communication perspective it is in fact not uncommon in nature. Many animals (including humans) can use scent to attract or repel other individuals and thus influence the future behaviour of those other individuals. Similarly, sound can be used, not as a language but rather as cries and screams. Although these may sometimes be directed at a specific individual in the real world sound can always have an effect on all individuals in hearing range. Other examples of indirect communication in the real world include road traffic signs or the state of some terrain. Changing a sign or modifying the terrain will have an effect on all other individuals in that vicinity. A consequence of prohibiting direct communication is that negotiations regarding task sharing cannot occur. This ensures a flat organizational structure that excludes a hierarchy of command or master-slave type relationships.

In summary, all the individuals of any particular type we use are identical in form, function and status. The number of different types we use is generally very small, most commonly either one or two.

5.3 Optimisation

5.3.1 Optimisation in General

For our purposes optimisation can be considered as finding the set of values that gives the best possible result for a given task. The values could be a collection of settings (numbers) for a process, the order in which a sequence of steps should be taken, the minimum number of items needed to perform a task or some combination of these. In order to find how good a particular set of values is we must be able to have a process (often a simulation) that performs a one-to-one mapping between that set of values and the quality of the result they would produce.

Having a process that goes from a set of values to a result quality does not necessarily mean that you can go from a desired result back to the set of values that would give that result quality. In fact, for most real world processes, this return relationship may well be 'one-to-many' and/or the interactions between the input values so strong that the reverse relationship is too complicated to be solved algebraically. In the real world we might be forced to a process of trial-and-error (perhaps augmented by a 'feeling' for parameter values that have previously proved to be good), which if tried systematically and exhaustively would often take far too long to be useful.

Under such conditions we would like an algorithm that efficiently and effectively searches even if it does not always find the very best result. This can be done by not exploring combinations of values in the vicinity of values already explored that have been found to produce poor results. If done well, this may reduce the time taken to search for an optimal set with only a slight risk of missing the optimum set.

5.3.2 Shades of Optimisation

Human nature being what it is, we commonly imagine that the term 'optimisation' implies a search for the very best. However, there are times when, although we might like the very best, we would be prepared to settle for something less provided that it is still very good. We usually do this because of time constraints. In general you can only know that you have the very best solution to something if you have exhaustively tried all possible solutions. For many real world problems there is just not enough time to do this and still get the answer in a meaningful time – there is no point in being told in five hours time what the optimum production process should be when you have to start the process in one hour.

When time is limited, algorithms are acceptable which regularly find very good optima orders of magnitude faster than it would take an exhaustive search algorithm to find the global optimum. It might not be the actual final result obtained that is important, rather the rate at which the best known solution improves with time. When time is really limited we might want an algorithm that has a rapid rate of improvement (in the current best known solution) for at least the first part of the process, even if later this rate will slow significantly and the final converged solution if we run it long enough be relatively poor. We would use this type of algorithm when the time we have to run it is so small that we can never get past this first part. We might gratefully accept what we obtained in the time available, even though we know that it was almost certainly sub-optimal.

5.3.3 Exploitation versus Exploration

To be really efficient optimisation algorithms require a balance between the exploration of new possibilities and the exploitation of prior experience. Excessive exploration can result in unsystematic and repetitive search, while extreme excessive exploitation can result in endless re-evaluation of the previously explored results. Managing the balance between these two factors is a critical part of algorithm design.

5.3.4 Examples of Common Optimisation Problems

Some results obtained using optimisation by collective intelligence will be given later in this chapter. Each of these optimisation problems will be briefly described now to familiarise the reader before discussing how collective intelligence can be applied to them.

Minimum Path Length

The most common example of this type of problem is the Travelling SalesPerson (TSP) problem, which requires us to find the shortest path that visits each of a number of locations exactly once and then returns to the origin point. The size of a path may be described either in terms of the total physical distance covered or the total time taken to complete the path. Whilst trivial for a small number of locations the number of possible solutions grows dramatically with the number of locations. If the distance or time between location-A and another location-B is the same no matter the direction of travel between the two locations for all the locations in the problem, the resulting TSP is called symmetrical, otherwise the TSP is called asymmetric. As a complete path is the same no matter which of the locations on it is specified as the starting point, the number of possible solutions to an N location asymmetric TSP is $(N-1)!$. For a symmetric TSP the direction of travel does not matter either so the number of possible solutions is $((N-1)!/2/)$. For a 14-location problem (considered to be an almost trivial TSP) the total number of different symmetrical paths is just over $3.1*10^9$. For a 52-location problem this has climbed to just over $7*10^{65}$. Exhaustive search, which might be marginally feasible for a 14-location problem, is almost certainly out of the question for a 52-location TSP. Practical TSPs involving hundreds of locations are not uncommon.

Many other problems can be cast into the form of a path minimisation problem. All that is required is that any solution to the problem can be built from a sequential series of decisions, each of which requires one choice to be selected from a known finite set of possibilities, and that a way exists of monotonically describing the quality of the solution as a simple number.

Function Optimisation

This class of problem involves finding a maximum or minimum value of a function. While fairly trivial for functions that involve a few variables and operators, the complexity rapidly increases as the number of variables increases, especially if the effects of variations in the parameter values are interdependent. Many techniques, such as hill climbing, can find a 'good' result quickly. However, there may be many 'good' results; the problem is to have a high probability of finding the best result. Only exhaustive search can guarantee to find the very best result, but for complex functions this might take an unreasonable time.

Any problem that can be cast in the form of a 'black box' (that given a set of real number inputs returns a real number output) can be transformed (in principle) into a function optimisation problem. The 'black box' can be a mathematical function or any process (or simulation) for which a given set of inputs always gives the same output. For example, it could be finding the set of input values that give the largest (or smallest) output from a previously trained neural network.

Sorting

In this type of problem the purpose is to sort a number of objects so that the distance between two objects is related to how similar they are (by some measure). Problems of this type are fairly trivial when there are a number of clearly delineated classes, but rapidly become harder as the spectrum of objects moves towards a continuum.

Multi Component Optimisation

The optimisation problems previously mentioned are neatly defined and clearly distinguishable from each other. Real life is rarely so obliging. Consider, for example, a schedule for a manufacturing process. A good schedule has many facets: it produces the maximum output, it keeps machinery as fully occupied as possible, it minimises waste and operating costs, it minimises the number of late deliveries and it may even provide time to perform maintenance on some machine. It may involve deciding whether to run a particular machine at all this shift or how many staff to allocate to various jobs. As such it may require the ability to find a composite optimum that involves deciding orders of things to do, how many things are to be used and when they are to be used. This is obviously harder than optimising any one of the previously mentioned problems in isolation.

5.3 Ant Colony Optimisation

Ant Colony Optimisation (ACO) algorithms are examples of algorithms that find good solutions without exhaustive search [6]. Each of these algorithms minimises the total cost of performing a fixed set of actions exactly once and returns the action order with a small (ideally the smallest) associated cost. It is most commonly described in the context of the Travelling Sales Person problem in which the actions are visits to locations and the cost is the total distance travelled, the sum of the path segments that join the various cities. Many papers have solved the TSP using ACO methods including the seminal [7], together with [8-10, 25]. This last paper contains a summary of all ACO applications to the TSP reported up to publication date.

ACO algorithms use an analogue of the chemical pheromone used by real ants. This pheromone is attractive to and tends to be followed by ants but is not the only factor influencing their path choices. Pheromone evaporates over time (diminishing the influence of old information) but is reinforced by ants laying more pheromone on each path they travel. The pheromone trails build up a collectively derived estimate of the worth of the various possible path segments.

A simple description, applicable to the whole set of ant colony optimisation algorithms, is first given below. Then the detail as to how individual algorithms differ from this is described.

1. The pheromone on every path segment is initialised to an initial value (τ_0) and N ants are randomly distributed among C cities. The initial city, although part of the path, is not considered as having been moved to.
2. Each ant decides which city to move to next. The exact way they do this is algorithm dependent (see later), however ants are always prohibited from returning to a city previously visited. Each ant moves to its new chosen city and then considers the city it should move to next, repeating this process until it is back at the city it started from. The ant has completed one tour when it has visited all the cities.
3. Each ant calculates the length of its tour, and updates the information about the best (shortest) tour found so far, if necessary.
4. The pheromone levels on each path segment are then updated. The exact update strategy is algorithm-dependent but all involve evaporating existing pheromone and adding extra pheromone to selected path segments.
5. All ants that have completed their assigned maximum number of tours (typically one) die and are replaced by new ants at randomly chosen cities.
6. The algorithm now returns to step 2 and the process continues until some stopping criteria is met, for example the best path length being below a threshold value or a maximum total number of tours having been completed.

It is in the detail of steps 2 and 4 that the various algorithms differ. Two of the most common algorithms, AS and ACS, are described below in Sects

5.4.1 and 5.4.2, with two later algorithms – AMTS and Max-Min – being described in sections 5.4.3 and 5.4.4, respectively. The term 'ACO' is used as an umbrella term that could mean any one of the four algorithms described in Sects. 4.1 through 4.4.

5.4.1 Ant Systems — the Basic Algorithm

The original of these algorithms – the Ant System (AS) – algorithm was inspired by the behaviour of Argentinean ants [6].

In AS an ant at city i calculates which city to visit next by first calculating the attractiveness A_{ij} of each possible city j. $A_{ij} = 0$ if the city has been visited before or $A_{ij} = P^{\alpha}{}_{ij} = P^{\beta}{}_{ij} D^{\alpha}{}_{ij}$ if it has not. P_{ij} is the pheromone level on the path segment from city i to city j, D_{ij} is the distance from city i to city j and α and β are two user chosen parameters (typically 2 and –2 respectively). The probability that an ant moves from city i to city j is $A_{ij} / \sum_{j=1}^{C} A_{ij}$,

and the decision as to which city to visit next is solely dependent on this set of probabilities.

In AS the pheromone levels are updated simultaneously once all ants have completed a tour. The pheromone on every path segment is first decreased by being multiplied by $(1 - \rho)$, where ρ is less than unity. Each ant then augments the pheromone on all the path segments it used on its tour by adding ρ_u/L units of pheromone to each path segment, where ρ_u is another parameter and L is the length of the tour.

Deciding which city to visit next involves at least one random number and this produces the exploration aspect of the algorithm. However, the balance between exploration and exploitation is stochastic and cannot be accurately predicted.

The Problem With AS

The basic ant system formula relies on the two random numbers to set the balance between exploration and exploitation. While such an approach is simple and ensures that each has a turn at being the most important factor (at least once in a while), it would be naïve to assume that the balance is likely to be anywhere near optimal. While this is not important on small problems (ones with a relatively small number of possibilities to choose from), on large problems it causes either slow exploration and hence a long time to find a stable situation *or* no further progress being made after obtaining a relatively poor result *or both*. For this reason researchers have developing modified algorithms that take at least some direct control over the balance between exploration and exploitation. The original algorithm is still the basis for these new algorithms and so should be understood before considering these later variants.

5.4.2 Ant Colony Systems

The Ant Colony System (ACS) [5] differs from AS in two important ways: the method of deciding the next city to visit and the pheromone update procedure. These changes make the algorithm more complex but generally lead to better performance than AS.

The decision process starts with the calculation of the attractiveness A_{ij} of each city j using the same formula as AS except that the value of α is unity. A random number is then compared to a preset threshold, the greedy probability[3]. If the random number is below this threshold the city with the highest attractiveness is chosen, otherwise the probabilistic selection process from AS is used. This change increases the emphasis on exploitation compared with AS and has the effect of ensuring that many of the path segments each ant follows are in common with the currently best known path, thus encouraging such exploration as occurs to occur around this best known path.

The second change is that two different types of pheromone update occur. The best-found path is reinforced even if no ant has exactly followed that path this tour. This update mechanism does not alter the pheromone levels on any path segment not involved in the best-known path. The pheromone on every path segment of the best known path is first decreased by being multiplied by $(1 - \rho)$ where ρ is a user chosen parameter (less than unity) and then has ρ_u/L^+ units of pheromone added to it, where L^+ is the length of this best known path.

The second update mechanism updates the pheromone on every path segment actually travelled by an ant, but in such a way as to *reduce* the value of the pheromone on these segments. Each ant updates all the path segments it used by first multiplying the current pheromone level by $(1 - \rho)$ and then adding $\rho * \tau_0$ units of pheromone where τ_0 is the initial value of pheromone deposited on all path segments.

The pheromone levels on used segments that are not part of the current best path are moved asymptotically down towards $\rho * \tau_0$ while the path segments used in the currently best-known path have their pheromone

---

[3] Actually there is no reason why the threshold should not be modified as the algorithm progresses. While such an approach is attractive it is not discussed here as the algorithm for varying the threshold will introduce yet extra complication and parameter values.

levels increased. Note that the pheromone levels on path segments not involved in the best-known path nor travelled by any ant this tour are left unchanged.

As a result, ants that do not travel on a best path segment are more likely – in time – to choose a less used segment. These two pheromone update techniques simultaneously encourage exploration of little used path segments and the exploitation of the segments that are part of the current best-known path.

5.4.3 Ant Multi-Tour System (AMTS)

Both AS and ACS share one particular common feature – after an ant has completed a tour it dies and is replaced by a new ant at a randomly chosen city. In the Ant Multi-Tour System (AMTS) ants do not die until they have completed a user chosen number of tours. After their first tour each ant carries some information from its previous tours; in particular how many times it has previously moved from city i to city j. This information is used to encourage the ant not to follow (exactly) the paths that it travelled previously but to explore variations to these paths.

Each ant builds up a list of the prior path segments it has followed from each city. It uses this information to reduce the probability of repeating a choice previously made. The only difference to the AS algorithm is the modification made to the attractiveness A_{ij} of city j to an ant at city i, which is now $A_{ij} = P^{\alpha}_{ij} D^{\beta}_{ij} / F_{ij}$. The extra term F_{ij} is a factor that is derived from the number of times that this ant has chosen to move from city i to city j in previous tours ($prior_{ij}$). A suitable formula for calculating F_{ij} is $F_{ij} = 1 + \sqrt{prior_{ij}}$ which produces the values shown in Table 5.1.

Reducing the probability of repeating choices that have already been made encourages an ant to increase exploration at the cost of exploitation on all tours after the first.

Table 5.1. The relationship between F_{ij} and prior

$Prior_{ij}$	0	1	2	3	4	5	6
$F_{ij} = 1 + \sqrt{prior_{ij}}$	1	2	2.4142	2.7321	3	3.2361	3.4495

5.4.4 Limiting the Pheromone Density – the Max-Min Ant System

If the pheromone level on the most densely travelled path becomes high, many ants will take this path further reinforcing the high pheromone level.

This will have the effect of decreasing the amount of exploration. At the same time, the pheromone levels on infrequently travelled path segments can drop so as to be insignificant, again decreasing the probability that they will ever be explored. Just normalising the pheromone densities on the paths to and from a city does not help as this does not alter the relative levels on the frequently and infrequently travelled path segments.

One way of addressing these problems is to limit both the maximum and minimum permissible pheromone levels. One algorithm for doing this is the Max-Min algorithm [26]. In this algorithm the pheromone on all path segments is initialised to the maximum density that is to be allowed, τ_{max}, which is equal to $1/\rho$ where ρ is the evaporation rate. At the end of an iteration when all ants have completed their tour, all paths suffer evaporation during which their pheromone is reduced by a factor of $(1 - \rho)$. A minimum pheromone level is specified for any path segment (τ_{min}). Any path segment whose pheromone level would drop below this as a result of the application of the $(1- \rho)$ factor has its pheromone level reset to τ_{min}. The path segments that make up the best path (only) have pheromone added to them. One unit of pheromone is added to their pheromone level but the level is then reduced to τ_{max} if necessary.

5.4.5 An Example - Using Ants to Solve a (Simple) TSP Problem

The Burma 14 data set (Table 5.2) is a relative small symmetric TSP problem. However, it is large enough to test optimisation TSP algorithms while simultaneously being small enough that the full set of possible solutions has been exhaustively generated. The problem is to find the shortest tour between the 14 cities, specified by their latitude and longitude in the data set, with each city being visited exactly once. There are 14 possible starting points for each closed tour and each tour can be travelled in two different directions; none of these choices affect the tour length. The number of unique tours is 13!/2, or approximately $3*10^9$.

Table 5.2. The Burma 14 data set

City	Latitude	Longitude	City	Latitude	Longitude
0	16.47	96.10	7	17.20	96.29
1	16.47	94.44	8	16.30	97.38
2	20.09	92.54	9	14.05	98.12
3	22.39	93.37	10	16.53	97.38
4	25.23	97.24	11	21.52	95.59
5	22.00	96.05	12	19.41	97.13
6	20.47	97.02	13	20.09	94.55

The length $dist_{ij}$ of a path segment between two cities i and j (latitudes la_i, la_j and longitudes lo_i, lo_j, respectively) is given by Eq. (5.1).

$$dist_{ij} = \text{int}(6378388 \arccos(0.5*((1+q_1)* q_2)-((1-q_1)* q_3)+1) \quad (5.1)$$

where $q_1 = \cos(long_1 - long_2)$
$q_2 = \cos(lat_1 - lat_2)$
$q_3 = \cos(lat_1 + lat_2)$
$lat_1 = \pi(\text{int}(la_i) + 5(la_1 - \text{int}(la_i))/3)/180$
$lat_2 = \pi(\text{int}(la_j) + 5(la_2 - \text{int}(la_j))/3)/180$
$long_1 = \pi(\text{int}(lo_i) + 5(lo_1 - \text{int}(lo_i))/3)/180$ and
$long_2 = \pi(\text{int}(lo_j) + 5(la_2 - \text{int}(lo_j))/3)/180$

Note that the length of each path segment is floored to an integer as prescribed in the TSLIB library of TSP problems [23].

Table 5.3. The top ten paths for Burma14

Length	City order
3323	0,1,13,2,3,4,5,11,6,12,7,10,8,9
3336	0,1,13,2,3,4,5,11,6,12,10,8,9,7
3346	0,1,13,2,3,4,11,5,6,12,7,10,8,9
3346	0,7,1,13,2,3,4,5,11,6,12,10,8,9
3359	0,1,13,2,3,4,5,11,6,12,10,9,8,7
3359	0,1,13,2,3,4,11,5,6,12,10,8,9,7
3369	0,7,1,13,2,3,4,11,5,6,12,10,8,9
3371	0,7,12,6,11,5,4,3,2,13,1,9,8,10
3381	0,1,13,2,3,4,5,11,6,12,7,9,8,10
3381	0,1,13,2,3,4,5,11,6,12,8,9,10,7

The distribution of the full set of tour lengths is shown in Fig. 5.3, together with the distribution of the ten shortest tours. The length and the actual city orders for the ten shortest tours are given in Table 5.3. The two shortest tours are shown in Fig. 5.4, and the number of path segments different between the top ten paths is shown in Table 5.4.

Each of the four algorithms (AS, AMTS ACS and Max-Min) was used to solve the Burma 14 TSP problem using the parameters shown in Table 5.5. The AS and AMTS values had been previously found by the author to be effective for the Burma 14 data set; the ACS values follow the recommendations in [10]. An iteration consists of all ants completing one tour. Runs were terminated after 100 iterations had occurred without a new best path being found.

Table 5.6 shows the results obtained using AS, AMTS (with different values for the maximum age for the ants), ACS and Max-Min. The results show the excess path length of the average result compared to the best possible path length of 3323.

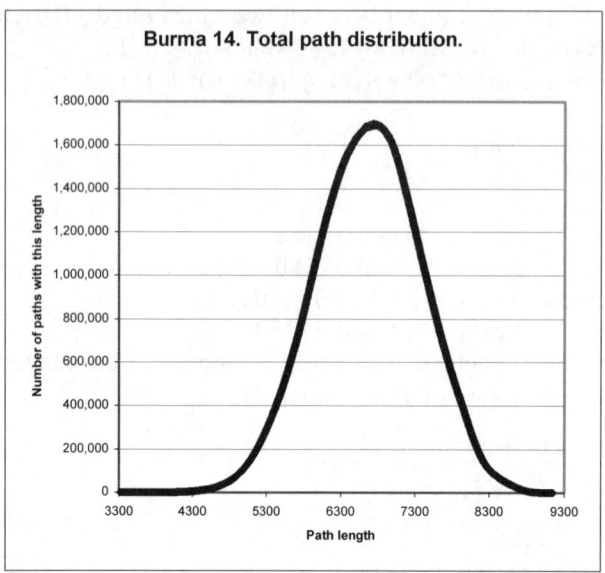

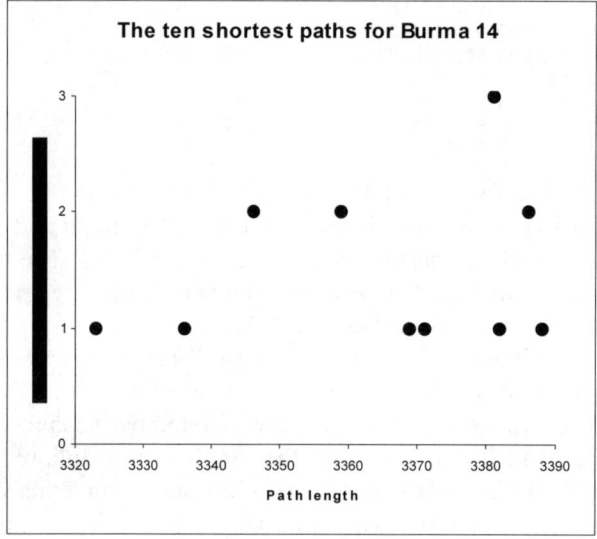

Fig. 5.3. The distribution of all Burma 14 path lengths (top), together with the path length distribution for the shortest ten paths 3323 to 3388 (bottom)

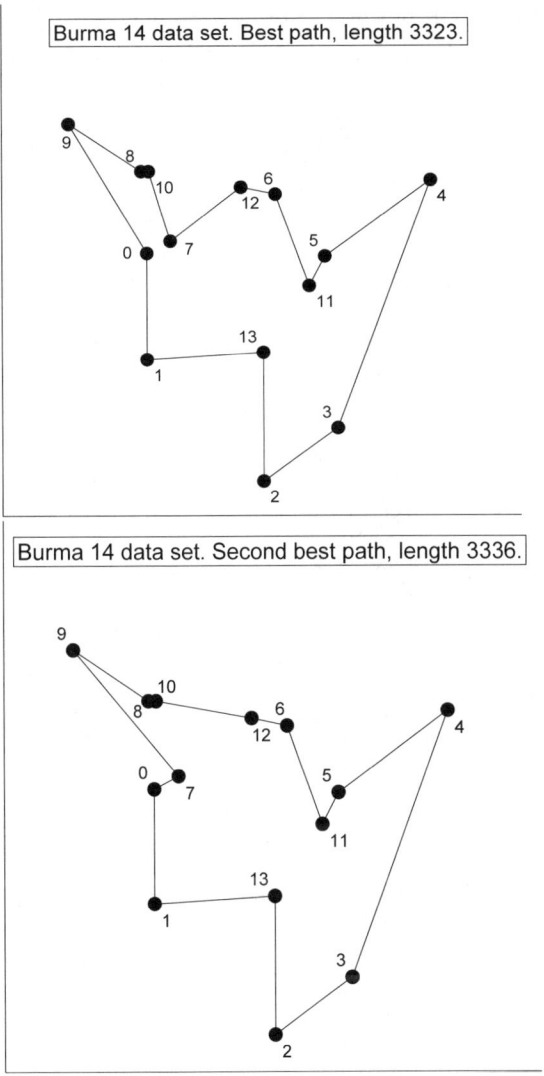

Fig. 5.4. The two best paths for the Burma 14 data set

It is clear that the AMTS algorithm outperforms all other algorithms on the Burma 14 TSP, in terms of the quality of the solutions that it finds. ACS takes the least number of iterations to reach its final path. The parameters used for the ACS algorithm are the generic values suggested by [10], while the parameter values used for AS have been optimised for this data set during prior work. The ACS algorithm produced the fastest results, although it found poorer solutions than those found using AMTS.

Table 5.4. The number of path segments different between the top 10 paths for Burma 14

	3323	3336	3346	3346	3359	3359	3369	3371	3381	3381
3323	-	3	2	3	5	3	6	5	2	3
3336	3	-	5	3	2	2	4	4	2	3
3346	2	5	-	3	6	3	3	5	4	5
3346	3	3	3	-	4	5	2	3	4	5
3359	5	2	6	4	-	2	6	5	3	2
3359	3	2	3	5	2	-	3	6	4	5
3369	6	4	3	2	6	3	-	5	6	7
3371	5	4	5	3	5	6	5	-	2	5
3381	2	2	4	4	3	4	6	2	-	4
3381	3	3	5	5	2	5	7	5	4	-

Table 5.5. Parameter values used for each of the algorithms

Parameters	AS and AMTS	ACS	Max-Min
Number of ants	10	10	10
Alpha	2	1	1
Beta	-2	-2	-2
Initial pheromone τ_0	0.1	$2.2*10^{-5}$	3.3
Pheromone decay ρ	0.1	0.1	0.3
Pheromone update ρ_u	0.5	0.1	1
Greedy probability	n.a.	0.9	n.a.
Pheromone maximum	n.a.	n.a.	3.3
Pheromone minimum	n.a.	n.a.	0.1

The AMTS algorithm typically ends at one of the two shortest paths. Consider Table 5.4 which shows the number of path segments that differ between each of the top ten paths. It can be seen that transitions from the majority of the poorer paths to the second best path involves one or more steps that generally require a smaller number of path segments to be changed than would be involved in moving to the best path. Moving directly from the second best to the best path requires three path segments to change with the result that both the two best paths of the Burma 14 TSP data set are fairly stable, as reflected in the percentage of times that one of them is the final solution found.

Table 5.6. AS, ACS and AMTS $F_{ij} = 1 + \sqrt{prior_{ij}}$ results (all results expressed as a percentage)

Algorithm	AS	AMTS $F_{ij} = 1 + \sqrt{prior_{ij}}$						ACS	Max-Min
Max age	n.a.	2	3	4	5	6	7	n.a.	n.a.
Path length = 3323	6	19	50	53	56	59	40	15	17
Path length = 3336	41	32	32	42	38	39	59	53	62
Path length = 3346	7	3	5		2				
Path length = 3359	15	16	10	3	2				
Path length = 3369	1	1							
Path length = 3371	4	8				1		22	10
Path length = 3381	23	20	3	2	2	1	1	9	6
Path length > 3381	3	1						1	5
Av excess path	30.29	27.16	10.65	7.7	7.28	6.13	8.25	24.08	22.09
Average iterations	47.15	51.51	48.6	41.49	38.4	29.99	22.02	9.45	30.54

As the maximum age of the ants increases in the AMTS algorithm a point is reached at which the quality of the solution starts to degrade, even though the speed with which they are found continues to fall. This happens as the value of F_{ij} becomes so large that the ant rarely follows any path segment of the current best path. Since it is likely any better path will share at least some (and probably many) path segments with the current best path, such a strong disinclination becomes counter productive.

The ACS and AMTS algorithms have a closer similarity in approach than might at first appear. The key difference is that ACS works on a global level with the best path found by anyone in the whole colony being reinforced and all path segments travelled by any ant being made less attractive to all other ants. In AMTS each ant reinforces the best path it has personally found so far while each ant is only influenced not to re-explore the path segments it has travelled in the past. This leads to a greater diversity of collective knowledge about good path segments in AMTS and reduces the emphasis on exploitation when compared to ACS. The faster convergence rate of ACS compared to AMTS is a result of this higher emphasis on exploitation.

The ant-specific nature of the re-exploration deterrent in AMTS, rather than the global deterrent of ACS, seems more appropriate. Each ant must evaluate the undesirability of each possible path segment based on the experience of and the paths already tried by this ant, as well as on the net experience gained from all the paths travelled before by any ant. Since AMTS always involves a probabilistic decision at each city, as opposed to often taking the most path with the highest desirability as in ACS, the quantity of exploration may be expected to be higher too but the convergence rate slower.

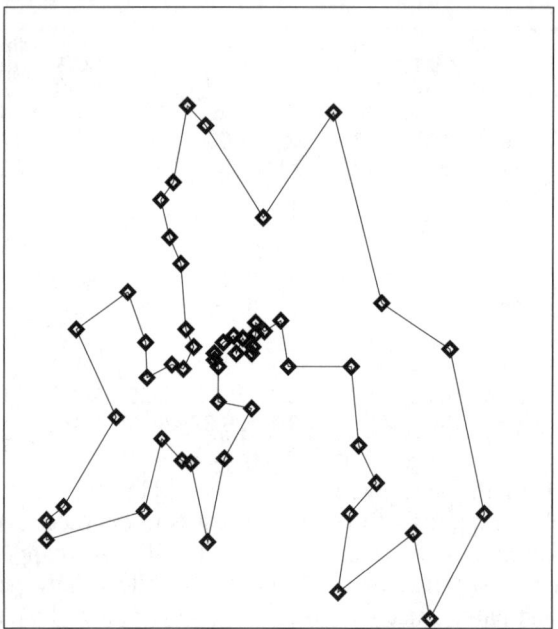

Fig. 5.5. The best solution for the Berlin 52 city TSP data set

Usually a new good path will include many path segments in common with previously best-known paths. An example of this can be seen from Fig.5.4 in which the difference between the best and second best path is restricted to the path segments between the cities in the top left corner of the plots. As a result, an over emphasis on exploration becomes counter productive as all paths tried are likely to contain too few of the already discovered necessary path segments. Thus there is likely to be an upper limit to the useful age of, and therefore the number of tours completed by, an ant. For the work described here the useful upper age limit appears to be about six. The useful upper age limit may well prove to be problem-dependent.

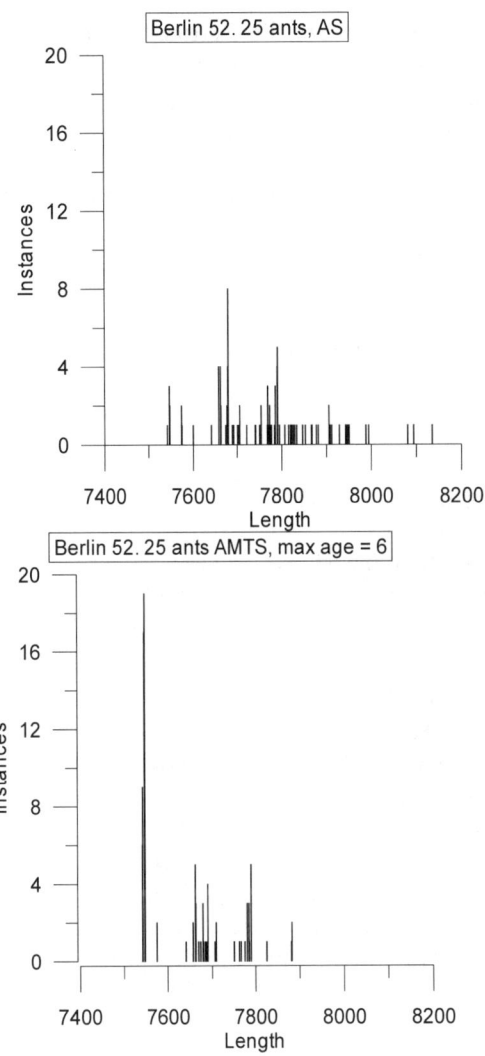

Fig. 5.6. The results of 100 repeats of the AS algorithm (top) and the AMTS algorithm (bottom) on the Berlin 52 TSP data set

The results for the Burma 14 data set, while significant, are not in themselves conclusive as there remains the possibility that it is some aspect of this particular data set that responds to the AMTS algorithm better than to the AS algorithm. Results have been obtained for the Berlin 52 data set and these are shown in Figs. 5.5 and 5.6. It should be noted that, apart from increasing the number of ants to 25, all the parameters used for both the AS and AMTS are identical to the values used for the Burma 14 data set and are unlikely to be optimum. It is not the absolute results that are important here, although the best solution (with a path length of 7542) was found (once by AS and nine times by AMTS in a 100 repeats). What is important is the improvement that accrues from allowing the ants to take multiple tours. Overall these results suggest that the AMTS algorithm may well be beneficial for TSP problems in general.

Table 5.7. The relative performance of the AS and AMTS algorithms on the Berlin 52 TSP data set

	AS	AMTS max age 6
Average excess path	216.4	90.8
Average iterations	66.0	33.1

5.4.6 Practical Considerations

For the basic AS algorithm, experience has shown suitable values of alpha and beta to be plus two and minus two respectively and to be remarkably stable across problems. Too many ants will lead to everywhere being swamped with pheromone, although of course the exact point at which this happens will also depend on the rates of deposition and evaporation of pheromone. General experience suggests that the number of ants should not exceed the number of cities (or equivalent) in the problem. The values for the initial level of pheromone, and the pheromone deposit and evaporation rates will need to be found by experimentation.

For ACO the value of alpha recommended is one with beta still at minus two. The initial pheromone level is recommended to be the reciprocal of the path length found by always selecting the nearest valid neighbour as the next city to visit (as if alpha was zero). Both the decay and deposit parameters need to be set to 0.1 and the probability of making a greedy choice to be 0.9.

For the Min-Max algorithm the pheromone decay parameter is recommended to be 0.3. This then sets the initial pheromone level and the maximum pheromone to both be the reciprocal of this (3.3). Alpha and beta are set as for AS and the minimum pheromone level low (for example 0.1).

Finally the AMTS system uses the same parameters as AS with the addition of a maximum age for each ant. While six seems to work well for a number of problems, this like most parameters given above should only be considered as being indicative − experimentation will be needed to optimise the values for any particular problem. Fortunately no parameter value appears to be overly critical.

5.4.7 Adding a Local Heuristic

Ant Colony Optimisation algorithms are broad search algorithms that consider a vast array of possible solutions quickly. They are not so good at local search. For this reason they are often coupled with a local search heuristic. This will be problem-dependent but which explores in the local vicinity around a good solution as found by ACO to see if there is an even better solution. Ideally this local search can be performed efficiently without adding excessively to the overall computational burden. Local search is done after all ants have completed a tour but before the pheromone is updated. This allows a better path, if found, to be included in the pheromone update process.

Taking the TSP as an example, a good local heuristic is to reverse the path between two randomly chosen points in the current best-known tour. Much of the path will not change, so the computational load will be small but the effect on the overall path length may be beneficial. Consider a trivial TSP with just 9 cities and let us assume that the best-known path at some time is:

$$3\ 0\ 2\ 6\ 9\ 7\ 4\ 1\ 5\ 8$$

Let two random numbers in the range from 1 to 9 be chosen, say 3 and 6, Then the section of the path between the third and sixth cities is reversed to give:

$$3\ 0\ 7\ 9\ 6\ 2\ 4\ 1\ 5\ 8$$

Provided we are considering a symmetrical TSP, it is not necessary to recalculate the whole path to find the new length, which may take considerable effort for a large TSP, as there will always be two path segments that are removed (in this case 0 to 2 and 7 to 4) being replaced by two others (0 to 7 and 2 to 4). All that is required is to deduct the lengths of deleted sections and add the lengths of the added path segments. Note that the direction of travel for the path segments between the two swapped cities has reversed, which makes no difference for a symmetrical TSP but would add to the recalculation necessary for an asymmetric TSP.

Of course all reversals do not result in a shorter path; if they do we keep them, if not we revert back to the best path. After we have performed a preset number of segment reversals we return to the ACO algorithm, update the pheromones and continue with new tours.

Despite the efficiency with which we can calculate the modified path length, the local heuristic can add significant computational load, especially if the number of repeats is high. Inspection of the magnitude of the path segment reversed and the number of times this is successful shows an interesting trend. Let N be the difference in city number between the two cities that mark the start and finish of a reversed path segment. The proportion of successful reversals dropped as N increased.

In 100-repeated solutions to the Berlin 52 symmetrical TSP, from random initial ant positions, an average of 109.4 cycles were taken before path stability was achieved (a cycle is one tour by all the ants). So, overall, the local heuristic was invoked on the best path 10,940 times. The local heuristic tried 100 random path segment reversals so that a total of 1,094,000 modified paths had to be evaluated. Of these only 6697 resulted in a path improvement (about 0.6%). But of those that were successful, 25% involved reversing the path between adjacent cities when N = 1 (in this case just swapping the two cities). A further 12% involved reversing a path segment of length three (N=2). N=3 and N=4 were responsible for about 5% each. So for efficiency it is advantageous to favour shorter paths by randomly selecting the city that marks the start of the segment to be reversed and then selecting the length using a non-linear relationship in which the probability of picking a particular number drops with the size of the number[4].

For example, the author has used a simple algorithm that starts by setting a parameter to one and choosing a random number in the range from zero to one. If this random number is less than 0.25 the parameter value is accepted to be the length of the path section to reverse. If not the parameter is incremented and the process repeated. This process continues until the parameter is selected or until the parameter value reaches one half of the number of cities in the problem, in which case it is always selected. The probabilities of selection lengths up to 10 are shown in the Table 5.8.

Table 5.8. Probabilities of Section Lengths up to ten

Length	1	2	3	4	5	6	7	8	9	10
Probability	0.25	0.19	0.14	0.11	0.08	0.06	0.04	0.03	0.03	0.02

The results shown previously for the four ant algorithms did not include a local heuristic in order to show the basic performance of the algorithms. When the local heuristic described above was used (100 times) each time all the ants had completed a tour of the Burma 14 data set, the average

---

[4] This may seem a bit of a surprise until one considers that the effect of reversing a path segment of length, say, 40 in the 52-city TSP produces the same effect as reversing the other 12, except that the ant direction of travel is reversed between the two cases – which has no effect for a symmetrical TSP.

excess length and the average number of iterations became those shown in Table 5.9 below.

Table 5.9. The performance on the Burma 14 data set by the four algorithms when a local heuristic is used (figures in brackets show the change produced by the addition of the local heuristic)

Algorithm	AS	AMTS (max age = 6)	ACS	Max-Min
Average excess length	14.2 (-16.09)	5.98 (-0.15)	24.4 (+0.32)	20.09 (-2.0)
Average no. of iterations	16.73 (-30.42)	16.05 (-13.94)	8.87 (-0.58)	29.97 (-0.57)

5.4.8 Other Uses For ACO

As well as the TSP, ACO algorithms have been applied to a number of other problems. What is common to these problems is that the pheromone analogy can be used to build up a collective view of the best way that they should be solved. These problems include assigning items to groups or resources, for example the optimal distribution of items among a number of containers (the Knapsack Problem KP). Here, pheromone is used to record the desirability of assigning an object to a particular container. Related to this are a number of other problems such as the Quadratic Assignment Problem (QAP), Generalized Assignment Problem (GAP) and Frequency Assignment Problem (FAP).

A further problem type that can be solved using ACO is the graph colouring problem. Here the aim is to colour a graph using a minimum number of colours such that no two adjacent regions have the same colour. Note that one has to find not only the minimum number of colours but also how these are to be distributed. Pheromone is used to reflect the desirability of assigning two regions the same colour. This is more effective than using pheromone to reflect the actual colour assigned to each region as the problem is really to group regions into sets that will have the same colour. The actual colour assigned to a particular set is not central to the solution.

Once one realises that pheromone can be used to reflect aspects of a problem (for example, which grouping a region should be assigned to) rather than the whole solution (such as which city should follow this), it becomes apparent that ACO has applications far outside the TSP with which it historically started. For a fuller discussion of problems ACO has been applied to and the pheromone structures used see, for example [20].

5.4.9 Using Ants to Sort

While pheromone is the most commonly used form of stigmergy, other forms can also be used. For example behaviour can be modelled on the way that termite ants sort eggs from faeces. The stigmergy is the way the terrain is left, which influences the ants that follow them.

Sorting information into appropriate (often fuzzy) categories is a difficult yet fundamental and frequent activity. Often the information is mapped, typically to a lower number of dimensions, so that the spacing between pieces of information reflects the strength of one or more relationships between these pieces. If the map is built by a process that is partly stochastic and partly procedural, the resulting map may at times reveal previously unrecognised relationships within the information. Self-organizing artificial neural networks have been used for some years for this purpose (see, for example, [18]).

Sorting in nature is also important and ants are known to be efficient at sorting, for example, sorting larvae as they mature. Models of how they do this have been applied to sorting information, to see if an ant-like efficiency can be reproduced.

Deneubourg proposed a model for this behaviour in which randomly moving unloaded ants pick up items of a particular type with a probability P_p of

$$P_p = \left(\frac{k_1}{k_1 + f}\right)^2 \quad (5.2)$$

and loaded ants drop what they are carrying with a probability P_d of

$$P_d = \left(\frac{f}{k_2 + f}\right)^2 \quad (5.3)$$

where k_1 and k_2 are constants and f is the average number of items encountered during some time period [5]. This algorithm allows you to sort clearly distinguished classes of items so that each type appears in a different place. A variant of this algorithm proposed by [21] discounts the influence of previously encountered objects by a time factor so that recently encountered objects have more weight than objects encountered long ago – this is to limit the size of object piles.

Lumer and Faieta have generalised this algorithm to allow for more complex sets of objects with A attributes in which various types may have degrees of similarity with each other [19]. The algorithm projects the A-

dimensional space onto a lower (usually two-) dimensioned grid in such a way as to provide *intracluster* distances smaller than *intercluster* distances while also preserving some of the *interobject* relationships that existed in the original dimension space. They define a local density function $f(o_i)$ of an object O_i within a square of size S. Let L be the list of all objects with that square, if L is empty $f(o_i)$ is zero. If not,

$$f(o_i) = \frac{1}{S^2} \sum_{o_j \in L} \left(1 - \frac{d(o_i, o_j)}{\alpha}\right) \qquad (5.4)$$

where $d(o_i, o_j)$ is the Euclidean distance between objects o_i and o_j in the original space, and α is a user chosen parameter.

This work can be extended in two ways to produce the SOMulANT algorithm. Firstly the ants have vision and, for any cell within their sight, can measure the local disorder index. This is similar to Lumer and Faieta's local density function, except that the further o_i is from o_j the less influence it has. Secondly, the ants move to some degree in a directed fashion. The probability that an ant continues to move in its previous direction is known as the momentum of the ant. Ants that do not continue to move in this way move towards the area within their sight in which their load will best fit (if loaded) or move to the area of highest local disorder within their sight (if unloaded).

SOMulANT is built on the concept of a measure of local disorder [13]. The value of this for any location reflects the distribution and variation of all the data examples visible to the ant, currently in the small region R, of radius O_{radius}, centered on the ant's position. While this measure can be defined for a map with any number of dimensions, the rest of this discussion concentrates on two-dimensional maps.

Let $Dist_{i,xy}$ be the distance of data example i from the center x,y of the region R and let $Diff_{i,j}$ be the difference between data example j located at x,y and another example i located somewhere within the region R. The local disorder is then:

$$\sum_{\substack{all \\ examples \\ j\, at\, xy}} \sum_{\substack{all \\ examples \\ i\, within \\ region\, R}} \left(\frac{Dist_{i,xy}}{1 + Diff_{i,j}}\right) \qquad (5.5)$$

An ant has a limited vision range. Within this range the ant can measure the local disorder at any point. If unloaded it has a certain probability (controlled by the ant's momentum) of continuing along its current path. If it does not continue it looks for the region with the highest local disorder visible to it and heads in that direction. At each step an ant may pick up an example from underneath it if it is currently unloaded. The probability of doing so is proportional to how much doing this would decrease the local disorder at the ant's current location.

An ant can only carry one data example at a time. When loaded there is again a certain probability (controlled by the ant's momentum) the ant will continue along its current path. However, now if it does not continue it looks for the region it can see whose local disorder would be least increased if the example it carries were to be dropped there and heads in that direction. At each step it may drop the load it carries with a probability inversely proportional to the increase this would cause in the local disorder.

Let the ant be at position x,y and let E be the example that the ant is either considering picking up or dropping. Let LD_{withE} be the local disorder with example E dropped at x,y and $LD_{withoutE}$ be the local disorder with example E removed from x,y.

The probability that example E will be picked up depends on the effect that such an action would have on the local disorder. The probability increases the more the local order would be decreased by such an action and is zero if the action would increase the local disorder. The pick up probability P_{PU} is given by:

$$P_{PU} = \alpha \frac{LD_{withE} - LD_{withoutE}}{1 + LD_{withE} - LD_{withoutE}} \text{ if } LD_{withoutE} \leq LD_{withE}, \quad (5.6)$$

$$= 0 \text{ otherwise}$$

The probability P_D that example E would be dropped at x,y is inversely proportional to the increase in local disorder that dropping the example would have:

$$P_D = \beta \frac{LD_{withoutE}}{LD_{withE}} \quad (5.7)$$

The parameters α and β in the equations above are a measure of how sensitive the ant is to the local disorder. With each step, during which an unloaded ant does not pick up a load, the value of α for that ant may be increased slightly, as soon as it loads α is returned to its original value. Similarly, with each step during which a loaded ant does not drop its load, the value of β for that ant may be increased slightly. As soon as it loads β is returned to its original value. This can be useful in allowing probabilistic map perturbations that may minimize any tendency of the map to settle in sub-optimal configurations. The full algorithm is:

Deposit the examples to be mapped in a pile at the center of the map space and randomly distribute the ants across that map space giving each a random direction of travel. Chose a probability that an unloaded ant will continue in its current direction (P_{UC}) and a probability that a loaded ant will continue in its current direction (P_{LC}).

```
Repeat
    For each ant in turn:
        If the ant is unloaded:
        1. select a random number
        2. If this is less than P_UC try to move
           the ant one unit in its current
           direction.
        3. If not, find the map position within
           the range of vision of this ant with
           the highest local disorder and try
           to move the ant is one unit towards
           this.
        4. If the place that the ant is to move
           to is unoccupied move the ant, else
           leave the ant where it was but
           change its direction to a new random
           direction.
        5. If the ant is over at least one
           example, calculate the highest
           pickup probability P_PU for the data
           examples under the ant. Choose a
           random number and pick up that
           example if the random number is less
           than this probability.
        6. If the ant loaded this move reset
           its sensitivity to the initial value
           else increase its sensitivity.
        If the ant is loaded:
        1. select a random number
        2. If this is less than P_LC try to move
           the ant one unit in its current
           direction.
        3. If not, find the map position within
           the range of vision of this ant
           whose local disorder would be least
```

increased should the data example carried by the ant be dropped there. Try to move the ant one unit towards this.
4. If the place that the ant is to move to is unoccupied move the ant, else leave the ant where it is but change its direction to a new random direction.
5. Calculate the probability P_D that the ant drops the data example it is carrying. Choose a random number and drop the data example being carried if the random number is less than this probability.
6. If the ant unloaded this move reset its sensitivity to the initial value else increase its sensitivity.

An Example of Sorting Using ACO

The test problem reported here is the sorting of the iris data set [1]. This common data set contains examples of iris flowers of varieties *setosa, versicolor* or *virginica*, giving the sepal length and width, petal length and width and class for each. Experience has shown that, given the first four pieces of information, it is relatively simple to identify all examples of the *setosa* class, but that some examples of the other two classes are very hard to tell apart. The ants were required to sort these classes onto a 10*10 grid, the minimum size on which a regular self-organising map algorithm is reliably able to sort them. Rather than randomly distribute the objects onto the grid, thus removing the large influence the initial distribution would have on performance, the problem was made more difficult by piling all the objects on top of each other in the very centre of the grid.

The ants were randomly positioned and the algorithm above was executed until completion was achieved. Completion required that no grid cell contain objects of more than one class and for five algorithm cycles to have passed during which no ant carried any item.

Fig. 5.7 shows a typical map produced. In this case the radius over which disorder was measured was three grid cells, the value of momentum was 0.25, and α and β were both 0.3. Note that the classes *versicolor* or *virginica* overlap slightly – which is consistent with experience building conventional self organizing maps using this data. The result shown above is typical of the completed maps built over a wide range of parameter values.

Momentum (and resetting the direction of blocked ants) is required else a traffic jam may occur especially during the early stage of map development. This is exacerbated if a large number of ants is used. The results described here were obtained with only five ants.

The majority of objects were only carried for a single cell before being dropped. As shown in Figs. 5.8 and 5.9, changing the value of α to be more than one and a half times the value of β increased both the average length of carry and the total number of cycles the algorithm required to complete the map. The longer the range of ant vision, the shorter time it took to complete a three cluster map, as shown in Fig. 5.10.

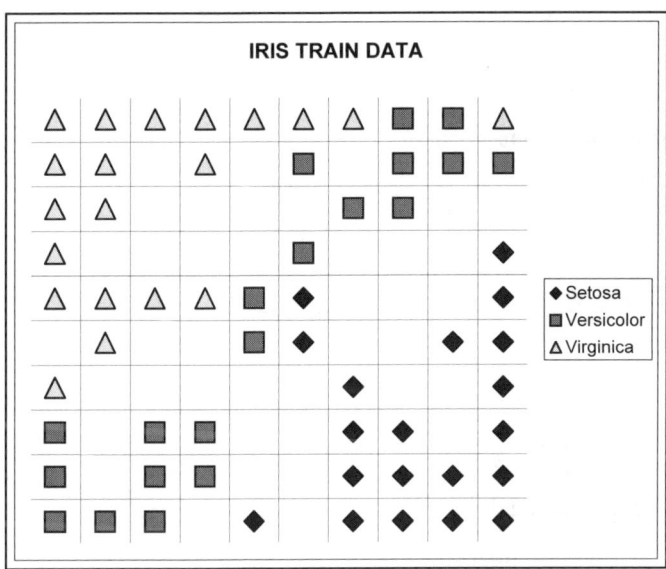

Fig. 5.7. A map produced by the SOMu1ANT algorithm

A vision range less than one half of the map dimensions (5 cells in the case of the maps being discussed here) resulted in maps with more clusters than necessary. For a vision range of 4 cells the maps were, however, complete in that no cell contained more than one type of object, definite clustering was observed and for a number of algorithm cycles no ant carried any object. Eventually the increases in α from being unloaded would result in the ants starting to modify the map again. For a vision range of 5 cells a three-cluster map was eventually produced after a number of completed maps with more than three clusters had been produced

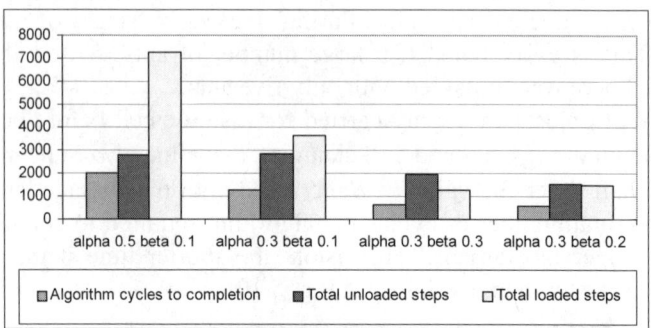

Fig. 5.8. Effect of varying α and β on Average Length of Carry

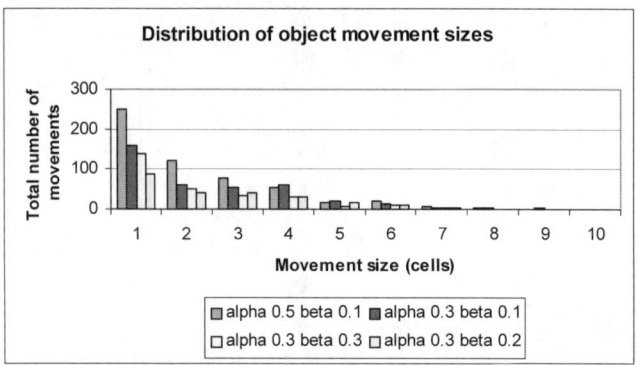

Fig. 5.9. Effect of varying α and β on Total Number of Cycles

For the vision ranges in Fig. 5.10, maps with three clusters would be produced, the longer the vision range the smaller the total number of cycles through the algorithm required to produce the map. Giving ants different visual ranges between 5 and 14 cells while building a map did not seem to change the resulting map but did increase the total number of cycles the algorithm required to produce a result.

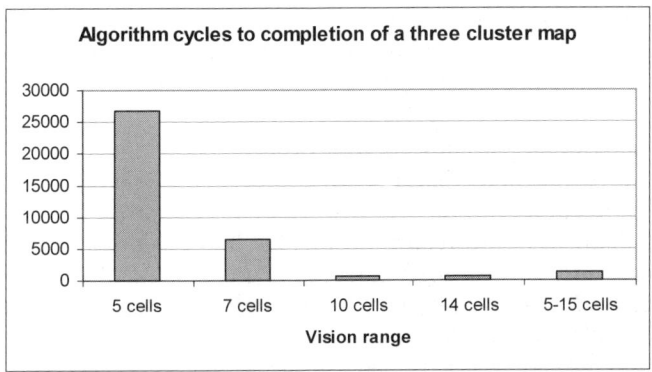

Fig. 5.10. Effect of Vision Range on Completion Time (3-cluster map).

As the visual range of an ant increases so does the number of cells it can see. Hence the amount of computation required to complete one algorithm cycle also increases as the ant had more possibilities to consider before moving. As a result the visual range to produce the map in the shortest time was not the same as the vision range to produce the map in the smallest number of algorithm cycles.

The SOMulANT algorithm has been effective at mapping a range of relatively simple data sets in addition to the data set described in this chapter. Unlike a conventional self-organising map it does not need a conscience applied to ensure that information is spread across the map. The inherently lower disorder of boundary cells ensures that information is spread right to the edge. Ant visual range is the main factor in deciding the map ordering scale, while the values of α and β largely determine the number of cycles through the algorithm needed to produce a completed map.

5.5 Particle Swarm Optimisation.

5.5.1 The Basic Particle Swarm Optimisation Algorithm.

Particle Swarm Optimisation (PSO) is a population-based algorithm that, unlike other population-based algorithms (for example a genetic or evolutionary algorithm), does not involve replacing individuals [4, 11, 12, 17]. Rather, the individuals move through the problem space with a velocity that is regularly updated as a consequence of the performance of the other individuals in the swarm. It is assumed that each individual broadcasts a message reflecting its current performance and that other individuals respond strongly to the messages from particles in their vicinity. Individuals also have a memory that contains the point of best performance found so far and this memory is updated as necessary as a result of the performance messages sent out by each of the swarm individuals.

All of the individual particles in the swarm update their velocities simultaneously at regular intervals. For each particle the updated velocity vector is a combination of the previous velocity vector ($\overline{V_T}$) of this individual, a component of the vector towards the best position ($\overline{B}$) found by any individual in the swarm so far, and a component of the vector towards a position ($\overline{S}$) derived by comparing the relative performance of this individual with the performances of a number of other swarm members. As each individual is assumed to have momentum, instantaneous changes in direction do not occur.

The movement towards the best position found so far provides the exploitation part of the algorithm, the influence of the performance of some sub-set of swarm members, together with the momentum of the individuals, provides the exploration. The balance between exploitation and exploration is essentially stochastic and controlled by two random numbers. Formally the new velocity at time $T+t$ is $\overline{V}_{T+t}$ and is given by:

$$\overline{V}_{T+t} = \chi(M\overline{V}_T + (1-M)(rand * P(\frac{\overline{X}-\overline{B}}{t}) + rand * G(\frac{\overline{X}-\overline{S}}{t}))) \quad (5.8)$$

where $\overline{V}_T$ is the velocity of this particle at time T, M is the momentum, $\overline{X}$ is the current position of the individual, *rand* returns a random number in the range from 0 to 1, and P and G set the relative attention to be placed on the positions $\overline{B}$ and $\overline{S}$.

Both M and G are bounded to the range from 0 to 1. The recently introduced factor χ is defined below and may be taken to be unity for the traditional algorithm. If M is large neither the $\overline{B}$ nor $\overline{S}$ positions have much effect. The influence of $\overline{B}$ increases with the value of G and P similarly influences the importance of $\overline{S}$. The parameter t is the time between updates and is required for dimensional consistency. It is usually taken to be one basic time interval and so is often omitted from the equation. However, it is important to realise that updates are made at regular intervals, not continuously. The effect this has will be discussed below.

5.5.2 Limitations of the Basic Algorithm.

The basic algorithm as descried in Section 5.5.1, while able to solve many problems, does have a number of limitations. The three main ones discussed below are:

1. how to decide on the best way to calculate the position $\overline{S}$,
2. how to detect a good, possibly optimum, position that a particle passes over in the finite time between updates,
3. and how to encourage the swarm to explore more aggressively as, unlike real life, our swarms only need one member to find the optimum rather than have the whole swarm at the same place.

5.5.3 Modifications to the Basic PSO Algorithm.

Choosing the Position $\overline{S}$

Without the component towards $\overline{S}$, the swarm tends to rush towards the first minimum found. An appropriately chosen $\overline{S}$ will cause the swarm to spread out and explore the problem space more thoroughly.

Each particle has its own position $\overline{S}$, which is often taken to be the position of the current best performing particle in its immediate neighbourhood (this could even be the position of the particle itself if it is performing better than any other particle in its neighbourhood). The neighbourhood could be either the N closest particles, or all particles within a distance R. In either case this involves a user chosen parameter which, while it does not seem to be very critical, is still somewhat problem dependent.

An alternate way of calculating the position $\overline{S}$ in an effective (but computationally modest) way that does not involve any user chosen parameter has been introduced in [14]. All other swarm members are considered to influence the position $\overline{S}$ for a given particle but the magnitude of the influence decreases with both fitness and distance from the given particle for a minimization problem (for a maximization problem it would increase with fitness but decrease with distance). Individuals within a unit distance are considered to be exploring the same region of space and make no contribution. An individual j will be attracted to the position $\overline{S}_j$ whose components S_{ji} are a function of D_{ni}, (the distance along the ith axis from this individual to some other individual n in the swarm) and F_n, the fitness of individual n. Each of the components of the position $\overline{S}_j$ for individual j is defined by:

$$S_{ji} = \sum_{n=1}^{N} W_{ji} \quad \text{where } W_{ji} = \frac{1}{D_{ni}^2(1+F_n)} \quad \text{if } D_{ni} \geq T, \quad (5.9)$$
$$= 0 \text{ otherwise}$$

if we are seeking a minimum. The parameter T is used so that the position of S is not heavily influenced by other particles so close that they are essentially just experiencing the same results as this particle. Should we be looking for a maximum, each component is defined by:

$$S_{ji} = \sum_{n=1}^{N} W_{ji} \quad \text{where} \quad W_{ji} = \frac{F_n}{D_{ni}^2(1+F_n)} \quad \text{if } D_{ni} \geq T, \quad (5.10)$$
$$= 0 \text{ otherwise}.$$

The Problem of a Finite t.

The performance of the swarm members is only noted at intervals of t during which a particle has travelled a distance of vt. Since the problem space is only sampled with a basic resolution of vt it is quite possible that useful minima can be missed. Since the units of t are arbitrary the sampling resolution can only be improved by deceasing v.

The allowable velocity is often limited to values between $\pm v_{max}$ thus setting an upper value on the speed a particle can attain.

A further enhancement introduced by Clerc involves a 'constriction factor' χ, a coefficient that weights the entire right side of the formula for the new velocity at time $T+t$ introduced in Section 5.5.1 [2, 3]. Clerc's generalised particle swarm model allows an infinite number of ways in which this factor can be constructed. The simplest of these is called 'Type 1' PSO (T1PSO), and recent work [12] showed that placing bounds on the maximum velocity of a particle in T1PSO "provides performance on the benchmark functions superior to any other published results known by the authors".

The equations defining the T1PSO constriction factor are:

$$\chi = \frac{2\kappa}{\left|2 - \phi - \sqrt{\phi^2 - 4\phi}\right|} \quad (5.11)$$

where $\phi = (P + G) > 4$, and $\kappa \in [0,1]$.

Aggressively Searching Swarms.

Knowledge of how a swarm is travelling can be used to influence how extended the swarm should be. When the swarm is centred upon a good solution, the particles will move with little common velocity component between members. In this case it is desirable for the swarm to converge. However, if the swarm is travelling as a structured entity in which all particles have a significant common velocity component, then the swarm should be made to spread out in the search space. This is achieved by an additional coherence velocity term *vc* that is added to the right hand side of the equation for V_{T+t} given above and which is defined as follows.

The coherence of the swarm's movement (*CSM*) can be defined by:

$$CSM = \frac{speed\_swarm\_centre}{average\_particle\_speed} \qquad (5.12)$$

where $speed\_swarms\_centre = \left\| \frac{\sum_{i=1}^{\#particles} \vec{v}}{\#particles} \right\|$

This *CSM* term is used as the basis with which to calculate *vc*. The desired effect of this term is that a swarm on the move will tend to be spread out, and when it encounters a local minimum, the 'momentum' term of the PSO will carry the swarm a small distance past the minimum. If the swarm has encountered the true minimum it will converge back on this point. If it has only encountered a local minimum the combination of the momentum and increased spread of the swarm is far more likely to cause the swarm to explore other surrounding minima, thus allow the swarm to move to the global minimum.

The coherence velocity for the ith particle in dimension *d* is given by:

$$vc_{id} = sh*Step(CSM, so) * average\_particle\_speed_d * \qquad (5.13)$$
$$CauchyDist()$$

where *sh* is a multiplier, *so* an 'offset' and

$$Step(CSM, so) = \begin{cases} 0 & CSM <= so \\ 1 & CSM > so \end{cases}$$

The Cauchy distribution random number is included to prevent the coherence velocity causing the swarm to accelerate in its current direction, without spreading out significantly. Because the random number can have either positive or negative values, the effect is to make the swarm to spread out around its current position. The average particle speed is included to scale the coherence velocity to the same order of magnitude as the velocity of particles within the swarm.

Adding Memory to Each Particle.

Another technique [15] adds memory to each swarm particle. The purpose of the added memory feature is to maintained spread (and therefore diversity) by providing particle specific alternate target points to be used at times instead of the current local best position.

To optimise this effect each particle in the swarm maintains its own memory. The user must specified the maximum size of the memory and the probability that one of the points it contains will be used instead of the current local optimal point. The current local optimum point will be added to the memory if the fitness of this point is better than the least fit stored point. The point may also be required to differ by at least a specified amount from the position of any other point already in the memory. If the memory is full the new point replaces the least fit point already in the memory. There is a certain probability that a point from the memory will be used instead of the position S described in Section 5.5.1. When a point from the particle memory is to be used, which point may be chosen randomly or with the probability of selection being fitness based (with better fitness producing a higher probability of selection).

Storing points with good fitness in the memory ensures that particles move back towards good areas when the history option is invoked. This continues the practice of always moving towards known good areas, but increases the range of known points still having influence. This acts as a counter to premature convergence of the swarm arising from a lack of influences encouraging the swarm to explore elsewhere.

5.5.4 Performance

The performance a swarm algorithm can achieve cannot (generally) be represented by a drawing on a page in the way that the best path found by an ACO algorithm can visually represent the quality of the solution it has found. Instead the performance has to be described. Like an Evolutionary Algorithm (EA) the progress of a swarm algorithm towards the solution can be divided into two parts, in the first deliberative progress is made – and in the second random events have to be relied on that happen to finish the task. For an EA, in the first part crossover has a good probability of producing steadily fitter individuals; in the second the available genetic material has been mixed to the best advantage possible and now only creep mutation can find the final solution[5].

For a swarm algorithm in the first stage the vector summation process leads the swarm to explore and finally circle one (or more) promising local optimum. The second stage can take a long time owing to the fact that the

---

[5] An EA is generally a good coarse search algorithm with poor local search capabilities. For this reason a local heuristic (a problem specific local search algorithm) is normally used as well.

performance of the particles is only sampled and in the time between samples particles may cross the optimum without detection. It is for this reason that the velocity constriction factor has been introduced, as the velocity is steadily reduced the distance travelled per time interval and therefore the resolution of the search is reduced. However, like an EA, PSO should not be considered as a fine scale search technique.

The main difference between the performance of an EA and PSO is speed. While this is hard to quantify, the PSO is often an order of magnitude faster than an EA when running both on the same computer. Part of this speed advantage comes from the use of particle momentum; a suitably chosen value will allow particles to move across local minima without becoming trapped. The PSO algorithm is less likely to revisit exactly the same position in problems space than the EA, especially when the diversity in the EA population has become low and mutation is not high. The EA can also use a history mechanism to try to reduce this effect but the PSO has the advantage that its search is inherently more efficient in this respect [22].

In real life the dominant effect on the time taken to reach a solution is the time to evaluate how good a particular particle (or individual) is. The use of estimated (rather than actually evaluated) fitness values has been shown not to impede the progress of an EA, provided that periodic real evaluations do occur. The use of fitness estimation can have a dramatic effect on convergence time when the fitness function takes a significant time to calculate [24]. Such an approach may also speed up PSO, although to the best on this author's knowledge it has not yet been tried.

One of the advantages of PSO is flexibility. Fundamentally it finds a good combination of a set of real numbers – how these values are used depends on the application. The sign of a particular number could be used to set a binary variable; conversion to an integer value is trivial. Indeed, mapping to any quantised set of values presents little difficulty. Even mapping the values to an order or path is not hard, as is shown in Section 5.5.5. This flexibility may be the greatest advantage of PSO when real life optimisation problems are concerned.

5.5.5 Solving TSP Problems Using PSO.

PSO optimises a set of real numbers; just how these numbers are interpreted is up to the user and is specified in the fitness calculation algorithm. It is quite possible to interpret them in a way consistent with a TSP problem.

All that is required is a way that can map a set of real numbers into a unique sequence of non-repeated values. It is important that, although the same order could result from many sequences of real numbers, the same sequence always decodes to the same order.

One suitable way is to associate each position in the sequence with a particular city (or equivalent) in the TSP. For example, the first number in the sequence is associated with city-one, the second with city-two, and so on. Given a particular sequence of numbers find the largest, if this is in position-n of the sequence then the city that it is associated with is the first in the order of cities we are decoding (city-n in this example). The position of the second largest number in the sequence defines the second city in the order of cities we are building. Continuing this, the position of the smallest number defines the last city in the order of cities. In the unlikely event that there are two numbers in the sequence that are exactly the same (very unlikely given that these are real numbers we are talking about) we make a random choice.

In the PSO each particle is at a definite position that is described by a number of coordinate values. Assembling these coordinate values in a predetermined sequence (such as coordinate-one first, coordinate-two second, and so on) and then interpreting this sequence in the way described maps each particle to an order. As a particle moves through problem space the order it represents will stay the same until one or more of its coordinate values changes sufficiently so that their position in the ordered list of values changes. Every point in problem space decodes to an order and using the fitness function this point can be associated with a fitness. To solve the TSP, PSO must place one (or more) particles in the region of space that decodes to the order with the highest fitness.

A fitness space so defined typically has many small-scale local minima and thus exploration is very important. Rather than random exploration, exploration in the light of prior experience is effective at reducing the solution time. It is for this type of problem that history was originally added to the basic PSO.

PSO Performance on a TSP.

Using the Burma 14 TSP described previously the following results were all obtained using a common particle population size of 30 and with the normal local point of influence derived from the ten closest neighbours (values that prior experimentation had shown to be suitable for this test set). For similar reasons the three parameters M, P and G in the equation for the new velocity at time $T+t$ (Eq. (5.8); Sect. 5.1) were set to 0.97, 0.04 and 0.96, respectively. The values that were varied were the memory depth

(0-20), the probability of using a memory point rather than the current local point (0 – 20%), whether the memory point to be used was based on the rank order of performance (rank) or just randomly selected (random) and the minimum difference (D) required between a potential memory point and all the points already in the memory before the potential memory point would be considered for storage.

Table 5.10. Performance variation with memory depth *(10% probability, rank, D=0)*

Memory depth	% ideal result	% in top $1*10^{-6}$%	% in top $5*10^{-6}$%	Average excess path length
No memory	2	9.1	24	390
5	3.2	11.6	28	353
10	4.8	17.2	32	313
20	6	20.4	38	296

In each of the Tables 5.10 through 5.13 one of these variables was altered, with the other three being held constant; the constant values are stated in each Table caption. The performance of PSO without the particle memory is also shown in each Table for easy reference.

Table 5.11. Variation with memory probability *(Depth = 10, rank, D=0)*

Memory use probability	% ideal result	% in top $1*10^{-6}$%	% in top $5*10^{-6}$%	Average excess path length
No memory	2	9.1	24	390
5	3.2	15.6	31.2	351
10	4.8	17.2	32	313
15	4.4	13.6	30.4	338
20	5.2	13.2	29.2	315

From Tables 5.10 – 5.13 it can be seen that, as might be expected, performance increases with memory depth. Excessive use of the memory contents appears to be counter-productive and for this problem a probability of 10% gives the best result. It also appears that, when choosing a location to be taken from the memory, a bias towards the better performing memory locations via rank selection outperformed random selection in almost every case (only two of which are shown in Table 5.12). Finally Table 5.13 shows that it is desirable that the locations stored in the memory are unique, as repeats decrease the effective memory size and also alter the selection process. However, insisting on too large a difference can also be counter productive as the memory may not actually fill. For the Burma 14 problem a difference of 100 to 150 seems to have been most suitable.

Table 5.12. Variation with memory choice *(Depth as noted, 10% probability, D=0)*

Memory choice	% ideal result	% in top $1*10^{-6}$%	% in top $5*10^{-6}$%	Average excess path length
No memory	2	9.1	24	390
10 deep rank	4.8	17.2	32	313
10 deep random	2	11.6	21.6	352
20 deep rank	5.2	13.2	29.2	314
20 deep random	2.8	13.6	29.6	340

Table 5.13. Variation with minimum separation *(Depth =10, 10% probability, rank)*

Minimum difference (D)	% ideal result	% in top $1*10^{-6}$%	% in top $5*10^{-6}$%	Average excess path length
No memory	2	9.1	24	390
0	4.8	17.2	32	313
50	10	22.4	44.4	224
100	16	29.6	47.6	214
150	19.6	38	50.8	198
200	10.8	25.6	43.2	242

Fig. 5.11 shows a plot of the average iterations taken by the algorithm (for a fixed set of parameter values) to find its final solution plotted against the number of times that the optimum solution was found. This shows that the improved performance comes at the cost of the number of iterations needed.

Table 5.14 shows the better performing parameter set values for each performance measure sorted into order with the performance without using particle memories shown for comparison. The order of merit is remarkably consistent, no matter which performance measure is used. The parameter set found by combining the individual best values from Tables 5.2-5.4 is actually only the second best performing of the parameter sets tried, possibly suggesting the parameters are not fully independent. The improvement obtained by using either of the top two-parameter sets is clearly significant compared to a swarm of particles without memory. The result obtained using the most basic ant colony optimization (AS) is also shown, these being taken from [16]. While a detailed comparison needs to take into account the values of the parameters used, the table is included to show that both algorithms obtain comparable results.

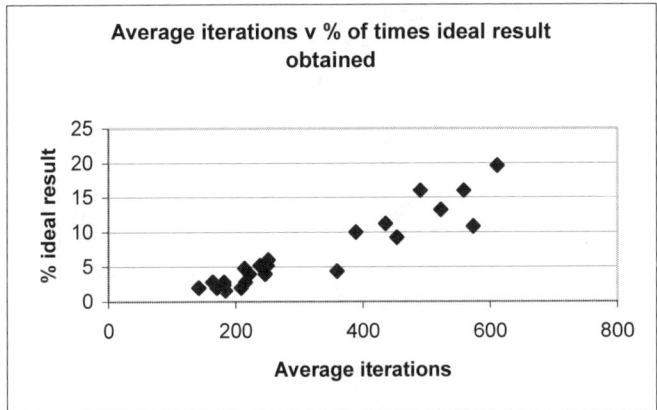

Fig. 5.11. The percentage of times that the ideal result was found versus the average iterations required to find the best result

Table 5.14. The top performing parameter sets for each performance measure sorted into order (Each parameter set is identified by memory depth/probability/selection method/minimum difference. The performance without using particle memories is shown for comparison. Also shown are results obtained using basic ant colony optimization (AS) for comparison)

Sorted by % ideal solution		Sorted by % in top 0.000001%		Sorted by top 0.000005%	
10/10/rank/150	19.6	10/10/rank/150	38	10/10/rank/150	50.8
20/15/rank/100	18.8	20/15/rank/100	35.2	20/15/rank/100	48.8
10/10/rank/100	16	10/10/rank/100	29.6	10/10/rank/100	47.6
10/10/random/200	16	10/10/random/150	27.6	10/10/rank/50	44.4
10/10/random/150	13.2	10/10/random/200	27.2	10/10/rank/200	43.2
10/10/rank/200	10.8	10/10/rank/200	25.6	10/10/random/150	41.6
10/10/rank/50	10	10/10/rank/50	22.4	10/10/random/200	40
10/10/random/100	9.2	10/10/random/100	22.4	20/10/rank/0	38
20/10/rank/0	6	20/20/rank/0	21.2	10/10/random/100	37.2
No memory	2	No memory	9.1	No memory	24
Ant System (AS)	2	Ant System (AS)	49	Ant System (AS)	67

5.5.6 Practical Considerations.

The first practical consideration must be deciding how to map the n real number associated with the position of each particle with the inputs to your problem. Of course, if all these problem inputs are themselves real numbers this is trivial. If some of the problem inputs are an order, the method described above can be used to map a chosen set of real numbers into the order. Another common problem input is a switch that can be in one of a fixed number of states. Assume for a moment that there are two states. Probably the simplest method is to associate one real number with the switch and chose a boundary value, if the real number value is greater than the boundary value assign one state to the switch, otherwise assign the other. While simple this can introduce very noisy behaviour as a particle moves close to the boundary. It is better to use the distance from the boundary as a probability that the switch is in one state. For example for a binary switch let the value -0.5 be the probability that the switch is in the 'on' state. If this 'probability' is less than zero then set it to zero, if above one set it to one. Now there is a narrow band outside which the state is consistent and within which the state need not always be mapped the same. This has the effect reducing the magnitude of the bounce effect around the boundary. Obviously this can be extended to a switch with any number of states by setting more than one boundary. The size of the probabilistic zone can be altered as required.

The number of particles need not be high – enough should be used to allow for realistic neighbourhoods to be formed but excessive numbers add little to the search while consuming computer resources. Adjusting the exploration/exploitation balance may take some experimentation, even when a system that attempts explicitly manage of this is used. The choice of momentum value can be important depending on the number of minima in the problem space. Increasing momentum may reduce the likelihood of getting caught in a local minimum but at the cost of generally increasing the time the swarm takes to converge.

5.5.7 Scalability and Adaptability.

Comparing ACO(AS) and PSO, ACO currently scales best as the size of the problem increases. For example, TSPs with hundreds of cities can be solved. PSO on the other hand becomes hard to use with many dimensions, partly because it is very hard to find a set of parameters that suit the problem. However, there is no escaping the fact that it is easier to map the PSO set of optimised numbers to a wider range of problems (including order-based problems) than ACO. In particular for problems that include combinations of parameter types to be optimised, including continuous parameter values, switched parameter values and sequences (orders) the PSO is the only realistic option. Evolution could also be used for such problems but one advantage that both ACO and PSO have over evolution is speed, usually finding solutions orders of magnitude faster.

Both ACO and PSO are relatively new technologies and no doubt the range of applicability will only increase with time. This may come as a result of refinement of these algorithms or perhaps as a result of developing new algorithms that have no real life parallel but which none-the-less exploit powerful emergent collective behaviour between numbers of similar entities, each with limited individual capability.

References

1. Batchelor BG (1974) *Practical Approach to Pattern Recognition*, Plenum Press, New York.
2. Clerc M (1998) Some Math about Particle Swarm Optimization, available online at: http://clerc.maurice.free.fr/PSO/PSO_math_stuff/PSO_math_stuff.htm (last accessed 28th January 2002).
3. Clerc M (1999) The Swarm and the Queen: Towards a Deterministic and Adaptive Particle Swarm Optimization, *Proc 1999 Congress on Evolutionary Computation*, Washington, DC: 1951-1957.
4. Corne D, Dorigo M and Glover F (Eds) *New Ideas in Optimization*, McGraw-Hill Publishing Co, London, Chapter 25: The Particle Swarm: Social Adaptation in Information-Processing Systems.
5. Deneubourg J-L, Goss S, Franks N, Sendova-Franks A, Detrain C and Chretien L (1992) The Dynamics of Collective Sorting: Robot-Like Ant and Ant-Like Robot, *Proc 1st European Conf Simulation of Adaptive Behaviour: from Animals to Animats,* Varela FJ and Bourgine P (Eds), MIT Press, Cambridge MA: 123-133.
6. Dorigo M (1992) Optimization, Learning and Natural Algorithms, PhD Thesis, Dipartimento di Elettronica, Politechico di Milano, Italy.
7. Dorigo M., Maniezzo V and Colorni A (1996) The Ant System: Optimization by a Colony of Cooperating Agents, *IEEE Trans Systems, Man and Cybernetics - Part B*, 26: 29-41.
8. Dorigo M and Gambardella L (1997) Ant Colony System: A Cooperative Learning Approach to the Traveling Salesman Problem, *IEEE Trans Evolutionary Computing*, 1: 53-66.
9. Dorigo M and Gambardella L (1997) Ant Colonies for the Traveling Salesman Problem, *Biosystems*, 43: 73-81.
10. Dorigo M and Di Caro G (1999) The Ant Colony Optimization Meta-heuristic, in Corne D, Dorigo M and Golver F (Eds) *New Ideas in Optimization*, McGraw-Hill Publishing Co, London: 11-32.
11. Eberhart RC, Dobbins P, and Simpson P (1996) *Computational Intelligence PC Tools,* Academic Press, Boston.

12. Eberhart RC and Shi Y (2000) Comparing Inertia Weights and Constriction Factors in Particle Swarm Optimisation, *Proc 2000 Congress Evolutionary Computation*, San Diego, CA: 84-88.
13. Hendtlass T (2000) SOMulANT: Organizing Information Using Multiple Agents, *Lecture Notes in Artificial Intelligence*, Springer Verlag, Berlin 1821: 322-327.
14. Hendtlass T (2001) A combined swarm differential evolution algorithm for optimization problems, *Lecture Notes in Artificial Intelligence*, Springer Verlag, Berlin, 2070: 374-382.
15. Hendtlass T (2003) Preserving Diversity in Particle Swarm Optimisation. *Lecture Notes in Computer Science*, Springer Verlag, Berlin, 2718: 31-40.
16. Hendtlass T and Angus D (2002) Ant Colony Optimisation Applied to a Dynamically Changing Problem, *Lecture Notes in Artificial Intelligence*, Springer-Verlag, Berlin, 2358: 618-627.
17. Kennedy J and Eberhart RC (1995) Particle Swarm Optimization, *Proc IEEE Intl Conf Neural Networks*, Perth, Australia, IEEE Service Centre, Piscataway NJ, IV:1942-1948.
18. Kohonen T (1988) *Self-Organisiation and Associative Memory*, Springer-Verlag, New York.
19. Lumer E and Faieta B (1994) Diversity and Adaption in Populations of Clustering Ants, *Proc 3rd Intl Conf Simulation of Adaptive Behaviour: from Animals to Animats*, MIT Press, Cambridge MA: 499-508.
20. Montgomery J, Randall M and Hendtlass T (2003) Automatic selection of appropriate pheromone representation in Ant Colony Optimisation, *Proc Australian Conf Artificial Life – ACAL*, Canberra, December.
21. Oprisan SA Holban V, and Moldoveanu B (1996) Functional Self-Organization Performing Wide-Sense Stochastic Processes, *Phys Lett A*, 216: 303-306.
22. Podlena J and Hendtlass T (1998) An Accelerated Genetic Algorithm, *Applied Intelligence*, 8(2).
23. Reinelt G TSPLIB95, available online from: http://www.iwr.uni-heidelberg.de/iwr/comopt/soft/TSPLIB95/TSPLIB95.html
24. Salami M and Hendtlass T (2002) A Fast Evaluation Strategy for Evolutionary Algorithms, *J Applied Soft Computing*, 2(3): 156.
25. Stutzle T and Dorigo M (1999) ACO Algorithms for the Traveling Salesman Problem, in Miettinen K, Makela M, Neittaanmaki P and Periaux J (Eds) *Evolutionary Algorithms in Engineering and Computer Science*, Wiley, New York.
26. Stützle T and Hoos HH (2000) Max-Min Ant Systems, *Future Generation Computer Systems*, 16(8): 889-914.

6 Where are all the mobile robots?

Phillip J McKerrow

1   School of Information Technology and Computer Science,
           University of Wollongong NSW 2522 Australia,
           phillip@uow.edu.au

6.1 Introduction

Fig. 6.1. Hertwig robot cleaning sugar from the floor, leaving a trail showing where it has been (*1st Intl Cleaning Robot Contest at IROS'02*)

The call for papers for the 1985 *IEEE International Conference on Robotics and Automation* in St. Louis painted a very bright future for mobile robots and encouraged submission of papers on research into mobile robotics. At a conference in December 2001, a senior mobile robotics researcher expressed the view to this author that all the mobile robot research had been done. Yet, at the *Intelligent Robots and Systems*

Conference in Lausanne in October 2002 [10], the robots entered in the 'First International Cleaning Robot Contest' performed poorly (Fig.6.1).

So if all the research has been done, where are the mobile robots? Why don't we see mobile robots performing everyday tasks in the home, in the office and in service industries? The only applications where mobile robots are in common use are swimming pool cleaning and in the delivery of bombs (Tomahawk missile). Pool cleaning occurs in a constrained, object-free environment; the pool is empty when cleaning is done.

The mobile robot that is best known to the public is the Mars rover Sojourner (Fig.6.2), probably the most expensive mobile robot ever made. Sojourner landed on Mars as part of the Pathfinder mission [13]. Sojourner was semi-autonomous. Each night programmers would load up the commands for the next day. They planned paths using images received the previous day. Then Sojourner would carry out these instructions autonomously and send sensor data back to earth.

Fig. 6.2. Sojourner rover on Mars Pathfinder mission examining a rock (tracks in the dust show where it has been)

Is the lack of application of mobile robot technology due to the lack of a 'killer' application, to limitations of the technology, or to robotics engineers not understanding how to develop a product for market? It is clear that current technology is unable to achieve the level of 'intelligence' required to robustly, repeatedly and reliably achieve many simple tasks. So contrary to the opinion that the research has been done, the evidence is that, in many areas, the research has only achieved partial solutions to problems.

Many problems need to be revisited in future research. In other cases, the solutions are too expensive for commercialization.

The claim by some researchers that they have solved all the problems in mobile robotics has lead to a reduction in funding for mobile robot research. It appears that by making excessive claims about their successes, robotics researchers have convinced the funding bodies that there is no need to do further research.

Roughly 45% of the papers at *IROS'02* in Lausane were on mobile robotics (Table.6.1) indicating that many researchers believe that there is much research still to be done.

In the following Sections, we look at mobile robots that can be purchased for applications in the home. We examine their navigation, planning and sensing capabilities, and their use of Artificial Intelligence. Then we look at a case study of developing a consumer robot. Finally, we ask: 'where is mobile robot research heading?'

Table 6.1. Distribution of the 500 papers presented at *IROS'02* in Lausane

Topic	%	Topic	%
Mobile Robots	11	Motion planning	5
Robot vision and sensing	10	Service and field robotics	4
Human-robot interaction	9	Tactile sensing and haptics	4
Robot Navigation	8	Visual servoing	3
Actuators and control	6	Industrial applications	3
Learning	6	Virtual reality and simulation	3
Multiple robot systems	6	Micro systems	2
Humanoid robots	6	Kinematics	2
Manipulation and assembly	5	Dynamics	1
Medical robotics	5	Entertainment robotics	1

6.2 Commercial applications

Over the last decade, many mobile robot companies have come and gone. Some aimed to produce general-purpose mobile robots, while others targeted specific application domains. Most general-purpose mobile robots have been sold to university researchers. Many of these companies are the offshoots of university research programs. They are owned by robotics enthusiasts and produce little economic return. One company developed a network for their robot. A communications company purchased the robot manufacturer and closed the mobile robot section, as they only wanted the network technology.

The companies that have targeted specific application domains have done so either by developing a mobile robot for that application or by modifying an existing manually operated machine. The companies that take the first approach are usually start-up companies spawned from a robotics research program. Many have struggled to get venture capital, to develop application expertise and to establish a market.

Established companies have tended to take the approach of modifying existing machines. They see robotics as the next step in their product development. The advantage of this approach is that the company already has both an established market and a knowledge base in the application. The disadvantage of this approach is that often the machine was not designed for automation. Consequently, there are no points for adding sensors, motors do not include encoders, and the motor controllers do not have computer interfaces. Thus, the machines have to be redesigned from the ground up with technology that is new to the company's design and engineering departments.

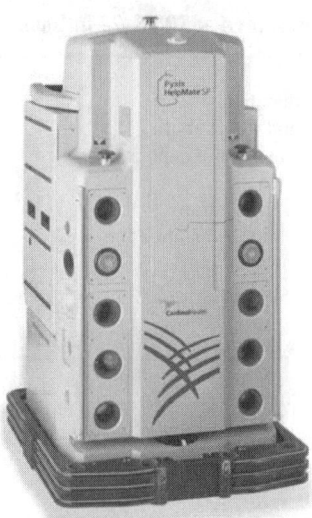

Fig. 6.3. Pyxis HelpMate - a robotic courier system

6.2.1 Robot Couriers

The longest surviving service robot is HelpMate (Fig.6.3), originally developed by Joe Engleberger (the father of the industrial robot), who wrote a book on service robots in an attempt to kick start the service robotics industry [3]. HelpMate is a robotic courier system designed to perform material transport tasks in hospitals. It transports pharmaceuticals, lab specimens, equipment, meals, medical records and radiology films back and forth between support departments and nursing wards. While the technology works and can be demonstrated to save a hospital money, sales have been slow. Pyxis is the third company to market it [15].

Helpmate uses light stripe vision for navigation. It uses sonar sensors for obstacle avoidance. When the hospital corridor it is traveling along is blocked, it stops and asks the object blocking the path to move. People soon move out of the way and the robot can continue on its mission. It can call an elevator with its infrared transmitter, enter the elevator and command it to ascend to the desired floor. A corridor map is digitized into the robot from the engineering drawings of the hospital.

Tasks are programmed using a language with high-level constructs that describe navigation in hallways. The Expert System that controls its operation has over 1470 rules. During early hospital trials it ran into objects with sloping surfaces, such as linen carts. The developers solved problems like this by adding extra sensors, which required additional rules to be added into the expert system.

Fig. 6.4. Electrolux Trilobite vacuum cleaner *(left)*, and close up of its ultrasonic transmitter *(right)*

6.2.2 Robot Vacuum Cleaners

Two companies that sell Robot Vacuum cleaners are Electrolux and iRobot. Both offer small units designed to clean lightly soiled areas. One of these was demonstrated at *IROS'02* but was not entered into the cleaning competition. When the Electrolux Trilobite (Fig.6.4) is started, it first goes to the nearest wall, and then follows the walls to circumnavigate the room [2]. It calculates the size of the room to determine the bounds of its navigation map. It does this for a minimum of 90 seconds and up to 15 minutes.

While vacuuming this path, it scans the room with ultrasonic sensors to create a map of the room: walls, furniture and open space. The Trilobite has a 180° sonar transmitter and 8 receivers, operating at 60Khz. After it has finished mapping the room, it starts cleaning. The wall phase forms the basis for calculating the time required to clean the open spaces.

The Trilobite's maximum speed is 0.4 meters per second. It avoids collision with objects on the floor – for example a dog's water bowl – by detecting them with the sonar. Then it calculates a new path and continues cleaning. Special magnetic strips are placed in doorways, near stairs and other openings. These act as a wall, keeping the Trilobite in the room.

The Trilobite is driven by two wheels with independent suspension so it can navigate over cables and the edges of rugs. Its ability to maneuver is further enhanced by its round shape and the fact that it is only 13 centimeters high and has a diameter of 35 centimeters. It is able to clean under very low furniture and beds. It is driven by four motors: one for each drive wheel, one for the roller that collects dirt and dust, and one for the fan.

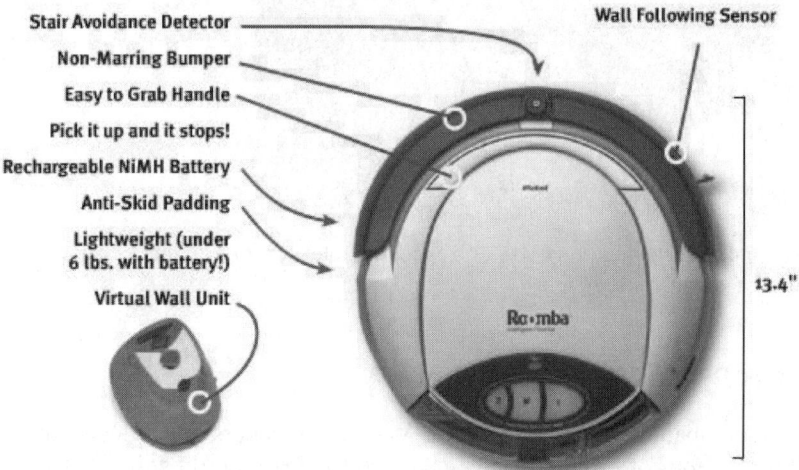

Fig. 6.5. iRobot Roomba vacuum cleaner and virtual wall unit

When the batteries run low (after 1 hour of cleaning) it automatically returns to the recharging station (which takes 2 hours). If it needs recharging before it has completed the cleaning, it will automatically resume cleaning once it is fully charged. The user can choose between three different cleaning programs: normal, quick and spot vacuuming. Dust is gathered using rubber lamellae and is accumulated in a reusable 1.2 litre dust box.

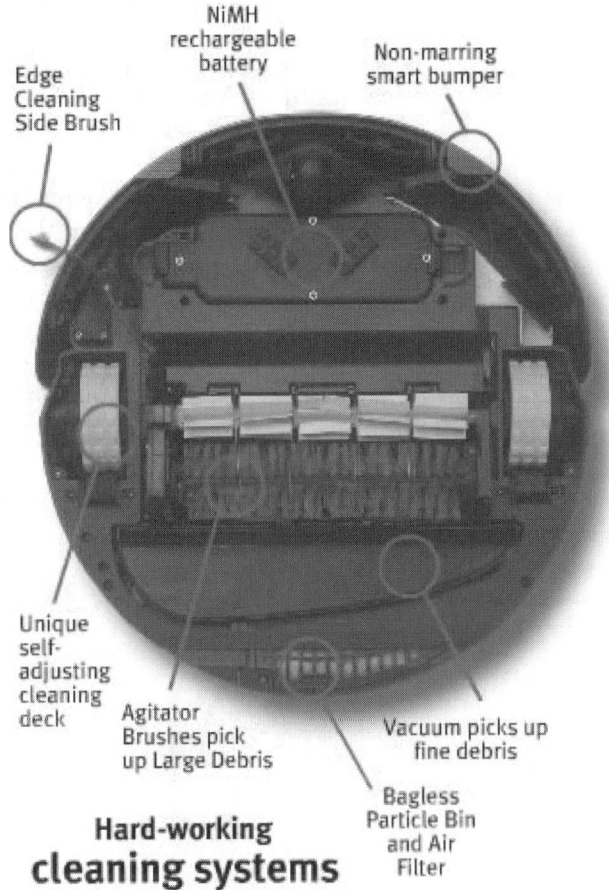

Fig. 6.6. Under side of Roomba

When the iRobot Roomba (Figs.6.5, 6.6) starts cleaning it travels around the floor in a spiral pattern (Fig.6.7). Its Non-Marring Bumper will contact a wall, or it may try to find a wall after spiraling for a while [9]. Upon first contact with an object, Roomba makes small, clockwise motions until its path becomes clear again. Once clear, these tiny bearing changes occur every time the bump sensor is activated and allow Roomba to track along a wall or object such as a sofa. Roomba follows the wall for a short period of time, utilizing its Spinning Side Brush to move objects away from the wall and into the vacuum's path.

After cleaning along a portion of the wall or other object, Roomba criss-crosses the room in straight lines (Fig.6.7.). Roomba repeats this cleaning pattern until its cleaning time has elapsed. This pattern is designed to provide maximum coverage of the room. Roomba uses a behavioural architecture. Its cleaning pattern is determined by its response to the objects with which it collides. As a result, it doesn't always look like it's doing what a person would think it should do. Also, it cannot find its way to a recharging station when the batteries are nearly flat.

To Roomba, a room is a space defined by walls, closed doors, and stairs. Roomba can also be directed to stay in or out of an area by a Virtual Wall Unit, and by furniture placement. Roomba automatically senses stairs with downward pointing infrared sensors and turns away from them. However, if the room to be cleaned contains a balcony, a physical barrier should be used to prevent access to the balcony to ensure safe operation. Although Roomba senses stairs on the vast majority of floor surfaces, there is the possibility that it will not detect rounded stair edges or light coloured steps.

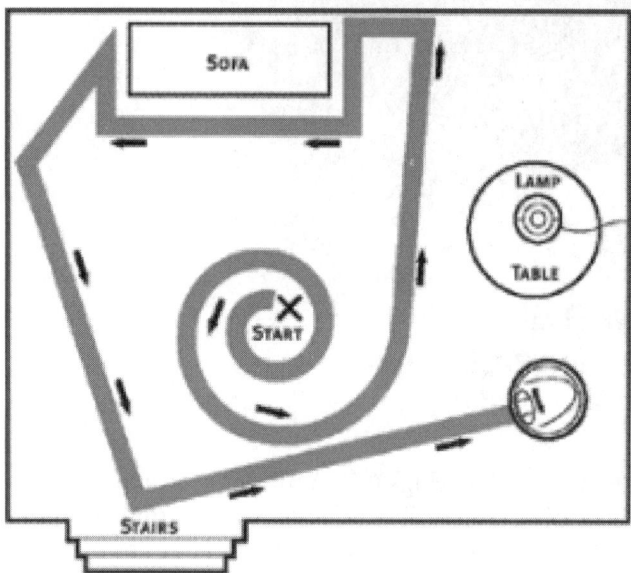

Fig. 6.7. Cleaning path followed by Roomba: short run along a wall, followed by straight legs away from the wall in random directions

Roomba uses a two-stage cleaning system. Two counter-rotating Cleaning Brushes sweep up larger particles, while a small vacuum sucks up the smaller particles. The Spinning Side Brush sweeps the dirt from next to walls and corners into the path of the main cleaning head. This allows Roomba to get impressive cleaning performance using only 30 Watts. For typical operation, Roomba cleans three 4m*5m rooms on a single charge.

Prager reviewed its operation on Tech TV[14]:

"To use Roomba, you need to remove miscellaneous items from the floor. For me, those items included clothes, CD cases, computer cables, and poker chips. Rugs with fringes and similar items must also be dealt with because they can get stuck in the agitator brushes.

Almost every square inch of floor space is covered, albeit some many more times than others. Roomba picks up small debris but would often miss things that tend to cling to a carpet's fibers, such as pillow feathers. Roomba was more effective in picking up the debris on the linoleum floor of my kitchen. I found plenty of debris collected inside the particle bins, but the carpet fuzz in the brushes can be a major pain to clean. After several uses, you learn how to arrange a room for Roomba so it can work most efficiently. Roomba can be a convenient and effortless way to quickly polish up any room with minimal effort. But for serious cleaning you'll need a more powerful vacuum."

iRobot has recently released a more expensive (US$250) version [18] that includes a remote control, spot cleaning mode and detection of objects stuck in the brushes. The remote control allows the user to drive the robot to the place where it is to clean rather than carrying it. The spot cleaning mode performs a series of small arcs. When an object jambs the brushes, Roomba stops and flashes a red light to request attention.

6.2.3 Robot Lawn Mowers

Robomower (Fig.6.8) from Friendly Robotics mows within a yard defined by a perimeter guide wire. To avoid mowing plants, trees, or other obstacles along the side of the lawn, a guide wire must be placed at a specified distance from the edge. Trees in the middle of the yard must also have wire around them. In edge mowing mode, Robomower follows this wire by sensing an electromagnetic field that emanates from it with a Hall effect sensor. Thus it avoids plants and other obstacles outside the perimeter.

After mowing around the perimeter it follows a forward-and-back V-shaped pattern from one side of the lawn to the other (Fig.6.9). Its navigational system is a floating compass that uses two settings for forward and backward direction. Once the lawn mower robot reaches the perimeter wire and knows it can't go any further, the on-board computer issues two new compass settings and the robot mows in the same V-shaped pattern against the previous path.

Fig. 6.8. Robomower – top view *(left)*; underside *(right)*

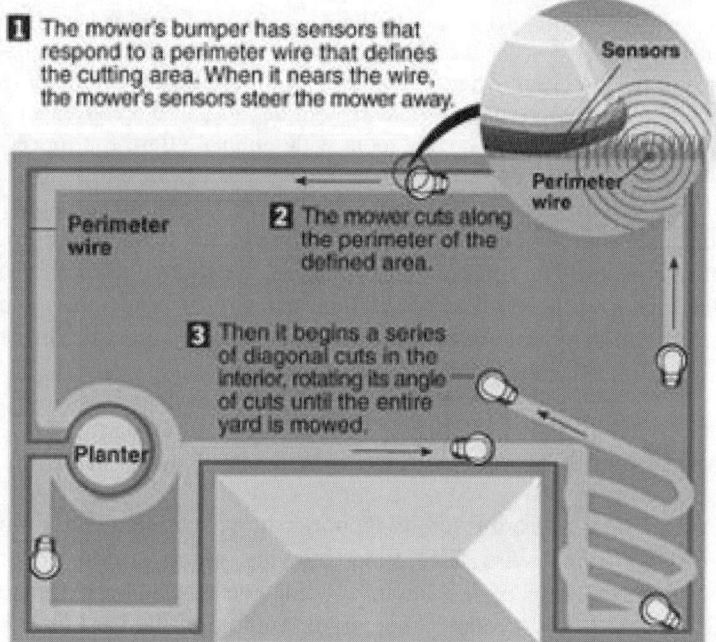

Fig. 6.9. Lawn mowing pattern of the Robomower

There are four pneumatic bump sensors on each side of the robot [7]. When it senses that it has bumped into an object it stops, backs up and changes direction. In addition, proximity sensors in the front and rear send out a sonar-like signal that tells the mower to slow down when it comes within one metre of an object. The robot also stops when it's tilted at an angle of over 20 degrees or when it is lifted. Before it begins the mowing process, the machine must be calibrated and mowing time set according to the size of lawn (100 square meters per hour).

6 Where Have All the Mobile Robots Gone?   189

Fig. 6.10. Husqvarna robot mower

Husqvarna [8] also make a robot mower (Fig.6.10). It can mow up to a 1500m^2 lawn, then find its recharger and automatically recharge its batteries. The boundary of the lawn is defined with a guide wire. It has 'collision sensors', which react to garden furniture, trees, rockeries and the like, and is equipped with anti-theft protection.

6.2.4 Robot Pool Cleaners

Companies that make robot pool cleaners include Aquabot (Fig.6.11), Dolphin (Fig.6.12) and Aquavac [17]. Models range in price from US$825 to US$1,500, depending on their features and options. The robot is self contained except for power. A 15-metre (typical) power lead is plugged into a (low voltage) power supply that is located away from the water. When it is time to clean the pool the operator puts the robot in the pool, sets the handle to indicate the forward cleaning direction, and turns the power on.

A typical robot contains two motors: one to drive it and one to operate the pump. The robot is driven by tracks on the side. The drive control can be set to clean the floor only or to clean walls and stairs as well. The drive motor also turns the front and rear brushes to sweep the dirt on the surface into the inflow of the suction system. The pump sucks the water through a filter, which has to be emptied manually.

Fig. 6.11. Aquabot robot pool cleaner used at the 2000 Olympic Games in Sydney

Robots for home use can clean a 6m*12m pool. More expensive models can clean up to Olympic pool size, depending on the length of the power tether and the ability of the drive motors to drag the tether through the water. A typical robot for a home pool is about 300mm*400mm*400mm in size and can pump about 16 Kilolitres per hour.

There appear to be two approaches to navigation. In one the robot is placed at one end of the pool, facing across the pool. It heads in the direction in which it is initially placed, following a compass bearing, until it bumps the other side of the pool. Then it backs away from the wall, turns 90°, moves its width, then turns a further 90° and heads back toward the other side. When it bumps a wall during the motion after the first 90° turn it knows that it has reached the other end of the pool and cleaning is finished. One problem with using a magnetic compass is that the iron reinforcement at the deep end of the pool can cause the compass to fail and the robot to loose its bearing.

The second approach involves the robot measuring the size of the pool, by traveling first across it and then along it, relying on the tracks to drive it in a straight line. Then it can calculate the size and an initial shape of the pool and plan an efficient cleaning pattern. As it follows this pattern, when it travels further than it expects, it updates the shape map.

Fig. 6.12. Dolphin robot pool cleaner cleaning a pool

6.2.5 Robotic People Transporter

The Segway Human Transporter is a robot with a human providing some of the intelligence (Fig.6.13). It represents the class of tele-operated vehicles where human and robot intelligence complement each other to produce a result that neither can alone. The robot balances on two wheels and responds to human navigation commands. The latter are communicated to the robot via a hand speed controller and by the human shifting their weight in the direction they want to travel.

The Segway is controlled by two processors operating in parallel, so that if one fails the other can continue to control the vehicle. Communication between the processors is performed optically to avoid electrical faults on one propagating to the other. The processors read the sensors 100 times per second, run the balance control and motor speed loops at 100 Hz, and pulse the motors at 1Khz.

The balance sensor assembly consists of five solid-state, vibrating-ring, angular-rate gyroscopes that use the Coriolis effect to measure rotation speed. These tiny rings are electromechanically vibrated in such a way that when they are rotated, a small force is generated that is measured by the electronics in the sensor. Each 'gyro' is placed at a unique angle that allows it to measure multiple directions. The measurement software compares the data from the five gyros to determine if any of the five is supplying faulty data and, when a failure occurs, compensates with data from the remaining sensors to continue balancing through a controlled safety shutdown. Two electrolyte-filled tilt sensors provide a gravity reference for achieving balance and keeping the control stork upright.

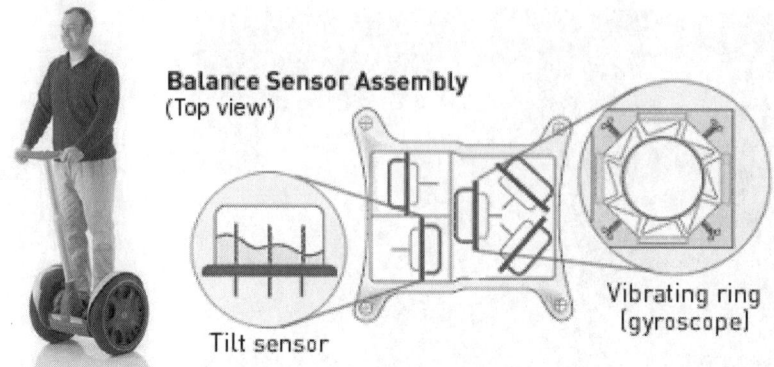

Fig. 6.13. The Segway Human Transporter and the balance sensor assembly

The 1.5 Killowatt brushless servo motors use neodymium-iron-boron rare-earth magnets to obtain the highest power-to-weight ratio of any motor in mass production. Each motor is constructed with two independent sets of windings, each driven by a separate control board. Under normal conditions, both sets of windings work in parallel, sharing the load. In the event of a failure, the motor is designed to instantly disable the faulty winding and use the remaining winding to maintain control of the Segway until it can be brought to a stop.

The motors are carefully balanced to operate up to 8,000 rpm, allowing them to produce very high power levels in a small package. Feedback from the motor to the Segway is provided by redundant, non-contact analog Hall sensors that sense the positions of magnets with no moving parts other than the motor shaft itself.

The motors drive the wheels through 24:1 ratio helical gearboxes. The meshes between the three gears in each box produce sounds that are two octaves apart, so the gearbox noise sounds musical. The tyres have been designed using a silica-based tread compound giving enhanced traction and minimized marking on indoor floors, and a specially engineered tubeless construction that allows low pressure for comfort and traction while minimizing rolling resistance for long distance travel. The tyres are mounted on wheels constructed from an engineering-grade thermoplastic that is lightweight, durable, and reduces the noise transmitted from the drive system.

The twin nickel-metal hydride battery packs deliver a nominal 72 volts. Each pack consists of an array of high-capacity cells and a custom-designed circuit board that constantly monitors the temperature and voltage of the pack in multiple locations. The internal electronics in the battery incorporate a 'smart' charging mode so the customer need only plug the Segway into the wall and the battery will choose the appropriate charge rate based on temperature, voltage, and level of charge. Under normal operation, the electronics monitors both batteries and automatically adjusts to drain the batteries evenly. In the unlikely event of a battery failure, the system is designed to use the second battery to operate the Segway and allow it to continue balancing until it is brought to a safe stop.

Fig. 6.14. Turtle-type Lego robot with odometry sensors on each drive wheel and touch sensors front and back.

6.2.6 Robot Toys

Possibly the only area of robotics that is turning in a profit is robot toys. The Lego Company found that their sales were dropping because children were playing with computer games rather than with Lego. They decided to make some computer games but lost money because they couldn't compete head on with games companies like Sony. So they returned to what they do best: toys that provide an open-ended learning experience. They developed a series of Lego toys that incorporated computers. One such toy is the Lego robot system known to consumers as 'Mindstorms' and to educators as 'Robolab' (Fig.6.14).

A second robot toy is Sony's AIBO robotic dog (Fig.6.15). AIBO uses 18 motors (four for the head, three for each leg and two in the tail) to walk, sit, sleep, beg, and perform a host of other dog-like actions [12]. A remote control emits tones that the robotic dog 'hears'. The remote puts the dog into one of three activity states: lying, sitting, or standing. Within each state, the dog performs one of 5 to 10 actions, based on the key pressed. It can walk as fast as 6m per minute.

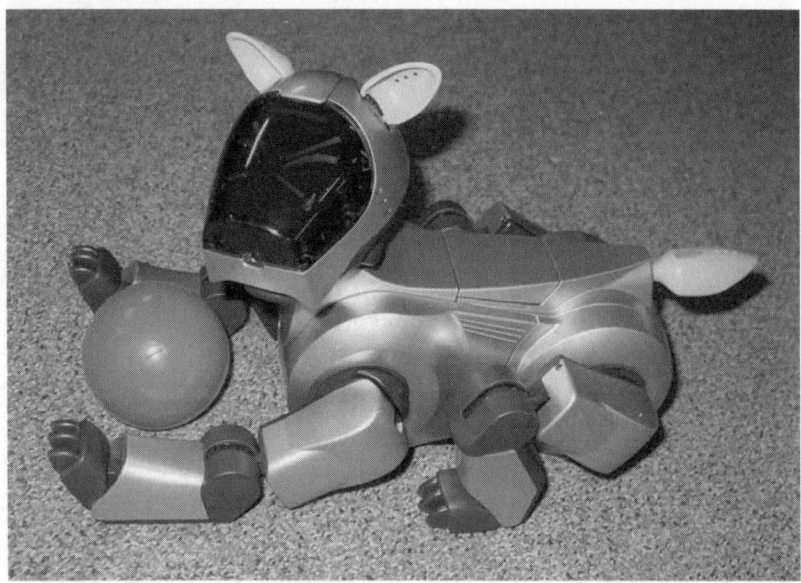

Fig. 6.15. AIBO – Sony's pet robot dog

AIBO includes a 180,000 pixel CCD camera in its head. AIBO also includes a pressure sensor so it can determine when it's being petted and LEDs that light up like eyes. The LEDs are a kind of doggy mood ring: green means AIBO is happy, while red means its angry. It also includes a range finder for detecting objects around it and motion sensors that let it regain its balance after falling. Sony claims it won't walk off the side of a table or into walls.

Sony also claims AIBO has 'instincts' and 'emotion', and that it will autonomously act on these capabilities. They claim that AIBO will learn from its behavior and environment, and move from being a puppy to an adult dog over time.

6.2.7 Other Applications

Mobile robots have been built for many applications. A few have made it to market. Most have not been a commercial success. Those currently on the market are listed on the web site hosted by the *IEEE Technical Committee on Service Robots* [6]. Some of the applications include robot pets, autonomous wheelchairs, robot tour guides, robot fire fighters, robot carriers of shipping containers, robot mining vehicles, robot security guards and robot guide dogs. In addition, some expensive tele-operated vehicles are being sold to police forces for bomb disposal operations.

6.2.8 Getting a Robot to Market

Most of the companies that have been started up to make and sell mobile robot-based appliances have failed. The reasons were varied, but most failed to sell sufficient robots to return a profit. The experience of iRobot demonstrates how daunting the task is. iRobot makes and markets the `Roomba` vacuum cleaning robot [14]. By August 2003, they had sold over 100,000 robots at US$200 each. By contrast, the `Trilobite` sells for US$1,600, the `Robomower` for US$700 and `AIBO` for US$2,500.

iRobot is known among roboticists as the company that makes and sells robots created by Rodney Brookes at MIT. For 13 years it built a reputation as a specialist robotics contractor. During this time iRobot had its fingers in lots of pies because they had no 'killer pie', as it were. The goal of such diversity was to mitigate risk; this diversity made iRobot flexible.

However, its founders had a dream: to place cheap robots in every home to labor together to clean carpets. To make the shift from making small volume high-cost robots to mass production iRobot had to learn some difficult lessons [1].

When developing a robot called `Grendel` for a customer, iRobot negotiated away their Intellectual Property (IP). When they developed subsequent robots they had to make sure that they used a different technology so that the customer who owned the IP wouldn't end up owning part of the new robot. Now iRobot retains the rights to all the technology it develops. Most customers are happy with that because they benefit from the IP developed on projects paid for by previous customers. Access to IP has become part of their marketing edge. For example, cleaning technology developed for Johnson Wax Professional was used in the development of `Roomba`. This intellectual capital made iRobot strong.

One of the hardest lessons for engineers used to making robots to learn is to make them cheap. Previous experience had taught iRobot that other companies were not going to take their expensive designs and make them cheap. From their customers they had learned how to source materials, ensure safety, manufacture cheaply, design for the mass market and work with companies in South East Asia. So they decided to go into production themselves. They had to learn that cost is priority one; every cent has to be justified. iRobot learned to make robots with plastics rather than with expensive machined aluminum.

To develop a robot that consumers would use they had their engineers observe focus groups of ordinary people who clean things. As engineers developed a sequence of 20 prototypes, they brought each iteration home for testing by spouses, friends, and neighbours. The marketing staff also used Roomba in their houses and filled out surveys reflecting the perspective of a non-technical user. Through this process it became clear that if the users had to be trained to operate the robot it would not succeed. The users wanted a very simple interface: a single button labeled 'Clean'.

The focus groups also reinforced the lesson that consumers don't like expensive products. To achieve the low price of US$200, the team developing Roomba had reduced the robotics to the minimum that would achieve the task. In contrast to most other robots, Roomba has very few sensors; its main sensor for navigation is a bump sensor. So it relies heavily on mechanical design and behavioural software. By contrast the robots that use more sophisticated sensing cost more.

A lot of the success of Roomba is due to marketing strategy. It was first released through gift shops in time for Christmas purchases. After Christmas it was released through specialty manchester stores which played videos in their store. They employed a marketing firm to push Roomba aggressively in major newspapers and magazines. This resulted in TV interviews, followed by reviews in magazines. iRobot raised US$27.5 million in venture capital to get Roomba to market.

6.2.9 Wheeled Mobile Robot Research

Mobile robot research began in an effort to solve some of the problems of Automated Guided Vehicles (AGVs), which are expensive and inflexible. They are used in heavy industry to move car bodies, steel coils, etc. from one processing plant to another. Their motion is restricted to paths defined by a guide technology, such as wires in the floor. People are kept out of these guide paths to minimize the possibility of collision. A central computer program plans and controls their motion. The guide system is expensive to install and alter.

The goal of mobile robotics is to make an autonomous mobile vehicle that is flexible and low cost. Flexibility requires the ability to move in an unconstrained environment so that the robot can travel from a start location to a goal location without collision. Once the robot has reached the goal location it may have to interact with (contact) an object in the environment to accomplish its given task. Both motion in an unconstrained environment and interacting with task objects requires the ability to sense the environment and rapidly change the planned path in response.

All the consumer robots described above require a defined workspace. Where the workspace is not defined by natural barriers such as walls and the sides of pools, human operator have to set them up before starting the robot. Also, the human has to remove objects from the robot's workspace that may cause the robot to malfunction.

These robots are a long way from the popular conception of what a mobile robot should be. Even when the results of current research is incorporated into them they will still require a high-level of human assistance. However, many research results have not transferred to consumer mobile robots because the solutions are too expensive, not robust, only work in a limited environment, and are hampered by poor perception of the environment.

6.3 Research Directions

The robots described above have taken a long time and a lot of detailed research to bring to market. Yet their use of sensing technology and Artificial Intelligence reflects research done a decade ago. Much of the research done in the last decade has not yet made it into product.

To get funding, European researchers have proposed a 'Beyond Robotics' initiative a part of the 6th European Framework Programme [5]. Their proposal is to incorporating AI into physical mobile artifacts. The proposal has three objectives:
1. *Cognitive robots as assistants to humans* – open ended conceptual learning of new tasks (ie. robust perception),
2. *Hybrid bionic systems* – augmenting human interaction with and perception of the environment, and
3. *Autonomous micro-robot groups (robot ecologies)* – able to organize to achieve an objective.

The proposal implies that classical robotics research has been done and reduces the place of mobile robotics to a technique for solving problems. The area most aligned to mobile robotics is the neuroinformatics initiative which aims to explore synergies between neuroscience and IT to enable the construction of hardware/software artifacts that live and grow.

It gives the impression that mobile robotics research has been done until you look at some of the proposed projects. They look very much like a continuation of the perception and learning projects that mobile robot researchers have been working on. For example, the central technological achievement of the AMOUSE project will be the construction of an artificial whisker system and test it on a mobile robot.

Other projects focus on machine perception - for example building an electronic bat - and on building serpentine robots with nano-technology. Research into improved sensing is essential to the development of mobile robot applications. However, having to make blue sky claims to get funding for mobile robot research will only make it more difficult to get the next round of funding.

6.4 Conclusion

We make tools to complement us – to do the things we find difficult, boring, tiresome or dirty – not to replace us. Our tools are good at doing the things we find difficult to do or don't want to do. In contrast, we find it difficult to make machines to do the things that we are good at, such as sensing and mobility. One reason is that we don't understand how we do these things.

These two issues – complementarity and understanding – make mobile robots difficult to sell and mobile robotics research challenging. Mobile robots that reliably execute tasks that people don't want to do will eventually find a market. Mobile robot research needs to focus on the problems stopping robots achieving these tasks.

The popular notion that because humans solve a task in a certain way then robots should too is questionable. People use vision for navigation. Only one of the robots described above uses vision, but it uses light stripe vision, not human-like vision. We are far from understanding human intelligence. The only one of these robots that uses human-like intelligence is the Segway which includes a human in the loop. The others use rule-based systems, behavioural architectures and control loops.

Mobile robot research need not be limited to solving task problems. Research in intelligence, sensing, perception, navigation and motion control will lead to improved theoretical understanding. This theoretical understanding will result in improved solutions to the practical problems that currently limit the application of mobile robotics in consumer appliances.

A Note on the Figures

All figures are reproduced herein with the kind permission of the copyright holders; the figure sources are as follows:

Fig.6.1	http://iros02.epfl.ch/gallery/view_album.php?set_album Name=Contest
Fig.6.2	http://mars.jpl.nasa.gov/MPF/ops/yogi-pres-col-2.jpg (courtesy of NASA/JPL/Caltech)
Fig.6.3	http://www.pyxis.com/multimedia/helpmate/HMflash.asp
Fig.6.4	http://trilobite.electrolux.se/presskit_en/node1281.asp
Figs. 6.5-6.7	http://www.roombavac.com/products/Roomba.asp
Figs. 6.8-6.9	http://www.robotic-lawnmower.com/index.htm
Fig.6.10	http://international.husqvarna.com/node1520.asp?id=23494&intLanguageId=1
Fig.6.11	http://www.smarthome.com/3246.html
Fig.6.12	http://dolphinautomaticpoolcleaner.com/Default.htm
Fig.6.13	http://www.segway.com/

References

1. Buchanan L (2003) Death to Cool, *Inc. Magazine*, July.
2. Electrolux (2003) http://trilobite.electrolux.se/presskit_en/ node1281.asp
3. Engleberger JF (1989) *Robotics in Service*, MIT Press, Cambridge, MA.
4. EURON (2003) European Robotics Network – http://www.euron.org/
5. FET (2003) Beyond robotic – http://www.cordis.lu/ist/fetro.htm
6. IEEE (2003) Robotics and Automation Society, Technical Committee on Service Robots, http://www.service-robots.org/companies.php
7. Friendly Robotics (2003) http://www.robotic-lawnmower.com/index.htm
8. Husqvarna (2003) – http://international.husqvarna.com/node1533.asp
9. iRobot (2003) http://www.roombavac.com/products/Roomba.asp
10. IROS (2002) – http://iros02.epfl.ch/finalStats.php
11. Kim JL (2000) Robomower From Friendly Robotics, *Tech TV - Fresh Gear*, June 2 http://www.techtv.com/freshgear/products/story/0,23008,2563453,00.html

12. Louderback J (1999) Sony AIBO Robotic Dog, *Tech TV – Fresh Gear*, May 11, http://www.techtv.com/freshgear/products/story/0,23008,2257334,00.html
13. NASA (2003) – http://mars.jpl.nasa.gov/MPF/ops/yogi-pres-col-2.jpg
14. Prager D (2003) First Look: iRobot Roomba Intelligent FloorVac, *Tech TV - Fresh Gear*, April 2, http://www.techtv.com/freshgear/products/story/0,23008,3408540,00.html
15. Pyxis (2003) – http://www.pyxis.com/multimedia/helpmate/HMflash.asp
16. Segway (2003) – http://www.segway.com/
17. Smarthome (2003) – http://www.smarthome.com/3246.html
18. Ulanoff L (2003) Roomba Returns, *PC Magazine*, August 27th.

7 Building Intelligent Legal Decision Support Systems: Past Practice and Future Challenges

John Zeleznikow[1,2]

1   Faculty of Law, University of Edinburgh, Old College, South
    Bridge, Edinburgh, EH8 9YL, Scotland, UK
    john.zeleznikow@ed.ac.uk

2   School of Information Systems, Victoria University, PO Box 14428,
    Melbourne City MC, VIC 8001, Australia
    john.zeleznikow@vu.edu.au

7.1 Introduction

In [91] Susskind outlines the past use of Information Technology (IT), and indicates probable future uses of IT by the legal profession. He indicates that until recently, there was only limited use of IT by legal professionals. Whilst the use of word processing, office, automation, case management tools, client and case databases, electronic data/document interchange tools and fax machines is now standard, only recently have legal firms commenced using knowledge management techniques. The use of applied legal decision support systems is in its infancy.

The development of intelligent systems in legal practice was investigated by [123]. They noted that most commercially successful systems have employed rules. The major reasons for this occurrence include that it is easy to model rules and there are many tools for building rule-based systems.

Although many commentators including [64] clearly express reservations about this approach for the majority of fields of law, rule-based reasoning is still the predominant basis for legal decision support systems. The fundamental limitation not addressed by this view of law can be reduced to two significant omissions: (a) the failure to model open texture, and (b) the failure to provide an analysis of how justification differs from the process used to arrive at decisions.

A decade later, in his description of commercial legal decision support systems build at the Donald Berman Laboratory for Information Technology and Law at La Trobe University, Zeleznikow notes that the majority of commercially available legal decision support systems model fields of law that are complex but not discretionary [121]. For example SoftLaw - an Australian software house that specializes in legal knowledge based systems for the Australian Government - has developed systems with tens of thousands of rules. It primarily builds systems with regard to entitlements, in domains such as social security legislation. There seems little doubt that the trend toward rule based systems to encode large and complex legislation will continue to a substantial extent, as claimed by [85]. This is assured by the increasing public demand for more transparency and consistency in government decision making alongside with the continuing enactment of increasingly complex legislation.

In this Chapter, we wish to examine theoretical issues and practical implications behind the development of intelligent legal decision support systems. To do so, we must consider why rule-based systems are inadequate for modeling legal reasoning, by first examining the nature of legal reasoning. After an examination of early legal knowledge-based systems, we consider how case-based reasoning, hybrid reasoning, machine learning and argumentation have been used to develop legal knowledge-based systems.

We conclude by examining how legal knowledge-based systems are being used on the World Wide Web, and how legal decision support systems can be used to support negotiation.

7.1.1 Benefits of Legal Decision Support Systems to the Legal Profession

The development of intelligent legal decision support systems is improving the training and skill of lawyers by:
- Supporting a more precise reading of legal materials;
- Supporting a more precise drafting of legal documents;
- Supporting a more rational management of risk; and
- Supporting a more efficient management of information.

Intelligent legal decision support systems will also help provide a more fair and efficient system of justice by
- Reducing the high transaction cost of legal services;
- Enhancing consistency in decision making – by replicating the manner in which decisions are made, decision support systems are encouraging the spread of consistency in legal decision-making. They ensure that legal decision-makers adhere to the principle that *like cases should be treated alike*;

- Transparency – by demonstrating how legal decisions are made, legal decision support systems are leading to a better community understanding of legal domains. This has the desired benefit of decreasing the level of public criticism of judicial decision making;
- Efficiency – one of the major benefits of decision support systems is to make firms more efficient; and
- Enhanced support for dispute resolution – users of legal decision support systems are aware of the likely outcome of litigation and thus are encouraged to avoid the costs and emotional stress of legal proceedings.

It is noted in [20] that increasing numbers of litigants represent themselves in court. This swelling tide of *pro se*[1] litigants constitutes a growing burden not only to the judiciary but to the entire legal process. Typically, unrepresented litigants:

1. Extend the time taken for litigation – due to their lack of understanding of the process,
2. Place themselves at a disadvantage compared to their opponent(s), and
3. Place the judicial decision-maker in the difficult position of deciding how much support and forbearance the decision-maker should offer to the *pro se* litigant.

A recent study conducted for the American Bar Association in the Supreme Court of Maricopa County, Arizona, USA, indicated that at least one of the parties was self-represented in over 88% of domestic relations cases, and both parties were self-represented in 52% of the cases. In [59] it is reported that in 1999, 24,416 of the 54,693 cases opened in the US Court of Appeals were filed by *pro se* appellants. This figure, whilst on first glance may not appear alarming, needs to be considered in light of the fact that many *pro se* appellants have neither the financial resources nor the legal skills to conduct their own appeals.

It is noted in [73] that there is a shortfall in legal systems for poor persons in the United States. It is claimed in [20] that domestic abuse victims are particularly likely to have few resources and little opportunity to obtain the services of a lawyer. He states that the growth of the consumer movement has increased the trend for *pro se* litigation. The growing availability of books, document kits and computerized forms, together with the increasing availability of legal materials on the World Wide Web, has increased the opportunities for pro se litigants to organize their own litigation.

---

[1] A *pro se* litigant is one who does not retain a lawyer and appears for them self in court.

When considering decision making as a knowledge-manufacturing process, the purpose of a decision support system is to help the user manage knowledge. A decision support system fulfils this purpose by enhancing the user's competence in representing and processing knowledge. It supplements human knowledge management skills with a computer-based means for managing knowledge. A decision support system accepts, stores, uses, receives and presents knowledge pertinent to the decisions being made. Its capabilities are defined by the types of knowledge with which it can work, the ways in which it can represent these various types of knowledge, and its capabilities for processing these representations.

In [119] it is claimed that whilst the construction of legal decision support systems will not have a drastic effect on improving access to justice, he makes the argument that the construction of such systems for community legal centres will improve their efficiency and increase the volume of advice they can offer. Until recently, most legal decision supports systems were rule-based and developed to run on personal computers. Whilst personal computer based tools are fine for lawyers, they may not be easily accessible to *pro se* litigants. Reasons for this difficulty include their lack of awareness of such systems, and the high cost of purchasing relevant software. Currently, very few legal decision support systems are available on the World Wide Web. The development of legal knowledge-based systems on the World Wide Web will help improve access to justice.

We should stress that any legal decision support systems should be used as tools for legal decision makers; it would be most inappropriate for them to make legal decisions. We will address this issue in this Chapter.

7.1.2 Current Research in Artificial Intelligence and Law

There are many important research issues (some specific to the legal domain, others of interest to the general Artificial Intelligence community), which arise from research in artificial intelligence and law. These include:
1. Representing legislation for both inference and maintenance – such as the work performed by SoftLaw.
2. Representing and reasoning with open-textured concepts – an issue we will discuss in Sect.7.2.
3. Representing and reasoning with normative concepts – early systems of law were based on a single morality – often based on religious beliefs. Such law is referred to as *natural law*. Legal positivism arose as a reaction to natural law. It assumes that law is based upon explicit norms

separate from the ideas of law and morality held by natural lawyers. These norms may have arisen from legislation, code, cases or doctrine. An ontology as an explicit conceptualization of a domain is defined in [38]. In Sect.7.4.5 we investigate the development of legal ontologies. Such ontologies represent legal norms and are very significant for developing legal knowledge-based systems on the World Wide Web.

4. Simulating the process of expert legal prediction/advising - In the Split-Up system [88, 125], a hybrid of rules, neural networks and argumentation is developed to provide advice upon the distribution of marital property following divorce in Australia. The development of legal advisory systems has led to an investigation of user interface issues (including for whom is the system designed). It is noted in [126] that the initial Split-Up system was designed to be used by lawyers and judges. However, following testing by practitioners, it turned out the system was of a greater benefit to mediators. It would be ideal if divorcees could use the system, but unfortunately the legal knowledge required, as well as their inability to make objective decisions, made it impractical for divorcees to use the system. Reich's theory of evaluation [74] was extended by [86] to the domain of legal knowledge based systems. [40] have developed a method for evaluating legal knowledge-based systems based upon the Criteria, Context Contingency-guidelines Framework of [42].

5. Reasoning and arguing using examples as well as rules – we will discuss the use of rule-based reasoning in Sect.7.3.1, case-based reasoning in Sect.7.3.2 and learning from cases in Sect.7.3.3.

6. Computerized statutes and regulations – including Legal Drafting, Tools to support drafting in normalized form and Document assembly/generation systems. LEDA [105] provides semi-intelligent document drafting assistance for the Dutch ministry of Justice, whilst OBW provides similar advice for the Dutch Ministry of Education and Science. In the 1980s, the Dutch government became worried about the quality and effectiveness of legislation. A set of measures aimed at the lasting improvement of legislative quality was introduced. LEDA and OBW were an attempt to provide automated support in document drafting.

7. Legal Decision Support and Advisory Systems - as noted above, Legal Decision Support and Advisory Systems have proved particularly successful in improving the efficiency and quality of advice given by legal aid organizations. Zeleznikow discusses the GetAid system [121]. When an applicant approaches Victoria Legal Aid (VLA) for assistance, his/her application is assessed to determine whether he/she

should receive legal aid. This task chews up 60% of VLA's operating budget, yet provides no services to its clients. After passing a financial test, applicants for legal aid must pass a merit test. The merit test involves a prediction about the likely outcome of the case if it were to be decided by a Court. VLA grants officers, who have extensive experience in the practices of Victorian Courts, assess the merit test. This assessment involves the integration of procedural knowledge found in regulatory guidelines with expert lawyer knowledge that involves a considerable degree of discretion. Since experts could not readily represent knowledge about an applicant's prospects for acquittal as a decision tree, we decided to model the process as a tree of Toulmin arguments. It was difficult for the principal domain expert to articulate many of the arguments. She could not express her heuristic as rules because the way in which the factors combine is rarely made explicit. Her expertise was primarily a result of the experience she had gained in the domain. Although it is feasible to attempt to derive heuristics, the approach used was to present a panel of experts with an exhaustive list of all combinations of data items as hypothetical cases and prompt the panel for a decision on acquittal prospects. Six experts and the knowledge engineer were able to record their decision in all of the exhaustive hypothetical cases (for that argument) in approximately 40 minutes. The decisions from each rater were merged to form a dataset of 600 records that were used to train neural networks. The PHP program that implements the argument based inferences is a small and relatively simple program that executes on the server side very quickly and is not memory intensive. GetAid was tested by VLA experts and developed in conjunction with web-based lodgment of applications for legal aid [41]. Commencing the middle of 2003, VLA clients have used the GetAid system.

8. Legal document management and retrieval systems – the significance of research in text retrieval for developing tools for legal practitioners is discussed in [97], in which it is noted that text retrieval models include: (a) Boolean, (b) Vector Space, (c) probabilistic, (d) rule-based, and (e) knowledge-based. WIN (Westlaw is Natural) uses a Bayesian network to compute the probabilities that documents in a database are relevant to a query, and returns the relevant documents in order of their significance to the relevant query. Bayesian Networks have been used to analyze criminal evidence (see for example [1] and [80]). Flexlaw [84] used a Vector Space Model for matching, whilst Datalex [39] used hypertext, rules and queries. [63] notes that current commercial retrieval systems either rely on a manual indexing of case texts or upon

a full text search. The disadvantage of the former is the tremendous cost, whilst the latter leads to a lack of reliable retrieval costs. The occurrence of a word or phrase in a text is no guarantee of the text's relevance for the search request. Whereas [63] used statistics to detect the features of cases, [77] used neural networks. The development of Legal Ontologies (see the E-court project of [21]) offers great opportunities.

9. Legal Argumentation and Negotiation – we discuss argumentation in Sect.7.3.4 and legal negotiation in Sect.7.4.

10. Intelligent tutoring systems – five case-based reasoning paradigms are identified in [6], namely: (a) statistically oriented; (b) model based; (c) planning or design oriented; exemplar based and precedent based. The most significant paradigm for legal case based reasoning is not surprisingly precedent based case based reasoning. GREBE [18] and HYPO [5] are two such systems. Ashley claims that case based reasoning contributes to the cognitive modeling of legal reasoning and other fields [6]. Following up this work on HYPO, [2] and [3] showed how case based reasoning can provide explanations through the development of CATO, an intelligent tutoring system. CATO presents students with exercises, including the facts of a problem, a set of online cases, and instructions on how to make or respond to legal arguments about the problem. The student is provided with a set of tools to analyze the problem and prepare an answer by comparing and contrasting it to other cases. The program responds to the student's arguments by citing counterexamples and providing feedback on a student's problem solving activities with explanations, examples and follow up assignments.

11. E-commerce – the development of electronic commerce has led to a variety of fascinating legal problems, including privacy, intellectual property, free speech, taxation, consumer protection, internet gambling and fraud. The regulation of activities in an information-based society is performed by a combination of legal, economic, social and technological means. A useful way to identify pressing issues and their solutions is to conceive of mechanisms that combine all four types of regulatory sources. The technological regulatory sources will mainly derive from intelligent systems of one sort or another. The ease with which created works can be copied and transmitted to a global audience with new cyber-space technologies presents serious challenges for the regulation of copyright. A number of commentators argue that new economic models that include a weakening of copyright protections are needed while others argue that the copyright law needs only minor

modification in order to accommodate new technology. Others claim that the most appropriate way to regulate the use of created works is to combine existing copyright law with technological measures such as encryption software. Although not yet widespread, the use of encryption software in conjunction with the modification of copyright legislation seems to be emerging as the main mechanism for the regulation of copyright. However, the main disadvantage with this solution is that the free availability of works for public benefit purposes such as study, research, news or review is threatened as encryption software can all too easily lock all unauthorized access to works regardless of public benefit purposes. A proposal for the regulation of copyright with the use of knowledge-based systems is advanced in [87]. They believe that regulation of copyright will be largely realized with a combination of software, statutory and other measures. They further propose a framework that involves the development of five knowledge based systems that can protect author's works using a variety of technological measures such as encryption without restricting access to works for public benefit purposes. Though not as simple as exclusive locking systems, the knowledge-based systems are likely to be used if users have sufficient trust in their performance and if appropriate incentives are made available. The systems are designed in such a manner that the user can easily vary the extent to which actions such as encoding information regarded desirable rights for a work, are performed automatically. Authors can use systems in minimal autonomy mode until they have sufficient trust that systems are performing in accordance with expectations. When this occurs, authors can incrementally direct the system to be increasingly autonomous. The principle of keeping the user in control will engender sufficient trust in the systems. This, together with appropriate copyright legislation and economic incentives will encourage the use of these systems. The architecture of the systems is agent based. This facilitates the operation performance of the knowledge-based systems in what is an open textured and complex domain.

7.2 Jurisprudential Principles for Developing Intelligent Legal Knowledge-Based Systems

In [114] it is argued that there is a science of reasoning underlying law. Whilst Wigmore's focus was upon the law of evidence, and in particular legal evidence, we believe this principle is true for all legal decision-making. Walton claims that in the past most books on legal logic – with the exception of [4] and [68] – have concentrated almost exclusively on deductive or inductive logic in modeling legal argumentation [110]. Whilst much current research in artificial intelligence and law focuses on argumentation, current commercial systems are primarily rule-based.

This is because most legal decision support systems built for commercial use model fields of law that are complex but not discretionary. For example, SoftLaw [46] has developed systems with tens of thousands of rules. Most of these systems have advised upon the determination of pension benefits. The Social Security Act is difficult to master because it is large and complex yet few decisions based on the Act involve the exercise of discretion.

7.2.1 Reasoning with Open Texture

Legal reasoning is characteristically indeterminate. [11] view the indeterminacy in law as a specific consequence of the prevalence of open textured terms. Open texture was a concept first introduced by [107] to assert that empirical concepts are necessarily indeterminate. To use his example, we may define gold as that substance which has spectral emission lines X, and is coloured deep yellow. However, because we cannot rule out the possibility that a substance with the same spectral emission as gold but without the colour of gold will confront us in the future, we are compelled to admit that the concept we have developed for gold is open textured.

It is indicated in [14] that legal reasoning is essentially indeterminate because it is open textured. [11] view the indeterminacy in law as a specific consequence of the prevalence of open textured terms. They define an open textured term as one whose extension or use cannot be determined in advance of its application. The term 'vehicle' in an ordinance invented by [43] can be seen to be an open textured term because its use in any particular case cannot be determined prior to that case. The substantial artificial intelligence literature on open texture has been collated and analyzed by [72] to point out that situations that characterize law as open textured include reasoning which involves defeasible rules, vague terms or classification ambiguities. This analysis of open texture is central to our discussion because we argue that the existence of judicial discretion is a form of open texture that is indistinct from the situations considered by [72]. The distinct types of situations that [Prakken 1997] notes are difficult to resolve because of the open textured nature of law are:
- *Classification difficulties.* A local government ordinance that prohibits vehicles from entering a municipal park is presented in [43]. He argues

that there can be expected to be little disagreement that the statute applies to automobiles. However, there are a number of situations for which the application of the statute is debatable. What of roller blades, for instance? [35], in a response to [43] posed the situation of a military truck mounted in the park as a statute. Considerable open texture surrounds the use of the term 'vehicle' in this case even though there is no question that the truck is a vehicle.

- *Defensible rules.* Another type of open texture arises from the defeasibility of legal concepts and rules. Any concept or rule, no matter how well defined, is always open to rebuke. Rarely do premises or consequents exist in law that are universally accepted. Driving whilst drunk is definitively prohibited by a Victorian statute, though few courts would convict a person who was forced to drive at gunpoint. The rule in this case is defeated in the context of exceptional circumstances.
- *Vague terms.* Legal tasks are often open textured because some terms or the connection between terms is vague. A judge finds the various interpretations of terms such as 'reasonable' or 'sufficient' stems from the vagueness of these terms and not from classification dilemmas or defeasibility requirements. [22] labels this a gradation of totality of terms which he claims is one reason that deduction is an inappropriate inferencing procedure for many problems in law.

The existence of judicial discretion contributes to the open textured nature of law. Yet situations that involve discretion cannot be described as instances of classification difficulties, defeasible rules or the presence of vague terms. In [122] it is argued that the existence of discretion is a distinct form of open texture.

7.2.2 The Inadequacies of Modeling Law as a Series of Rules

Rule- and logic-based systems handle open texture poorly, and generally rely on the user to resolve the open textured predicate without assistance. To model reasoning in open-textured and discretionary domains, we investigate the use of cases. One of the real benefits of using inductive and analogical reasoning as well as Knowledge Discovery from Databases (KDD) is the ability of these techniques to assist in resolving open texture.

Nonetheless, deduction is a powerful form of reasoning, particularly in areas which are based on legislation and which have little case law, customary law, implied doctrines, discretionary provisions and so on. Examples of appropriate domains for using deduction include civil law jurisdictions where case law provides only some assistance in interpreting the code, or in common law countries, rarely litigated statutes, new statutes and codifications of the case law. Lawyers form and use rules constantly – a feature described and commented upon by [98] where they discuss how lawyers reason with rules. Notwithstanding the legal philosophical issues, the practical difficulty with deductive reasoning in legal decision support systems is its poor handling of variations and amendments to the law, in particular statutory amendment and case law.

Given that we must include cases in our decision support system the question naturally arises: how should we do this? The most obvious answer would be to turn our cases into production rules or a logic programming representation. This simply means that we use a domain expert to review cases, statutes, and heuristics, and from this material formulate a number of rules. These rules then would be represented in the decision support system's rule base.

Case based reasoning is the catch all term for a number of techniques of representing and reasoning with prior experience to analyze or solve a new problem. It may include explanations of why previous experiences are or are not similar to the present problem and includes techniques of adapting past solutions to meet the requirements of the present problem. Case-based reasoning is very useful in resolving open-texture. The distinction between landmark and commonplace cases was introduced by [124] to help differentiate between those cases that are used in case-based reasoners and those that are used in a machine learning environment.

7.2.3 Landmark and Commonplace Cases

Kolodner incorporates context in her definition of a case for case based reasoning systems [50]. She states that 'a case is a contextualised piece of knowledge representing an experience that teaches a lesson fundamental to achieving the goals of the reasoner'. It is noted in [124] however that even in non-contentious areas, Kolodner's definition provides scope for considerable problems. They disagree with Kolodner that a case necessarily '…teaches a lesson fundamental to…the reasoner'. Certainly some cases do fit this description. Most notably within law, those decisions from appellate courts which form the basis of later decision and provide guidance to lower courts do provide a fundamental lesson, or normative structure for subsequent reasoning. [71] considers such cases to be formal, binding and inviolate prescriptions for future decision making, whilst [55] sees them as beacons from which inferior or merely subsequent courts navigate their way through new fact situations. The common name for such cases is landmark cases.

However, most decisions in any jurisdiction are not landmark cases. Most decisions are commonplace, and deal with relatively minor matters such as vehicle accidents, small civil actions, petty crime, divorce, and the like. These cases are rarely, if ever, reported upon by court reporting services, nor are they often made the subject of learned comment or analysis. More importantly, each case does not have the same consequences as the landmark cases.

Landmark cases are therefore of a fundamentally different character to commonplace cases. Landmark cases will individually have a profound effect on the subsequent disposition of all cases in that domain, whereas commonplace cases will only have a cumulative effect, and that effect will only be apparent over time.

Take, for example, the case of *Mabo v Queensland (No.2)*[2]. Prior to *Mabo* the indigenous people of Australia, the aborigines, had few, if any, proprietary rights to Australian land. Under British colonial rule, their laws were held to be inchoate and Australia itself was held to be *terra nullius*, 'empty land' at the time of white settlement. Hence, the only property laws applicable were those stemming from the introduction of white rule, laws which were less than generous in their grant of land to Aborigines. In *Mabo*, the High Court held that previous decisions holding that Australia was *terra nullius* at settlement, and decisions holding that Aborigines had no property laws affecting land, were simply wrong at law. Hence, the High Court said, Aborigines had sovereignty over parts of Australia under certain conditions. Whether one agrees with the High Court's interpretive technique, it is indisputable that this is the landmark case in the area, and has formed the basis of future decisions in the area. Indeed, *Mabo*, like many other leading cases, was the spur for political action and we soon saw the introduction of the Federal *Native Title Act*. Thus, landmark cases have the dual effect of determining (to some degree) the interpretation of subsequent fact situations as well as influencing the invocation of normative legislative processes.

As with rule-based systems, landmark cases often require interpretation. For example, following Mabo, most politicians believed Australian Aboriginals' rights to native title had been extinguished if farmers were given pastoral leases over land. In *the Wik Peoples v The State of Queensland* (1996), the High Court of Australia, in a 4—3 ruling, determined pastoral leases and native title could co-exist. This ruling has been the basis of many political debates. In late 1997, the new conservative Australian government introduced legislation to further interpret the Wik

---

[2] (1992) 175 CLR 1

decision (Federal Native Title Act, 1997). The Act was heavily amended in the Senate. Many Australian lawyers claim the Federal Native Title Act (1997) conflicts with the Racial Discrimination Act (1975) and the referendum of 1967 that allowed the Commonwealth of Australia to enact legislation for the benefit of indigenous Australians. There is debate as to whether the constitutional amendment allows the Parliament to enact any legislation relevant to indigenous Australians or only legislation beneficial to indigenous Australians.

Two leading United States landmark cases *Plesy v. Ferguson*[3] and *Brown v. Board of Education. of Topeka*[4], deal with the issue of segregated schools. The Fourteenth Amendment to the US constitution declared that *no state might deny any person the equal protection of the laws*. In 1896, in the case of *Plesy v. Ferguson* the United States Supreme Court ruled that the demands of the Fourteenth Amendment were satisfied if the states provided separate but equal facilities and the fact of segregation alone did not make facilities automatically unequal. In 1954, in *Brown v. Board of Education of Topeka* the Supreme Court seemingly overturned the decision made in *Plesy v. Ferguson*. In *Brown v. Board of Education of Topeka*, in the opinion of [15], the Supreme Court declared racial segregation in public schools to be in violation of the equal protection clause of the Fourteenth Amendment. They did so, not by overturning *Plesy v. Ferguson* but by using sociological evidence to show that racially segregated schools could not be equal [28].

To further indicate the similarity between landmark cases and rules we note that in *Miranda v Arizona* [5] the United States Supreme Court ruled that prior to any custodial interrogation the accused must be warned:
1. That he has a right to remain silent;
2. That any statement he does make may be used in evidence against him;
3. That he has the right to the presence of an attorney; and
4. That if he cannot afford an attorney, one will be appointed for him prior to any questioning if he so desires.

Unless and until these warnings or a waiver of these rights are demonstrated at the trial, no evidence obtained in the interrogation may be used against the accused. Miranda v Arizona is a landmark case with regards to the rights of the accused in a United States criminal trial. This case has assumed such significance that its findings are known as the Miranda rule.

---

[3] 163 US 537 (1896)
[4] 347 U.S. (1954)
[5] 384 U.S. 436 (1966)

Landmark cases rarely occur in common practice and are reported and discussed widely. These cases set a precedent that alters the way in which subsequent cases are decided. As an example of the preponderance of landmark versus commonplace cases, in the last two decades, the number of landmark cases in the Family Court of Australia is in the order of hundreds while the number of commonplace cases is in the order of multiple tens of thousands.

It should be noted that or notion of a landmark or commonplace case, is relevant to the system in which it is used. HYPO [5] has a case knowledge base consisting of thirty legal cases in the domain of trade secrets. None of these cases have particular legal significance, and thus legal scholars may consider these cases as commonplace cases. However, because HYPO reasons from these cases, for the purpose of the HYPO system, these cases can be considered as landmark cases.

Some critics believe the use of legal case-based reasoners is limited. Berman believed legal case-based systems must by necessity simulate rule-based systems and that factors emulate rules [13]. He said: 'For developers, as contrasted to researchers, the issue is not whether the resulting rule base is complete or even accurate or self-modifying – but whether the rule base is sufficiently accurate to be useful'. We believe that jurisprudes and developers of legal decision support systems use landmark cases as norms or rules. Commonplace cases can be used to learn how judges exercise discretion.

7.3 Early Legal Decision Support Systems

7.3.1 Rule Based Reasoning

There are many knowledge representation techniques. Logic is particularly useful in the domain of automated theorem proving, which can trace its roots to the work of Newell and Simon in the early 1960's [65]. The earliest legal knowledge based systems were developed in the 1970's; they were primarily rule- or logic-based.

The JUDITH system [69] used rules to represent part of the German Civil Code. Their rules were very similar to those developed in the Mycin system [82].

Logistic regression and basic nearest neighbour methods were used in [56] to support case based retrieval to predict judicial decisions. They did not develop a model of legal reasoning; their domain was that of Canadian capital gain cases in the decade 1958-1968.

Two different kinds of rules were used in [60]: general rules which define the elements of the claim, and specific rules extracted from cases. Things and relations are used to represent the *everyday world of human affairs* and are classified hierarchically into categories. A fact comprises two things and a relation between them; facts are assembled into situations. These situations are compared with the situation of the instant case, and the system determines the extent to which the instant case falls within or near the law of intentional torts (for example assault and battery).

TAXMAN was a logic-based deductive reasoner concerned with the taxation of corporate organizations. McCarty chose that domain because he believed the corporate tax domain is primarily *a tidy world of formal financial rights and obligations*. TAXMAN I [53] used an entirely rule based model. TAXMAN II [54] proceeded beyond the scope of rule-based systems by attempting to deal with open-textured concepts such as *continuity of interest*, *business purpose* and *step transactions*. It represented legal arguments as a sequence of mappings from a prototypical case to a contested case, in an attempt to perform analogical reasoning. Instead of adding cases to the knowledge base, open textured concepts were represented using a prototype – a concrete description expressed in the lower level representation language – together with a sequence of deformations or transformations of one concrete description into another.

In developing TAXMAN II, McCarty noted:
- Legal concepts are open textured;
- Legal rules are dynamic – as they are applied to new situations they are constantly modified to *fit the new facts*;
- In the process of theory construction there are plausible arguments of varying degrees of persuasiveness for each alternative version of the rule in each new fact situation, rather than in a single correct answer.

TAXADVISOR [62] used EMYCIN to assist lawyers in estate tax planning. It collected data about clients and suggested strategic plans about various aspects such as life insurance, retirement schemes, wills and making gifts and purchases. Rather than provide statutory interpretation, TAXADVISOR uses lawyers' experience and strategies to produce plans.

The British Nationality Act as a Logic Program [81] uses logic programming to perform statutory interpretation upon the British Nationality Act of 1981. The data needed for individual cases is stored in the APES shell. The answers produced by APES are the logical consequences of the rules together with supplied information. The knowledge in the rules is represented in and/or graphs.

Whilst the system is an interesting application of logic, the paper is jurisprudentially flawed, because it believes that law is straightforward and ambiguous. For example, the authors claim that a statement as to whether *an infant was born in the United Kingdom* is a readily verifiable fact. But is this statement true?

The boundaries of the United Kingdom are both constantly changing an in dispute. When the system was developed in 1986, if a child was born in Hong Kong, was she born in the United Kingdom under the Act?[6] At that time, Hong Kong citizens had British citizenship, but not the right of abode in the United Kingdom. Are the Falkland Islands (or Malvinas to others), part of the United Kingdom? These issues cannot be determined by reference to the Act or precedents. They depend on International Treaties, and even more significantly, delicate negotiations.

ExperTAX [83] was developed by Coopers and Lybrand to provide advice to clients of United States' certified public accountants on how to conduct the tax accrual and tax planning functions. The system improves staff accountants' productivity, the quality of information provided to them and accelerates their training process.

Ernst and Young UK developed three legal expert systems: VATIA [92], Latent Damage Adviser [23] and THUMPER [93]. VATIA (*V*alue *A*dded *T*ax *I*ntelligent *A*ssistant) placed specialist Value Added Tax expertise in the hands of auditors. VATIA enabled auditors to carry out overviews of clients' VAT affairs.

The Latent Damage Adviser modelled the Latent Damage Act 1986. The problems solved by the Latent Damage Adviser presented few difficulties for latent damage experts, but proved difficult for non-experts because they are not familiar with the complex web of inter-related rules that constitute this area of law. Susskind claims the statute was poorly drafted, complex and *barely intelligible*.

THUMPER was developed for use by corporate tax practitioners at Ernst and Young who give advice about tax liability and planning in respect of stamp duty. THUMPER has a three-layer conceptual model:
- Outermost level – users' view of the problem;
- Middle level – expert's interpretation of the principles and legislation of stamp duties; and
- Innermost layer – represents the legislation and the case law. The case law is stored in the form of rules induced by the experts from the cases.

The Rand Corporation built numerous Expert Systems in the early 1980's [112] to advise upon risk assessment. One of their early systems – LDS – assisted legal experts in settling product liability cases. LDS's knowledge

---

[6] On June 30 1997, Hong Kong was returned to China

consisted of legislation, case law and, importantly, informal principles and strategies used by lawyers and claims adjustors in settling cases.

Another Rand Corporation decision support system – SAL [113] – also dealt with claims settlement. SAL helped insurance claims adjusters evaluate claims related to asbestos exposure. SAL used knowledge about damages, defendant liability, plaintiff responsibility and case characteristics such as the type of litigants and skill of the opposing lawyers.

These two systems are important for they represent early first steps in recognizing the virtue of settlement-oriented decision support systems.

As we note from the work reviewed in this section, most legal decision support systems advise about risks and entitlements rather than predicting the results of litigation.

The Credit Act Advisory System (CAAS) is a strictly rule-based legal expert system which advises in relation to a small part of the Credit Act 1984 (Vic) [106]. CAAS does not argue directly with either statutes or precedents. Instead, it is a production rule system where the production rules forming the knowledge base of CAAS are heuristic rules supplied by experts from Allan Moore & Co. It was prototyped using NExpert Object and then compiled into C++ under Windows 3.1. In developing CAAS, Vossos developed a framework for building commercial legal expert systems using C++, rather than the more expensive expert system shells. The system was commercially marketed to organizations involved in the provision of credit. Back-office employees at the Bank of Melbourne commercially used the system. CAAS provided advice as to whether a transaction is regulated, not regulated or exempt under the Act.

CAAS was essentially the rule-based component of IKBALSIII [128], which is a deductive and analogical system that:
1. Includes induction as the basis for its case based retrieval function; and
2. Relies on distributed artificial intelligence techniques and the object oriented paradigm, rather than a blackboard architecture.

Whilst CAAS – the rule-based part of IKBALS III – covered the whole domain of the Victorian Credit Act (and was commercially exploited), IKBALS III only dealt with one possible open-textured predicate – *was the transaction for a business purpose*. IKBALSIII's novel technique for resolution of open texture had little commercial benefit.

SoftLaw Corporation Limited is an Australian company that provides software solutions for the administration of complex legislation, policy and procedure. SoftLaw's product – STATUTE Expert – is a knowledge base management system specifically designed for *n* of administrative rules.

SoftLaw has a comprehensive software project management methodology, which provides the following tools to software teams:

1. A document process model, which outlines all procedures and the products to be developed during the life of a project;
2. Templates for producing documentation on all issues to be considered at each step in the process; and
3. A team model, which ensures representation of all perspectives in the team.

Many Government agencies administer complicated legislation, policy or processes. Agencies structure their work organization, budget and level of client service around managing this complexity. The traditional management approach uses high numbers of specialized in-house staff, trained in individual aspects of an agency's work.

SoftLaw can create a rule base model of any complex legislation, policy or process. This makes the source material accessible to generalist users. STATUTE Expert guides a user through the rule base, and advises on the right course of action.

Law firms interpret and apply legislation to give advice to businesses and individuals. STATUTE Expert models complex legislation and rules and removes the experience of complexity for the user. With STATUTE Expert, lawyers can provide online advice on procedural law, supported by comprehensive legal reasoning. They can work effectively and quickly with unfamiliar legislation. Routine work can be done by non-specialists and generalists. Costs to a firm and to clients can be reduced.

Government regulation affects every business. Regulatory regimes are often complex, costly and burdensome for businesses, which want to meet their obligations as simply and cheaply as possible. Regulatory agencies have conflicting interests. Their policy is to target their regulations precisely and maximize the level of compliance. They also want to meet the needs of the businesses they regulate and the industry groups that pressure them to reduce the burden of compliance.

Using rulebase technology, Softlaw has created tools that remove most of the complexity from regulations, helping industry to comply quickly, easily and reliably. The use of rulebase technology has several benefits:

- reduced complexity,
- ease of compliance,
- reduced compliance costs, and
- improved levels of compliance.

7.3.2 Case Based Reasoning and Hybrid Systems

As stated in [70], although several developers have used rule-based systems to model statute law and case law, rules are fundamentally inadequate for representing cases. The FINDER system – a case-based system which advises upon the rights of finders of lost chattels – was developed by [99]. This area of law is based entirely on cases. Finder has a database of leading cases and a set of attributes which were significant in these cases. FINDER assigns a weight to each attribute equal to the inverse of the variance of the values of that attribute across all the cases. FINDER uses these weights to find the weighted Euclidean distance between the instant case and each of the leading cases. It uses the nearest case and the nearest case with the opposite result to build an argument about the likely result in the instant case. SHYSTER adopts and expands upon FINDER's approach to case law.

HYPO [5] operated in the domain of US trade secrets – where there is no legislation and all reasoning is performed with cases. It analyses problem situations dealing with trade secret disputes, retrieves relevant legal cases from its database and fashions them into legal reasonable arguments as to who should win. The arguments demonstrate the program's ability to reason symbolically with past cases, draw factual analogies between cases, cite them in arguments, distinguish them and pose counter examples and hypotheticals based upon past cases.

In HYPO, cases are represented at two levels:
1. *Factual feature level* – here a case is represented as a structured hierarchical set of frames with slots containing the values of the factual features;
2. *Abstract level* – a case is presented in terms of features computed from the concrete factual features input into the system.

The legal concepts of relevant similarities and differences, most relevant cases, distinguishing amongst cases and counter examples are defined in terms of factors. Factors are a kind of expert legal knowledge that tends to strengthen or weaken a plaintiff's argument in favour of a legal claim. The factors that affect trade secrets misappropriation claims were gleaned from legal treatises and law review articles.

In HYPO, factors are indexed with dimensions. A dimension is a general framework for recording information about a factor for the program to manipulate. HYPO has a library of dimensions for the claim of trade secrets misappropriation, each associated with a factor that identifies a common strength or weakness in trade secrets claim. Each dimension has a set of prerequisites used to test if the factor applies to a case and a range of values over which the magnitude of the factor in a particular case may vary. Dimensions encode domain knowledge that certain groups of operative facts and features derived from them enable us to make specific arguments or approach a case in a certain way. They summarise lines of argument or address cases in a specific manner. Metrics for assessing relevancy or on-pointedness are based on the intersection between the sets of dimensions applicable to the current case and the cases in the case knowledge base.

HYPO's notion of factors and dimensions and its ability to argue both sides of a case, made it a very significant legal case based reasoner, and one of the earliest legal argumentation systems.

Tom Gordon in his tutorial for the *9th International Conference on Artificial Intelligence and Law* noted that the following argument schemes use case-based reasoning.

Argument from Theory Construction (Taxman [53]): if a theory can be constructed which explains a line of precedents and hypotheticals about the same issue as the current case, then the theory should be applied to the current case.

Argument from Shared Factors (HYPO [5]): if a prior case – the precedent – has factors in common with the current case, then the current case should be decided in the same way as the precedent.

Argument from Criterial Facts (GREBE [18] and CABARET [76]): if an open textured predicate in a statute was decided to have been satisfied by the criterial facts of a precedent and these criterial facts are similar to the facts of the current case, then the open textured predicate is also satisfied in this case.

Popple states that the law in most common-law countries is based on both statutes and cases and hence legal expert systems must take into account both statute and case law [70].

GREBE [18] is an integrated deductive and case based reasoner which operates in the domain of Texas workers' compensation law. It uses legal and commonsense rules and precedent cases to construct explanations of conclusions about cases. Unlike the blackboard approach of PROLEXS and CABARET, GREBE does not have an agenda mechanism to mediate between the two paradigms. Instead it always tries both rules and cases. The choice between the two is evaluated retrospectively, and measures the strength of the explanations generated by both approaches.

PROLEXS [108] operates in the domain of Dutch landlord-tenant law. It operates with four knowledge groups, each of which have their own knowledge representation language and dedicated inference engine:
- legislation – a rule-based system;
- legal knowledge;

- expert knowledge; and
- case law – uses case-based retrieval.

PROLEXS uses neural networks in case selection, case abstraction and credit assignment.

The inference engine of a knowledge group may contain several reasoning groups. Conclusions derived in one knowledge group are written on a blackboard in a standard format and therefore readable to all inference engines.

CABARET (CAse Based Reasoning Tool) [76] performs legal interpretation by integrating case-based and rule-based reasoning in the domain of US income tax law. It deals with the circumstances under which a taxpayer may legitimately deduct expenses relating to a home office maintained at the taxpayer's expense. CABARET operates by:

1. The user inputs a case, a point of view regarding it and an overall goal for the system;
2. The system analyses the fact situation by using rules and cases, creating the trace of the necessary reasoning tasks and a determination of the relevancy of known rules and cases; and
3. The system generates arguments and/or explanation as to the advantages and disadvantages of a certain interpretation of the input case.

CABARET also has an inductive learning component.

IKBALS [118] used the object-oriented approach to build a hybrid rule-based/case-based system to advise upon open texture in the domain of Workers Compensation. IKBALSI and IKBALSII both deal with statutory interpretation of the *Accident Compensation (General Amendment) Act* 1989 (Vic). The Act allows a worker who has been injured during employment to gain compensation for injuries suffered. These compensation payments are called WorkCare entitlements, and IKBALS focuses on elements giving rise to an entitlement.

The original prototype IKBALSI was a hybrid/object oriented rule-based system. Its descendant – IKBALSII – added case-based reasoning and intelligent information retrieval to the rule-based reasoner, through the use of a blackboard architecture.

IKBALSIII [128] was an integrated deductive and analogical system working in the domain of Victoria (Australia) Credit Law. It
- Includes induction as the basis for its case based retrieval function; and
- Relies on distributed artificial intelligence techniques and the object oriented paradigm, rather than a blackboard architecture.

Induction is used in IKBALSIII to generate the indices into the cases. Thus, the developer can specify a number of cases, including the relevant factors and the outcome, and the induction algorithm will generate the indices automatically.

7.3.3 Knowledge Discovery in Legal Databases

According to [34], Knowledge Discovery from Databases (KDD) is the 'non-trivial extraction of implicit, previously unknown and potentially useful information from data.' Data mining is a problem-solving methodology that finds a logical or mathematical description, eventually of a complex nature, of patterns and regularities in a set of data [31].

Law has not yet been characterized with data collected in structured formats in the quantities apparent in other fields. However, knowledge discovery techniques have begun to be applied to legal domains in useful ways. Existing attempts in this direction provide important insights into the benefits for the practice of law in the future and also illustrate problems for KDD unique to the legal domain.

According to [31] KDD techniques, in general can be grouped into four categories:
1. *Classification.* The aim of classification techniques is to group data into predefined categories. For example, data representing important case facts from many cases may be used to classify a new case into one of the pre-defined categories, 'pro-plaintiff' or 'pro-defendant'.
2. *Clustering.* The aim of clustering techniques is to analyze the data in order to group the data into groups of similar data. For example, a clustering technique may group cases into six main clusters that which an analyst would interpret in order to learn something about the cases.
3. *Series Analysis.* The aim of series analysis is to discover sequences within the data. Sequences typically sought are time series. For example, past cases over a time period may be analyzed in order to discover important changes in the way a core concept is interpreted by Courts.
4. *Association.* The objective of association techniques is to discover ways in data elements are associated with other data elements. For example, an association between the gender of litigants and the outcome of their cases may surprise analysts and stimulate hypotheses to explain the phenomena.

Although KDD with data from law is not prevalent, important examples of classification, clustering, series analysis and association have been performed.

Rule induction was used by [75] to analyze a domain in order to detect a change in the way a legal concept is used by Courts. Whilst it is accepted that legal concepts change over time, developing computer techniques to detect such change is non-trivial.

Decision Trees that represented US bankruptcy law cases at various time intervals over a ten-year period were induced by [75]. They developed a structural instability metric to compare decision trees. This allowed them to measure the change in the concept of 'good faith' in bankruptcy proceedings by plotting the change over time of Rissland and Friedmann's structural instability index.

Knowledge discovery techniques were used by [67] to identify a prototypical exemplar of cases within a domain. An exemplar pro-defendant case has features that are most like the cases in which the defendant won and most unlike the cases the defendant lost. Genetic Algorithms were used in this application because the search space for a prototypical exemplar is typically enormous.

Large numbers of cases were examined by [116] in order to estimate the number of days that are likely to elapse between the arrest of an offender and the final disposition of the case. The time to disposition depends on variables such as the charge, the offenders age, and the county where the arrest was made. Values on more than 30 variables from over 700,000 records from 12 US states were used. Rules were automatically extracted using the `ID3` rule induction algorithm. Although Wilkins & Pillaipakkamnatt themselves had hoped for rule sets that predicted the time to disposition more accurately than their results indicate, this study remains an impressive demonstration of the potential for KDD techniques to contribute to the effective delivery of legal services.

There has been much recent research on the use of machine learning to improve information retrieval. The `SALOMAN` project [63] aims to generate a summary of a judgment. This is done by combining text matching using information retrieval algorithms with expert knowledge about the structure of judgments.

The `INQUERY` text retrieval engine was used by [26] to build a hybrid case-based reasoning and information retrieval system called `SPIRE`. `SPIRE` locates passages likely to contain information about legally relevant features of cases found in full-text court opinions.

A case study of legal document classification was performed by [61]. The core task of classification is performed by a non-standard neural network model with a layered architecture consisting of mutually independent unsupervised neural networks. The resulting system has a remarkably fast training time and explicit cluster representation.

It has been illustrated that association rule generators can highlight interesting associations in a small dataset in family law [90]. In an example of KDD in law that aims to analyze a legal domain rather than making specific predictions regarding judicial outcomes, association rules were generated by [45] from over 300,000 records drawn from a database of applicants for government funded legal aid in Australia. In that country, applicants for legal aid must not only pass an income and assets test but must also demonstrate that their case has merits for success. Consequently considerable data is recorded about the applicant and the case. The purpose of the association rules study performed by [45] was to determine whether this data mining technique could automatically analyze the data in order to identify hypotheses that would not otherwise have been considered. For example, as a result of this study, an association between the applicant's age and categories of legal aid applied requested, was discovered. It can be summarized as follows: 89% of applicants between 18 and 21 applied for legal aid for criminal offences, whereas 57% of applicants between 40 and 50 applied for aid for criminal offences. This result surprised experts in the field who did not expect young applicants to be so highly represented in criminal law matters. This result is not, of itself used to explain much, but is advanced to assist in the formulation of hypotheses to explain the associations observed.

A State Supreme Court Judge in Brazil (V. Feu Rosa Pedro) has initiated a program for the resolution of traffic accident disputes [32]. His 'Judges on Wheels' program involves the transportation of a judge, police officer, insurance assessor, mechanical and support staff to the scene of minor motor vehicle accidents. The team collects evidence, the mechanic assess the damage, and the judge makes a decision and drafts a judgement with the help of a program called the `Electronic Judge` before leaving the scene of the accident. The `Electronic Judge` software uses a KDD approach that involves data mining tools called neural networks. Although the judge is not obliged to follow the suggestion offered by the `Electronic Judge`, the software is used in 68% of traffic accidents by judges in the state of Espirito Santo. The system plays an important role in enhancing the consistency of judicial decision-making.

`Split-Up` provides advice on property distribution following divorce [126]. The aim of the approach used in developing `Split-Up` was to identify, with domain experts, relevant factors in the distribution of property under Australian family law. They wanted to assemble a dataset of values on these factors from past cases that can be fed to machine learning programs such as neural networks. In this way, the manner that judges weighed factors in past cases could be learnt without the need to advance rules. The legal realist jurisprudence movement inspired this approach.

In the Split-Up system, the relevant variables were structured as data and claim items in 35 separate arguments. The claim items of some arguments were the data items of others, resulting in a tree that culminated in the ultimate claim that indicated the percentage split of assets a judge would likely to award the husband. The tree of variables is illustrated in Fig.7.1. In 15 of the 35 arguments, claim values were inferred from data items with the use of heuristics, whereas neural networks were used to infer claim values in the remaining 20 arguments. The neural networks were trained with data from only 103 cases. This was feasible because each argument involved a small number of data items due to the decomposition described in Fig.7.1.

Ninety-four variables were identified as relevant for a determination in consultation with experts. The way the factors combine was not elicited from experts as rules or complex formulas. Rather, values on the 94 variables were to be extracted from cases previously decided, so that a neural network could learn to mimic the way in which judges had combined variables.

However, according to neural network rules of thumb, the number of cases needed to identify useful patterns given 94 relevant variables is in the many tens of thousands. Data from this number of cases is rarely available in any legal domain. Furthermore, few cases involve all 94 variables. For example, childless marriages have no values for all variables associated with children so a training set would be replete with missing values. In addition to this, it became obvious that the 94 variables were in no way independent.

In the Split-Up system, the relevant variables were structured as separate arguments following the argument structure advanced by [96]. Toulmin concluded that all arguments, regardless of the domain, have a structure that consists of six basic invariants: claim, data, modality, rebuttal, warrant and backing. Every argument makes an assertion based on some data. The assertion of an argument stands as the claim of the argument. Knowing the data and the claim does not necessarily convince us that the claim follows from the data. A mechanism is required to act as a justification for the claim. This justification is known as the warrant. The backing supports the warrant and in a legal argument is typically a reference to a statute or a precedent case. The rebuttal component specifies an exception or condition that obviates the claim.

A survey of applications of the Toulmin Structure has revealed that the majority of researchers does not apply the original structure but vary it in one way or another. Fig.7.3 illustrates the structure representing the variation used in Split-Up. The rationale for the variations applied are described in [89].

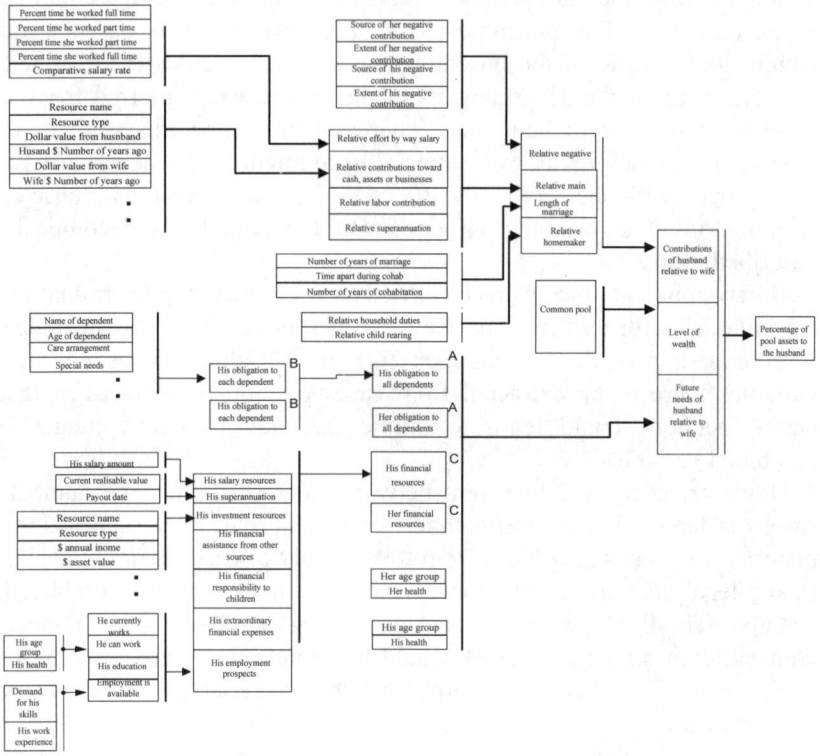

Fig. 7.1. Tree of actual arguments for Split-Up

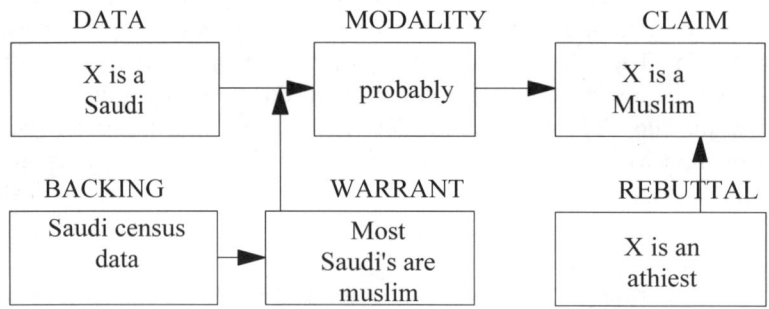

Fig. 7.2. Toulmin Argument Structure

7 Building Intelligent Legal Decision Support Systems   227

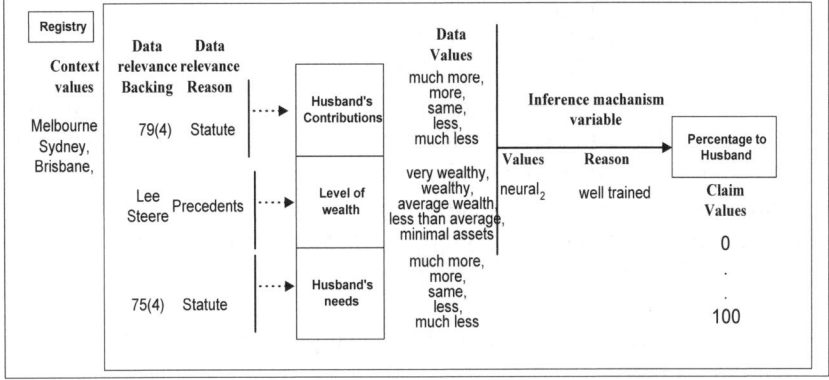

Fig. 7.3. Generic Argument for Percentage Split of Assets to the Husband

Fig.7.3 illustrates one argument from the Split-Up system. We see from that figure that there are three data items. Each of these is the claim item of other arguments leading to a tree of arguments where the ultimate claim of the system is the root of the tree.

A key difference in our variation from the original is the specification of an inference mechanism variable. In the argument in Fig.7.3, the inference mechanism is a neural network. The network, once trained with appropriate past cases, will output a claim value (% split of assets) given values of the three data items.

In twenty of the thirty-five arguments in Split Up, claim values were inferred from data items with the use of neural networks whereas heuristics were used to infer claim values in the remaining arguments. The neural networks were trained from data from only 103 commonplace cases. This was possible because each argument involved a small number of data items due to the argument-based decomposition.

The Split-Up system produces an inference by the invocation of inference mechanisms stored in each argument. However, an explanation for an inference is generated after the event, in legal realist traditions by first invoking the data items that led to the claim. Additional explanatory text is supplied by reasons for relevance and backings. If the user questions either data item value, she is taken to the argument that generated that value as its claim.

The Split-Up system performed favorably on evaluation, despite the small number of samples. Because the law is constantly changing, it is important to update legal decision support systems. The original hybrid rule-based/neural network version of Split-Up was constructed in 1996. Currently, the tree of arguments is being modified in conjunction with

domain experts from Victoria Legal Aid to accommodate recent changes in legislation. In particular
1. The recent tendency by Family Court judges to view domestic violence as a negative financial contribution to a marriage.
2. The re-introduction of spousal maintenance as a benefit to one of the partners. Under the *clean-break philosophy*, Family Court judges were reluctant to award spousal maintenance, since it would mean one partner would continue to be financially dependant on his/her ex-partner. However the increasing number of short, asset-poor, income-rich marriages has led to a re-consideration of the issue of spousal maintenance.
3. The need to consider superannuation and pensions separately from other marital property.

The argument-based representation facilitates the localization of changes and makes maintenance feasible. The use of the argument-based representation of knowledge enables Machine Learning techniques to be applied to model a field of law widely regarded as discretionary. The legal realist jurisprudence provided a justification for the separation of explanation from inference.

With the provision of domain expertise and financial support from VLA, we are currently developing a web-based version of Split-Up using the web-based shell ArgShell and the knowledge management tool JustReason developed by an Australian start up company JUSTSYS (www.justsys.com.au). As a web-based system, Split-Up will inform divorcees of their rights and support them to commence negotiations pertaining to their divorce.

7.3.4 Evaluation of Legal Knowledge Based Systems

KDD is a statistical endeavour. The results derived from KDD rely on empirical work, including:
1. *Learning the application domain* – this includes developing relevant prior knowledge and identifying the goal and the initial purpose of the KDD process from the user's viewpoint.
2. *Creating a target data set* – including selecting a data set or focusing on a set of variables or data samples on which the discovery is to be performed.
3. *Data cleaning and pre-processing* – includes operations such as removing noise or outliers if appropriate, collecting the necessary information to model or account for noise and deciding on strategies for handling missing data fields.

4. *Interpretation* – this step can also involve visualization of the extracted patterns and models or visualization of the data given the extracted models;
5. *Evaluation of KDD purpose* – newly discovered knowledge is often used to formulate new hypotheses; also new questions may be raised using the enlarged knowledge base.

80% of the KDD effort, especially in law, is involved in steps 1-3, the non-technical issues.

As is the case with conventional computer systems, Legal Knowledge-Based Systems (KBS) need to be evaluated in terms of verification, validation, usefulness and usability. Given the statistical nature of KDD, it is essential that all KDD based legal decision support systems be appropriately evaluated. In the case-based research of HYPO [5], CATO [3], GREBE [18] and the KDD work on Split-Up [88] there has been an focus upon evaluation of Legal Knowledge Based Systems. At the 9th *International Conference on Artificial Intelligence and Law* there was an important workshop on the evaluation of legal knowledge based systems (www.cfslr.ed.ac.uk/icail03).

The Criteria, Context, Contingency-guidelines framework (CCCF) was introduced by [40]. It is designed for general use, but can be specifically tailored to framing (planning) the evaluation of LKBS. CCCF assists an evaluator plan an evaluation in three distinct but interdependent ways. Initially it provides guidance to assess the context of the evaluation – in other words, both of the system to be evaluated and of the evaluation itself – and to determine the goals of the evaluation. CCCF also contains a hierarchical four-quadrant model of potential evaluation criteria and offers contingency-guidelines to select appropriate criteria to satisfy the goals of the evaluation and any arising contingency.

CCCF subsumes and integrates material sourced from the international standards and other frameworks. CCCF seeks to be wide-ranging in scope and facilitate ordered access for the inexperienced legal evaluator to the multitude of information available. The experienced software engineer should also find it a useful reference to the many general evaluation criteria available, a means of organizing their own criteria and a framework upon which to plan, standardize, report and compare their evaluations. It will also facilitate the inclusion of any legal domain specific evaluation requirements.

Oskamp & Tragter consider evaluation of LKBS to be a broad based process that should not be restricted to V (Validity) and V (Verification). They suggest that benefits of the system to all stakeholders should be evaluated in particular the impact of the LKBS on its environment [66].

Their suggestion is accommodated by the CCCF, which supports evaluation activities beyond the 'in the small' aspects of usability and V and V. Where the environmental impact of such systems is significant, an evaluation module using qualitative methods can be supported. Where LKBS systems have been developed in a research environment, with the intention of eventual migration to an operational environment, much can be gained by considering how they might fit in with the technical infrastructure required to maintain the proposed operational environment.

CCCF was itself evaluated on:

1. GetAid – a LKBS which advises solicitors and their clients whether the client is eligible for legal aid provided by Victoria Legal Aid (VLA) [41]. The evaluation of GetAid was framed by a committee of three using a modified DELPHI methodology supported by the CCCF: a VLA domain expert charged with representing management in the evaluation (the Domain Expert); the researcher and developer of the CCCF (the Evaluator); and GetAid's designer, knowledge engineer and developer (the Developer). The GetAid evaluation involved five distinct stages: framing, laboratory based evaluation, field testing, technical and impact assessment.

2. FamilyWinner – an automated system designed to support the negotiation process. It proposes settlements in a variety of negotiation settings making significant use of trade-off capabilities to mimic the dynamics of negotiation [9]. FamilyWinner was developed with minimal resources in a research laboratory, using Artificial Intelligence and game theory techniques. The evaluation of FamilyWinner was framed using the combined domain expertise of the developer and evaluation expertise of the evaluator supported by the CCCF. A structured interview technique, based on properties canvassed from the CCCF, was used to gain an understanding of the system and evaluation contexts. The research nature of the project and the minimal resources available afforded an over-riding contingency, so evaluation criteria were manually selected. The developer commented that she found the CCCF a useful tool to support her choice of evaluation criteria. She had not previously considered the possibility of evaluating some of the areas suggested. Importantly, groups of criteria that would not be included in evaluation modules, often due to lack of resources, were explicitly considered before being discarded and the resulting limits to the evaluation were well understood.

7.3.5 Explanation and Argumentation in Legal Knowledge Based Systems

Many rule-based systems provide an explanation for the advice offered by the system through a trace of the rules leading to the decision. This is not possible when the legal knowledge based system uses KDD.

As previously indicated, the Split-Up system produces an inference by the invocation of inference mechanisms stored in each argument. However, an explanation for an inference is generated after the event, in legal realist traditions by first invoking the data items that led to the claim. Additional explanatory text is supplied by reasons for relevance and backings. If the user questions either data item value, she is taken to the argument that generated that value as its claim.

Tom Gordon claims there are two views on the validity of arguments:
1. *Deductive Logic View* – An argument is valid only if the conclusion is a necessary implication of the premises. Validity is a property of a *relation* between premises and conclusion.
2. *Dialectical View* – An argument is valid only if it furthers the goals of the dialog in which it is used. Validity depends on how the argument is used in a *process*.

Problems with the deductive view include:
- Abductive arguments – for example: it is wet outside this morning, therefore it rained last night. There may be other reasons why the grass is wet, such as the sprinklers were left on all night.
- Inductive Arguments – because it rained last night it will rain tomorrow. There is a probability attached to this inference.
- Exceptions – the train leaves every day at 9:15, therefore it will leave today at 9:15. But what happens if today is a holiday.
- Open Textured Concepts.

Defendants of the logical view note that non-deductive arguments (often) can be reconstructed as deductively valid arguments (for example, by making all premises explicit). But
1. When should the proponent of an argument have to do this? (In which types of dialogs? In which roles? Who has the burden of proof?), and
2. If invalid arguments can be reconstructed as deductively valid arguments, what is the utility of this conception of validity in practice?

Dialectical approaches typically automate the construction of an argument and counter arguments normally with the use of a non-monotonic logic where operators are defined to implement discursive primitives such as *attack, rebut,* or *accept*.

Dialectical models have been advanced by amongst others [10, 37, 72, 109] on discussing the new dialectic noted that validity can depend on the context of the argument in a dialog. Issues to be considered are:
1. Who made the argument?
2. In what role did they make the argument?
3. When was the argument made?
4. In what kind of dialog was the argument made?

Prakken [72] has a four level model for argumentation:
1. Dialectical Level for evaluating and comparing arguments,
2. A Procedural Level,
3. A Logic Level. in which there is a formal language for representing knowledge to construct or generate arguments (the dialog is regulated at this level), and
4. At the Strategic Level Knowledge about how to 'play the game' well is stored.

Many legal decision support systems have used Toulmin Arguments. [27] illustrates how relevant cases for an information retrieval query can be retrieved despite sharing no surface features if the arguments used in case judgements are represented as Toulmin structures. Hypertext based computer implementations have been built by [52] that draw on knowledge organized as Toulmin arguments. Hypertext links connect an argument's assertions with the warrants, backing and data of the same argument and also link the data of one argument with the assertion of other arguments. Complex reasoning can be represented succinctly enabling convenient search and retrieval of relevant information.

User tasks have been represented by [57] as Toulmin arguments and associated a list of keywords to the structure. These keywords were used as information retrieval queries into a range of databases. Results indicate considerable advantages in precision and recall of documents as a result of this approach compared with approaches that require the user to invent queries.

Branting expands Toulmin Argument Structure warrants as a model of the legal concept of ratio decidendi [19]. The Split-Up project [88] used Toulmin Argument Structures to represent family law knowledge in a manner that facilitated rule/neural hybrid development. [89] discusses how Toulmin Argument Structures have been used to represent and reason with legal knowledge about:
- copyright (RightCopy);
- refugee law (EMBRACE), and
- eligibility for legal aid (GetAid).

7.4 Legal Decision Support on the World Wide Web

7.4.1 Legal Knowledge on the WWW

Over the past decade, there has been increasing use of the World Wide Web to provide legal knowledge and legal decision support. The first and still a significant provider of legal advice is the *A*ustralian *L*egal *I*nformation *I*nstitute (AustLII – www.austlii.edu.au). AustLII provides free internet access to Australian legal materials. AustLII's broad public policy agenda is to improve access to justice through better access to information. To that end, AustLII has become one of the largest sources of legal materials on the net, with over seven gigabytes of raw text materials and over 1.5 million searchable documents. AustLII publishes public legal information – that is, primary legal materials (legislation, treaties and decisions of courts and tribunals) and secondary legal materials created by public bodies for purposes of public access (law reform and royal commission reports for example). It does not have any decision support systems on its internet site.

The British and Irish Legal Information Institute – BAILII – (www.bailii.org) provides access to the most comprehensive set of British and Irish primary legal materials that are available for free and in one place on the internet. CanLII (www.canlii.org), now a permanent resource in Canadian Law, was initially built as a prototype site in the field of public and free distribution of Canadian primary legal material. Following the establishment of AUSTLII, BAILII, and CANLII there is now a WordLII web site: www.worldlii.org

JUSTSYS (www.justsys.com.au) is an Australian start up company which develops legal knowledge based systems for the World Wide Web. Its rationale for developing such tools came from the extended use of Toulmin Argument Structures for represented legal knowledge [89]. JUSTSYS is focusing upon: (i) Tools for the rapid development of decision support; (ii) On line dispute resolution tools and (iii) Data Mining.

JUSTSYS has recently launched GetAid – a web-based decision support system for determining eligibility for legal aid. Clients of Victoria Legal Aid (www.vla.vic.gov.au) use the GetAid system to determine their eligibility for Legal Aid. After passing a financial test (using Sequenced Transition Networks – a variant of decision trees), applicants for legal aid must pass a merit test. This assessment involves the integration of procedural knowledge found in regulatory guidelines with expert lawyer knowledge that involves a considerable degree of discretion. Knowledge

Discovery from Databases (KDD) was used to model the discretionary task. `GetAid` was developed in conjunction with web-based lodgement of applications for legal aid and has recently gone on-line.

`JUSTSYS` have developed `WebShell`, a knowledge based system shell that enables knowledge-based systems to be developed and executed on the World Wide Web. `WebShell` models knowledge using two distinct techniques: (i) decision trees for procedural type tasks, and (ii) argument trees for tasks that are more complex, ambiguous or uncertain. `JUSTSYS` have used Toulmin's [96] theory of argumentation to structure knowledge in a specific manner and to build `JustReason`, a knowledge management tool for structuring legal knowledge in the form of Toulmin Arguments.

SoftLaw (www.softlaw.com.au) is now building web-based systems for governments, lawyers and regulators. It has opened offices in Europe and the United States. JNANA (www.jnana.com) realized that there was a large commercial need for decision support systems that advise upon risk assessment. Such systems are not made available to the public.

JNANA currently focuses upon building a software platform to enable advice to be deployed over the Internet and Intranet. JNANA is now being used broadly in many industries, such as financial services, health care, customer relationship management, legal, and regulatory compliance.

Incorporator (www.incorporator.com.au) helps individuals with the legal requirements of incorporating their company. The company provides a product involving very specific legal knowledge in Australia, the United Kingdom and the United States.

7.4.2 Legal Ontologies

Breuker *et al* claim that unlike engineering, medicine or psychology, law is not ontologically founded [21]. They claim law is concerned with constraining and controlling social activities using documented norms. They have developed a core upper level ontology LRI-core. This ontology has over 200 concepts and has definitions for most of the anchors that connect the major categories used in law – person, role, action, process, procedure, time, space, document, information, intention, and so on. The main intended use is supporting knowledge acquisition for legal domains, but a real test of its semantics is whether it enables natural language understanding of common sense descriptions of simple events, as in the description of events in a legal case documentation. This is of course the core principle of the Semantic Web initiative of WC3.

An ontology is defined by [38] as an explicit conceptualization of a domain. The development of legal ontologies has been examined by [12]. Ontologies have benefits for:
- Knowledge sharing;
- Verification of a knowledge base;
- Knowledge acquisition; and
- Knowledge reuse.

A formal legal ontology was built by [104] by developing a formal specification language that is tailored in the appropriate legal domain. Visser commenced by using Van Kralingen's theory of frame-based conceptual models of statute law [102]. Visser uses the terms ontology and specification language interchangeably. He claims an ontology must be:
1. epistemologically adequate;
2. operational;
3. expressive;
4. reusable; and
5. extensible

Visser chose to model the Dutch Unemployed Benefits Act of 1986. He created a CommonKADS expertise model [79]. Specifying domain knowledge is performed by:
1. Determining the universe of discourse by carving up the knowledge into ontological primitives. A domain ontology is created with which the knowledge from the legal domain can be specified.
2. Domain specification is created by specifying a set of domain models using the domain ontology.

A legal ontology based on a functional perspective of the legal system was developed by [103]. He considered the legal system as an instrument to change or influence society in specific directions by reacting to social behaviour. The main functions can be decomposed into six primitive functions each of which corresponds to a category of primitive legal knowledge:
1. *Normative knowledge* – which describes states of affairs which have a normative status (such as forbidden or obligatory);
2. *World knowledge* – which describes the world that is being regulated, in terms that are used in the normative knowledge, and so can be considered as an interface between common-sense and normative knowledge;
3. *Responsibility knowledge* – the knowledge which enables responsibility for the violation of norms to be ascribed to particular agents;
4. *Reactive knowledge* – which describes the sanctions that can be taken against those who are responsible for the violation of norms;

5. *Meta-legal knowledge* – which describes how to reason with other legal knowledge; and
6. *Creative knowledge* – which states how items of legal knowledge are created and destroyed.

Valente's ontology, which he described as a Legal Information Server, allows for the storage of legal knowledge as both text and an executable analysis system interconnected through a common expression within the terms of the functional ontology. The key thrust of his conceptualization is to act as a principle for organizing and relating knowledge, particularly with a view to conceptual retrieval.

Many organizations are now building legal ontologies to provide legal knowledge on the World Wide Web. The Dutch Tax and Customs Administration have developed the POWER (Program for an Ontology-based working environment for rules and regulations) research project [100, 101]. POWER develops a method and supporting tools for the whole chain of processes from legislation drafting to executing the law by government employees. The POWER program improves legislation quality by the use of formal methods and verification techniques.

CLIME, e-COURT and FFPOIROT are all legal ontology projects funded by the European Union. Because of the plethora of legal systems, there is a great need to develop legal ontologies which are applicable across the European Union.

In the CLIME project, a large-scale ontology was developed for the purpose of a web-based legal advice system MILE (*M*aratime *I*nformation and *L*egal *E*xplanation). The system features both extended conceptual retrieval and normative assessment on international rules and regulations regarding ship classification (Bureau Veritas) and maritime pollution (MARPOL). The user can formulate a case using a structured natural language interface. The interface uses only the terms available in the ontology, which ensures that the user formulates a query on a topic the system knows about. The ontology also provides a means for adequate knowledge management of the rules and regulations.

The KDE (Knowledge worker Desktop Environment) project reused the CLIME ontology in a knowledge and workflow-management environment. In the KDE system, the CLIME ontology functioned as a domain ontology for the work of those associated with ship classification within the Bureau Veritas organization.

The CLIME knowledge base has two separate components:
1. *Domain* – A domain ontology of the design, construction, maintenance, repair, operation and construction of ships. The domain ontology

incorporates a small abstract top ontology, distinguishing things like artifacts, substances, agents and functions;
2. *Norms* –A knowledge base of norms: mappings from rules in legal documents to deontic constraints that allow or disallow certain types of cases. These norms are often limited and incomplete interpretations of the norms expressed in the rules.

The knowledge acquisition for the CLIME ontology can be split into two phases: (a) the conceptual retrieval phase in which the concepts and their relations are identified, created and defined, and (b) the phase in which knowledge acquisition for normative assessment takes place.

The e-COURT project [21] is a European project that aims at developing an integrated system for the acquisition of audio/video depositions within courtrooms, the archiving of legal documents, information retrieval and synchronized audio/video/text consultation. The focus of the project is to process, archive and retrieve legal documents of criminal courtroom sessions. In principle, these documents should be accessible via the World Wide Web. The system has the following major functions:

- Audio/Video/Text synchronization of data from court trials and hearings;
- Advanced Information Retrieval – multilingual, tolerant to vagueness. Statistical techniques are combined with ontology based indexing and search;
- Database management – multimedia documents support retrieval;
- Workflow management defines and manages rules for sharing relevant information and events among judicial actors; and
- Security management plays an important role to protect privacy information and to comply with national and international regulations about the interchange of criminal information.

The project is aimed at the semi-automated information management of documents produced during a criminal trial: in particular, the transcription of hearings. The structure of this type of document is determined by the debate/dialogue nature of the hearings, as well as by specific local court procedures. The developers identify and annotate content topics of a document. These annotations can vary from case descriptions in oral testimony to indictments in criminal law. Their first completed task was an ontology for Dutch criminal law, which served as a framework for ontologies on Italian and Polish criminal law.

FF-POIROT is a multi-million euro-dollar venture to develop European standards for the prevention, detection and successful investigation of financial fraud. The goal of the project is to build a detailed ontology of European Law, preventive practices and knowledge of the processes of financial fraud. The FF-POIROT project aims at compiling for several languages (Dutch, Italian, French and English) a computationally tractable and sharable knowledge repository (a formally described combination of concepts and their meaningful relationships) for the financial fraud domain.

This knowledge source is being constructed in three ways:
1. having human experts analyze and model the domain(s), in particular identifying the most abstract notions.
2. using computers to automatically find relevant notions (the most specific ones) from existing documents and semi-structured corpora including the Internet (text mining).
3. having humans validate the automatically generated suggestions to combine/merge already existing similar knowledge sources (semi-automatically aligning).

The resulting environment is useful to at least three different and EU-relevant types of user communities:

- Investigative and monitoring bodies: benefit from the strongly enriched information retrieval made possible by linking e.g. internet or database search facilities to the FF-POIROT ontology in order to detect or investigate instances of attempted or actual financial fraud. Species of fraud (typologies) have been identified so that macro and micro analysis can be undertaken then used as 'templates of fraud'. These templates can be stored, accessed and used to mine for new frauds across linguistic and jurisdictional boundaries.
- Financial professionals: Accountants, auditors, banks, insurance agencies, government departments, regulators and financial experts will benefit from an 'FF-POIROT-style' ontology using it as an authoritative concept base, extensively cross-linked (to other domains, systems and languages) and available for customized applications. Exploitation in this area could be as a high-tech service extending similar services and products (in particular with respect to accounting practices related to European VAT), currently already commercialized by at least two of FF-POIROT's users.
- Law enforcement: Police and other law enforcement agencies benefit by the availability of relevant parts of the FF-POIROT ontology – for example as an RDF-mapped Semantic Web resource, to support *future police-oriented query systems* – in a non-technical user-friendly, attractive, and comprehensive manner. Additionally, sharing of information with investigative bodies and understanding of related documents will be substantially enhanced if such communication and documents are hyperlinked to a shared ontology. Optimizing the

investigation, discovery, prevention and reduction of complex frauds is being made routine and efficient.

The partners in the FF-POIROT projects include universities (in Belgium, Romania and the United Kingdom), software houses (in Belgium and Italy) and two industry partners, CONSOB and VAT Applications who wish to commercialize the consortium's results.

CONSOB is the public authority responsible for regulating the Italian securities market. It is the competent authority for ensuring: transparency and correct behaviour by securities market participants; disclosure of complete and accurate information to the investing public by listed companies; accuracy of the facts represented in the prospectuses related to offerings of transferable securities to the investing public; compliance with regulations by auditors entered in the Special Register. CONSOB conducts investigations with respect to potential infringements of insider dealing and market manipulation law. Within the FF-POIROT project, CONSOB is particularly interested in detecting and prosecuting companies that spread fraudulent information on the internet.

VAT Applications NV is a Belgian software company developing automated software to deal with issues surrounding Value Added Tax at a European level. It has packages for all countries and in eleven languages. The recent decision to add ten new members to the European Union in 2004, will put further pressure on to develop software packages to help compliance with VAT requirements across the European Union and the identification, prevention and reduction of fraud across jurisdictions.

The University of Edinburgh's Joseph Bell Centre for Forensic Statistics and Legal Reasoning is performing the following tasks:
- Prepare for the construction and testing of the financial forensics repository using macro and micro analytical techniques;
- Gather information on how relevant authorities accumulate and analyze evidence of financial fraud, and analyze the tools auditors and accountants use to maintain up-to-date awareness of financial services regulations; and
- Collect requirements for the retained data, its validation and the applications needed to optimize the use of the information.

User requirements were collected by conducting structured interviews and using consortium expertise to accumulate necessary information for the construction and testing of a financial fraud ontology. Advice was obtained from end-users on how law enforcement agencies and investigative and regulatory bodies accumulate and analyze criminal evidence in domains of financial fraud, and analyze resources by which financial regulatory knowledge is available for auditing and accounting professionals.

7.4.3 Negotiation Support Systems

Ross states 'the principal institution of the law is not trial; it is settlement out of court' [78]. So what influence does judicial decision-making have over the outcome of negotiated settlements? The answer is a major one, since judicial decisions serve as the very basis from which negotiations commence [117].

Litigation can be damaging to both parties in a dispute. It is a zero-sum game; in that what one party wins the other loses.[7] Mediation can strive to reduce hostility between the parties, to fashion an agreement about tasks each party is willing to assume and to reach agreement on methods for ensuring certain tasks have been carried out. It can lead to a win-win result.[8]

Chung et al stress that although dispute resolution is a human problem, computers are already at the bargaining table to transform the negotiation process [24]. The Harvard Negotiation Project [33] introduced the concept of principled negotiation which advocates separating the problem from the people. Fundamental to the concept of principled negotiation is the notion of *know your Best Alternative to a Negotiated Agreement (BATNA)* – the reason you negotiate with someone is to produce better results than would otherwise occur. If you are unaware of what results you could obtain if the negotiations are unsuccessful, you run the risk of:

1. Entering into an agreement that you would be better off rejecting, or
2. Rejecting an agreement you would be better off entering into.

Sycara notes that in developing real world negotiation support systems one must assume bounded rationality and the presence of incomplete information [94]. Our model of legal negotiation assumes that all actors behave rationally. The model is predicated on economic bases – that is, it assumes that the protagonists act in their own economic best interests.

Traditional Negotiation Support Systems have been template-based with little attempt made to provide decision-making support. Little attention is given to the role the system should play in negotiations. Two template-based software systems which are available to help lawyers negotiate – NEGOTIATOR PRO and THE ART OF NEGOTIATING are discussed by [29]. INSPIRE [49] used utility functions to graph offers, while in

---

[7] It is actually worse than a zero-sum game and indeed can often lead to a lose-lose result. This is because of the large legal fees arising from litigation.

[8] For example if both parties value the list of items in dispute, it is not uncommon (as long as they do not value the items in an identical manner) for each party to receive 70% of their requested points.

DEUS [127] the goals of parties (and their offers) were set on screen side by side. The primary role of these systems is to provide users with a guide to how close (or far) they are from a negotiated settlement.

Fallback bargaining is discussed by [16]. Under fallback bargaining, bargainers begin by indicating their preference rankings over alternatives. They then fall back, in lockstep, to less and less preferred alternatives - starting with first choices, then adding second choices, and so on, until an alternative is found on which all bargainers agree.

MEDIATOR [51] used case retrieval and adaptation to propose solutions to international disputes. PERSUADER [94] integrated case-based reasoning and game theory to provide decision support with regard to United States' industrial disputes. NEGOPLAN was a logic based expert system shell for negotiation support [58]. GENIE [95] integrates rule based reasoning and multi-attribute analysis to advise upon international disputes.

Whilst no computer system can replace the human element, negotiation support systems can provide valuable support in structuring mediations, advising upon trade-offs and providing suggested compromises.

Katsh & Rifkin state that compared to litigation, Alternative Dispute Resolution has the following advantages [48]:
- Lower cost;
- Greater speed;
- More flexibility in outcomes;
- Less adversarial;
- More informal;
- Solution rather than blame-oriented;
- Private

To avoid the risks of extra costs and an unfavourable outcome, disputants often prefer to negotiate rather than litigate. Whilst investigating how disputants evaluate the risks of litigation researchers are faced with a basic hurdle – outcomes are often, indeed usually, kept secret. If the case is litigated, it could be used as a precedent for future cases, which may be a disincentive for one or more of the litigants [36]. Publicity of cases and the norms resulting from cases makes the public aware of the changing attitudes towards legal issues. The adjudication decision not only leads to the resolution of the dispute between the parties, but it also provides norms for changing community standards [30]. This latter facet is lost in negotiated settlements.

There has been much recent research on discourse and deliberation systems (especially for the domains of e-democracy and e-government). Recent work on negotiation concerns the area of discourse and

argumentation theory. The following six main types of dialogues are defined in [111], one of which is negotiation:
1. persuasion dialogue,
2. negotiation,
3. inquiry,
4. deliberation,
5. information seeking, and
6. eristic dialogue.

In this Section we focus upon negotiation support systems.

Alternative Dispute Resolution (ADR) has moved dispute resolution away from litigation and courts. On-line Dispute resolution (ODR) extends this trend [25]. As stated by [48], the trend toward non-legalistic systems of settling conflict will push mediation and arbitration to the foreground and litigation into the background.

In our desire to construct decision support systems to support legal negotiation, we realize how much the building of such systems depends upon the domain context. We chose as our domain to be modeled Australian family disputes. In most legal conflicts, once a settlement is reached the parties to the settlement are not required to have an on-going relationship. This is not the case in Australian Family Law. Family Law varies from other legal domains in that in general [44]:

1. There are no winners or losers – save for exceptional circumstances, following a divorce both parents receive a portion of the property and have defined access to any children;
2. Parties to a family law case often need to communicate after the litigation has concluded. Hence the Family Court encourages negotiation rather than litigation.

Zeleznikow has developed numerous systems to support negotiation in Australian family Law [120]. He claims all of the techniques developed can be used in other negotiation domains.

His first attempt at building negotiation support systems was to build a template-based system, DEUS [127]. In building DEUS, he developed a model of family law property negotiation, which relies upon building a goal for each of the litigants, with the goals being supported by their beliefs. Goals can only take real number values, because in simplifying the model it is assumed that the goal of each party is a monetary figure. Beliefs, which support the goals, are expressed in natural language. In the system, which has been implemented using this model, goals are used to indicate the differences between the parties at a given time. The beliefs provided are used to support the goals.

The model calculates the agreement and disagreement between the litigants' beliefs at any given time. The agreement and disagreement are only in relation to the beliefs and hence do not resolve the negotiation. In order to reach a negotiated settlement, it is essential to reduce the difference between the goals to nil. The system supports the negotiation process by representing the goals and beliefs of the opposing parties to a property conflict arising from a divorce application. It helps mediators understand what issues are in dispute and the extent of the dispute over these issues.

Split-Up can be used to determine one's BATNA for a negotiation. It first shows both litigants what they would be expected to be awarded by a court if their relative claims were accepted. It gives them relevant advice as to what would happen if some or all of their claims were rejected. Users are then able to have dialogues with the system to explore hypothetical situations to establish clear ideas about the strengths and weaknesses of their claims.

Suppose the disputants' goals are entered into the system to determine the asset distributions for both W and H in a hypothetical example. For the example taken from [9], the Split-Up system provided the following answers as to the percentages of the marital assets received by each party:

Table 7.1. Use of Split-Up to provide negotiation advice

	W's%	H's %
Given one accepts W's beliefs	65	35
Given one accepts H's beliefs	42	58
Given one accepts H's beliefs but gives W custody of the children	60	40

Clearly custody of the children is very significant in determining the husband's property distribution. If he were unlikely to win custody of the children, the husband would be well advised to accept 40% of the common pool (otherwise he would also risk paying large legal fees and having on-going conflict).

Whilst Split-Up is a decision support rather than negotiation support system, it does provide disputants with their respective BATNAs and hence provides an important starting point for negotiations. However, more is required of negotiation support systems. Namely they should model the structure of an argument and also provide useful advice on how to sequence the negotiation and propose solutions.

A generic framework for classifying and viewing automated negotiations has been developed by [47]. This framework was then used to analyze the three main methods of approach that have been adopted to automated negotiation, namely:
1. Game theory;
2. Heuristics;
3. Argumentation based approaches.

All three techniques have been used by [9] in building negotiation support systems. Family_Negotiator [7] is a hybrid rule-based and case-based system which attempts to model Australian Family Law. The system models the different stages of negotiation (according to Principled Negotiation Theory) by asking individuals for their positions and reasons behind these.

Game theoretic techniques and decision theory were the basis for AdjustedWinner [8], which implemented the procedure of [17]. AdjustedWinner is a point allocation procedure that distributes items or issues to people on the premise of whoever values the item or issue more. The two players are required to explicitly indicate how much they value each of the different issues by distributing 100 points across the range of issues in dispute. The AdjustedWinner paradigm is a fair and equitable procedure. At the end of allocation of assets, each party accrues the same number of points. It often leads to a win-win situation. Although the system suggests a suitable allocation of items or issues, it is up to the human negotiators to finalize the agreement acceptable to both parties.

Arising from our work on the AdjustedWinner algorithm, they noted that
- The more issues and sub-issues in dispute, the easier it is to form trade-offs and hence reach a negotiated agreement; and
- They chose as the first issue to resolve the issue on which the disputants are furthest apart - one wants it greatly, the other considerably less so.

Instead of using points as in AdjustedWinner, they used influence diagrams in Family_Winner. Family_Winner [9] uses both game theory and heuristics. It supports the process of negotiation by introducing importance values to indicate the degree to which each party desires to be awarded the issue being considered. The system uses this information to form trade-off rules. The trade-off rules are used to allocate issues according to the logrolling strategy. The system makes this analysis by transforming user input into trade-off values, used directly on trade-off maps, which show the effect of an issue's allocation on all unallocated issues.

Users of the `Family_Winner` system enter information such as the issues disputed, indications of their importance to the respective parties and how the issues relate to each other. An analysis of the aforementioned information is compiled, which is then translated into graphical trade-off maps. The maps illustrate the relevant issues, their importance to each party and trade-off capabilities of each issue. The system takes into account the dynamics of negotiation by representing the relations that exist between issues. Maps are developed by the system to show a negotiator's preferences and relation strengths between issues. It is from these maps that trade-offs and compromises can be enacted, resulting in changes to the initial values placed on issues.

The user is asked if the issues can be resolved in its current form. If so, the system then proceeds to allocate the issue as desired by the parties. Otherwise, the user is asked to decompose an issue chosen by the system as the least contentious. Essentially the issue on which there is the least disagreement (one party requires it greatly whilst the other party expresses little interest in the issue) is chosen to be the issue first considered. Users are asked to enter sub-issues. As issues are decomposed, they are stored in a decomposition hierarchy, with all links intact. This structure has been put in place to recognize there may be sub-issues within issues on which agreement can be attained. It is important to note that the greater the number of issues in dispute, the easier it may be to allocate issues, as the possibility of trade-offs increases. This may seem counter intuitive, but if only one issue needs to be resolved, then suggesting trade-offs is not possible.

This process of decomposition continues through the one issue, until the users decide the current level is the lowest decomposition possible. At this point, the system calculates which issue to allocate to which party, then removes this issue from the parties respective trade-off maps, and makes appropriate numerical adjustments to remaining issues linked to the issue just allocated. The resulting trade-off maps are displayed to the users, so they can see what trade-offs are made in the allocation of issues. When all issues are allocated at the one level, then decomposition of issues continues, re-commencing from the top level in a sequential manner.

The algorithms implemented in the system support the process of negotiation by introducing importance values to indicate the degree to which each party desires to be awarded each issue. It is assumed that the importance value of an issue is directly related to how much the disputant wants the issue to be awarded to her. The system uses this information to form trade-off rules. Systems such as Family_Winner are offer far more negotiation support than decision support systems that advise upon BATNAs.

7.5 Conclusion

In this Chapter we have attempted to discuss current and possible future trends in the development of intelligent legal decision support systems. To be able to do this, we need to carefully investigate the history of the development of such systems. But even before undertaking such a task, we need to investigate the nature of legal reasoning, focusing upon the notion of open texture and discussing why rules are inadequate for modeling legal reasoning. If we are to use cases in developing intelligent legal decision support systems, we need to distinguish between landmark cases (used for case-based reasoning) and commonplace cases (used for knowledge discovery from databases).

We then investigated the development of a number or legal decision support systems created in the 1970's and 1980s. These included:

- Rule-based reasoners – TAXMAN, British Nationality Act as a logic program, Ernst and Young and Rand Corporation systems, CAAS and the work of the SoftLaw company;
- Case-based and hybrid systems – FINDER, HYPO, GREBE, PROLEXS and IKBALS.

Systems using KDD were developed in the 1990s. These include INQUERY and Split-Up. Because KDD is an empirical exercise, we need to investigate issues of evaluation, explanation and argumentation.

The World Wide Web is becoming an important source of knowledge, even in the legal community. We discuss the provision of legal knowledge, legal ontologies and legal decision support systems on the World Wide Web.

Given that the principal institution of the law is not trial; it is settlement out of court, we discuss the development of legal negotiation support systems. The development of our legal decision support systems has led to increases in consistency, efficiency, transparency and increased support for alternative dispute resolution.

Acknowledgements

Much of the ideas in this chapter were developed in preparing a tutorial for the *9th International Conference on Artificial Intelligence and Law*, held at the University of Edinburgh in June 2003. My co-presenters were Tom Gordon and Carole Hafner. As always, I am grateful for their thoughts.

References

1. Aitken C (1995) *Statistics and the Evaluation of Evidence for Forensic Scientists*. John Wiley & Sons, Chichester, UK.
2. Aleven V and Ashley K (1997) Evaluating a Learning Environment for Case-Based Argumentation Skills, *Proc 6th Intl Conf Artificial Intelligence & Law,* Melbourne, Australia, June 30 – July 4, ACM Press, New York: 170 – 179.
3. Aleven V (1997) Teaching Case-Based Argumentation through a model and examples, PhD dissertation, University of Pittsburgh, PA.
4. Alexy R (1989) *A Theory of Legal Argumentation*, Clarendon Press, Oxford, UK.
5. Ashley KD (1991) *Modeling legal argument: Reasoning with cases and hypotheticals*, MIT Press, Cambridge, MA.
6. Ashley KD (1992) Case-based reasoning and its implications for legal expert systems, *Artificial Intelligence and Law*, 1(2): 113-208.
7. Bellucci E and Zeleznikow J (1997) Family—Negotiator: an intelligent decision support system for negotiation in Australian Family Law, *Proc 4th Conf Intl Soc for Decision Support Systems,* Lausanne: 359-373.
8. Bellucci E and Zeleznikow J (1998) A comparative study of negotiation decision support systems, *Proc 31st Hawaii Intl Conf System Sciences*, Maui, Hawaii, IEEE Computer Society, Los Alamitos, CA: 254-262.
9. Bellucci E and Zeleznikow J (2001) Representations of Decision-Making Support in Negotiation, *J Decision Systems*, 10(3-4): 449-479.
10. Bench-Capon TJ.M (1997) Argument in artificial intelligence and law, *Artificial Intelligence and Law*, 5:249-261.
11. Bench-Capon TJM and Sergot MJ (1988) Towards a rule-based representation of open texture in law, in Walter C (ed) *Computer Power and Legal Language,* Quorum Books, New York, NY: 39-61.
12. Bench-Capon TJM and Visser PRS (1997) Ontologies in Legal Information Systems: the Need for Explicit Specifications of Domain Conceptualizations, *Proc 6th Intl Conf Artificial Intelligence and Law*, Melbourne, Australia, June 30 – July 4, ACM Press, New York: 132-141.
13. Berman DH (1991) Developer's choice in the legal domain: the Sisyphean journey with case-based reasoning or down hill with rules, *Proc 3rd Intl Conf*

Artificial Intelligence and Law, Oxford, UK, June 25-28, ACM Press, New York:307-309.
14. Berman DH and Hafner CD (1988) Obstacles to the development of logic-based models of legal reasoning, in Walter C (ed) *Computer Power and Legal Reasoning*, Quorum Books, New York, NY: 183-214.
15. Black HC (1990) *Black's Law Dictionary*, West Publishing Company, St. Paul, Minnesota.
16. Brams SJ and Kilgour DM (2001) Fallback Bargaining, *Group Decision and Negotiation*, 10(4): 287-316.
17. Brams SJ and Taylor AD (1996) *Fair Division, From cake cutting to dispute resolution*, Cambridge University Press, UK.
18. Branting LK (1991) Building explanations from rules and structured cases, *Intl J Man Machine Studies*, 34(6): 797-838.
19. Branting KL (1994) A Computational Model of Ratio Decidendi, *Artificial Intelligence and Law*, 2: 1-31.
20. Branting LK (2001) Advisory Systems for Pro Se Litigants, *Proc 8th Intl Conf Artificial Intelligence and Law*, St Louis, Missouri, May 21-25, ACM Press, New York: 139-146.
21. Breuker J Elhag A, Petkov E and Winkels R (2002) Ontologies for Legal Information Serving and Knowledge Management, *Proc Jurix 2002: 15th Annual Conf Legal Knowledge and Information Systems*, London, UK, December 17-18, IOS Press, Amsterdam: 73-82.
22. Brkic J (1985) *Legal Reasoning: Semantic and Logical Analysis*, Peter Lang. New York.
23. Capper P and Susskind R (1988) *Latent Damage Adviser – The Expert System*, Butterworths, London.
24. Chung WWC and Pak JJF (1997) Implementing Negotiation Support Systems: Theory and Practice, *Proc 13th Annual Hawaii Intl Conf Systems Science (Mini track on Negotiation Support Systems)*, Hawaii.
25. Clark E and Hoyle A (2002) On-line Dispute Resolution: Present Realities and Future Prospects, *Proc 17th Bileta Conf*, Amsterdam, www.bileta.ac.uk/02papers/hoyle.html
26. Daniels J and Rissland E (1997) Finding legally relevant passages in case opinions, *Proc 6th Intl Conf Artificial Intelligence and Law*, Melbourne, Australia, June 30 – July 4, ACM Press, New York:39-46.
27. Dick JP (1991) A conceptual, case-relation representation of text for intelligent retrieval, PhD Thesis, University of Toronto, Canada.
28. Dworkin R (1986) *Law's Empire*, Duckworth, London.
29. Eidelman JA (1993) Software for Negotiations, *Law Practice Management*, 19(7): 50-55.
30. Eisenberg MA (1976) Private Ordering Through Negotiation: Dispute Settlement and Rulemaking, *Harvard Law Review*, 89: 637-681.
31. Fayyad U, Piatetsky-Shapiro G and Smyth P (1996) The KDD process for Extracting Useful knowledge from volumes of data, *Communications ACM*, 39(11):27-41.

32. FeuRosa PV (2000) The Electronic Judge, *Proc AISB'00 – Symp Artificial Intelligence & Legal Reasoning*, Birmingham, UK, April: 33-36.
33. Fisher R and Ury W (1981) *Getting to YES: Negotiating Agreement Without Giving In,* Haughton Mifflin, Boston, MA.
34. Frawley WJ, Piatetsky-Shapiro G and Matheus CJ (1991) Knowledge discovery in databases: an overview, *Knowledge discovery in databases*, AAAI/MIT Press: 1-27.
35. Fuller L (1958) Positivism and the separation of law and morals - a reply to Professor Hart, *Harvard Law Review*, 71: 630-672.
36. Goldring J (1976) Australian Law and International Commercial Arbitration, *Columbia J Transnational Law*, 15: 216-252.
37. Gordon TF (1995) The Pleadings Game: An exercise in computational dialectics, *Artificial Intelligence and Law*, 2(4): 239-292.
38. Gruber TR (1995) Towards principles for the Design of Ontologies used for Knowledge Sharing, *Intl J Human-Computer Studies*, 43: 907-928.
39. Greenleaf G, Mowbray A and Van Dijk P (1996) Representing and Using Legal Knowledge in Integrated Decision Support Systems: Datalex Workstations, *Artificial Intelligence and Law*, 4: 97-142.
40. Hall MJJ, Hall R and Zeleznikow J (2003) A method for evaluating legal knowledge-based systems based upon the Context Criteria Contingency-guidelines framework, *Proc 9th Intl Conf Artificial Intelligence and Law*, Edinburgh, Scotland, June 24-28, ACM Press, New York: 274-283.
41. Hall MJJ, Stranieri A and Zeleznikow J (2002) A Strategy for Evaluating Web-Based Discretionary Decision Support Systems, *Proc ADBIS2002 – 6th East-European Conf Advances in Databases and Information Systems*, Slovak University of Technology, Bratislava, Slovak Republic, 8-11 September: 108-120.
42. Hall MJ.J and Zeleznikow J (2002) The Context, Criteria, Contingency Evaluation Framework for Legal Knowledge-Based Systems, *Proc 5th Business Information Systems Conf*, April 24-25, Poznan, Poland, ACM Press, New York: 219-227.
43. Hart HLA (1958) Positivism and the separation of law and morals, *Harvard Law Review*, 71: 593-629.
44. Ingleby R (1993) *Family Law and Society*, Butterworths, Sydney.
45. Ivkovic S, Yearwood J and Stranieri A (2003) Visualising association rules for feedback within the Legal System, *Proc 9th Intl Conf Artificial Intelligence and Law*, Edinburgh, Scotland, June 24-28, ACM Press, New York: 214-223.
46. Johnson P and Mead D (1991) Legislative knowledge base systems for public administration - Some practical issues, *Proc 3rd Intl Conf Artificial Intelligence and Law*, Oxford, UK, June 25-28, ACM Press, New York:108-117.
47. Jennings, N.R., Faratin, P., Lomuscio, A.R., Parsons, S., Wooldridge, M.J. and Sierra, C. (2001) Automated Negotiation: Prospects, Methods and Challenges, *Group Decision and Negotiation*, 10 (2):199-215.

48. Katsh E and Rifkin J (2001) *Online Dispute Resolution: Resolving Conflicts in Cyberspace*, Jossey-Bass, San Francisco, CA.
49. Kersten GE (1997) Support for Group Decisions and Negotiations, in Climaco J (ed) *An Overview, in Multiple Criteria Decision Making and Support*, Springer Verlag, Heidelberg.
50. Kolodner J (1993) *Case based reasoning*, Morgan Kaufman. San Mateo, CA.
51. Kolodner JL and Simpson RL (1989) The Mediator: Analysis of an Early Case-Based Problem Solver, *Cognitive* Science, 13: 507-549.
52. Loui R, Norman J, Altepeter J, Pinkard D, Craven D, Lindsay J and Foltz M (1997) Progress on Room 5: a Testbed for Public Interactive Semi-Formal Legal Argumentation, *Proc 6th Intl Conf Artificial Intelligence and Law*, Melbourne, Australia, June 30 – July 4, ACM Press, New York: 207-214.
53. McCarty LT (1977) Reflections on TAXMAN: an experiment in artificial intelligence and legal reasoning, *Harvard Law Review*, 90: 837-893.
54. McCarty LT (1980) The TAXMAN project: Towards a cognitive theory of legal argument, in Niblett B (ed) *Computer Science and Law*, Cambridge University Press: 23-43.
55. McCormick DN (1978) *Legal Reasoning and Legal Theory*, Clarendon Press, Oxford.
56. McKaay E and Robilliard P (1974) Predicting judicial decisions: the nearest neighbour rule and visual representation of case patterns, *Datenverarbeitung im Recht*: 302-331.
57. Matthijssen LJ (1999) *Interfacing between lawyers and computers. An architecture for knowledge based interfaces to legal databases*, Kluwer Law International, The Netherlands.
58. Matwin S, Szpakowicz S, Koperczak Z, Kersten GE and Michalowski G (1989) NEGOPLAN: an Expert System Shell for Negotiation Support, *IEEE Expert*, 4: 50-62.
59. Meachem L (1999) *Judicial Business of the United States Courts*, Technical Report, Administrative Office of the United States Courts.
60. Meldman JA (1977) A structural model for computer-aided legal analysis, *Rutgers J Computers and Law*, 6: 27-71.
61. Merkl D and Schweighofer D (1997) The Exploration of Legal Text Corpora with Hierarchical Neural Networks: a Guided Tour in Public International Law, *Proc 6th Intl Conf Artificial Intelligence and Law*, Melbourne, Australia, June 30 – July 4, ACM Press, New York: 98-105.
62. Michaelsen RH and Michie D (1983) Expert systems in business, *Datamation*, 29(11): 240-246.
63. Moens MF (2000) *Automatic indexing and abstracting of document texts*, Kluwer International Series on Information Retrieval 6, Boston, MA.
64. Moles RN and Dayal S (1992) There is more to life than logic, *J Law and Information Science*, 3(2): 188-218.
65. Newell A and Simon H (1972) *Human problem solving*, Prentice Hall, Englewood Cliffs, NJ.

66. Oskamp A and Tragter MW (1997) Automated legal decision systems in practice: the mirror of reality, *Artificial Intelligence & Law*, 5(4), 291-322.
67. Pannu AS (1995) Using Genetic Algorithms to Inductively Reason with Cases in the Legal Domain, *Proc 5th Intl Conf Artificial Intelligence and Law*, College Park, Maryland, ACM Press, New York: 175-184.
68. Perelman C and Olbrechts-Tyteca L (1969) *The New Rhetoric*, translated by Wilkenson J and Weaver P, University of Notre Dame Press, Notre Dame, IN??, originally published in French as Perelman C and Olbrechts-Tycteca L (1958) *La Nouvelle Rhétorique: Traité de l'Argumentation*. Presses Universitaires de France.
69. Popp WG and Schlink B (1975) JUDITH, a computer program to advise lawyers in reasoning a case, *Jurimetrics*, 15(4): 303-314.
70. Popple J (1993) SHYSTER: A pragmatic legal expert system, PhD Dissertation, Australian National University.
71. Pound R (1908) Mechanical Jurisprudence, *Colombia Law Review*, 8: 605.
72. Prakken H (1997) *Logical Tools for Modelling Legal Argument. A Study in Defeasible Reasoning in Law*, Kluwer Academic Press, Dordrecht.
73. Quatrevaux E (1996) Increasing legal services delivery capacity through Information Technology, Technical Report LSC/OIG-95-035, Legal Services Corp, oig.lsc.gov/tech/techdown.htm
74. Reich Y (1995) Measuring the value of knowledge, *Intl J Human-Computer Studies*, 42(1): 3-30.
75. Rissland EL and Friedman MT (1995) Detecting change in legal Concepts, *Proc 5th Intl Conf Artificial Intelligence and Law*, Melbourne, Australia, June 30 – July 4, ACM Press, New York: 127-136.
76. Rissland EL and Skalak DB (1991) CABARET: Rule interpretation in a hybrid architecture, *Intl J Man-Machine Studies*, 34(6): 839-887.
77. Rose DE and Belew RK (1991) A connectionist and symbolic hybrid for improving legal research, *Intl J Man-Machine Studies*, 35(1): 1-33.
78. Ross HL (1980) *Settled Out of Court*, Aldine.
79. Schreiber JT, Akkermanis AM, Anjewierden AA, de Hoog R, Shadbolt A, Van de Velde W and Wielinga BJ (1999) *Knowledge Engineering and Management: The Common Kads Methodology*, MIT Press, Cambridge, MA.
80. Schum DA (1994) *The Evidential Foundation of Probabilistic Reasoning*, John Wiley & Sons, New York.
81. Sergot MJ, Sadri F, Kowalski RA, Kriwaczek F, Hammond P and Cory HT (1986) The British Nationality Act as a logic program, *Communications ACM*, 29: 370-386.
82. Shortliffe EH (1976) *Computer based medical consultations: MYCIN.*, Elsevier, New York.
83. Shpilberg D, Graham LE and Scatz (1986) ExperTAX: an expert system for corporate tax planning, *Expert Systems*, July.
84. Smith JC, Gelbart D, MacCrimmon K, Atherton B, McClean J, Shinehoft M, Quintana L, (1995) Artificial Intelligence and Legal Discourse: the Flexlaw Legal Text Management System, *Artificial Intelligence and Law*, 3: 55-95.

85. SoftLaw (2000) http://www.softlaw.com.au (accessed June 1, 2001).
86. Stranieri A and Zeleznikow J (1999) Evaluating legal expert systems, *Proc 7th Intl Conf Artificial Intelligence and Law*, Oslo, Norway, June 15-18, ACM Press, New York: 18-24.
87. Stranieri A and Zeleznikow J (2001) Copyright Regulation with Argumentation Agents, *Information & Communications Technology*, 10(1): 109-123.
88. Stranieri A, Zeleznikow J, Gawler M and Lewis B (1999) A hybrid-neural approach to the automation of legal reasoning in the discretionary domain of family law in Australia, *Artificial Intelligence and Law* 7(2-3):153-183.
89. Stranieri A, Zeleznikow J and Yearwood J (2001) Argumentation structures that integrate dialectical and monoletical reasoning, *Knowledge Engineering Review* 16(4): 331-348.
90. Stranieri A, Zeleznikow J and Turner H (2000) Data mining in law with association rules, *Proc IASTED Intl Conf Law and Technology (Lawtech2000)*, San Francisco, California, October 30 – November 1, ACTA Press, Anaheim, CA: 129-134.
91. Susskind R (2000) *Transforming the Law: Essays on Technology, Justice and the Legal Marketplace*, Oxford University Press.
92. Susskind R and Tindall C (1988) VATIA: Ernst and Whinney's VAT expert system, *Proc 4th Intl Conf Expert Systems*, Learned Information.
93. Swaffield G (1991) An expert system for stamp duty, *Proc 3rd Intl Conf Artificial Intelligence and Law*, Oxford, UK, June 25-28, ACM Press, New York: 266-271.
94. Sycara K (1990) Negotiation planning: an AI approach, *European J Operations Research*, 46: 216-234.
95. Sycara K (1998) Multiagent Systems, *AI Magazine*, 19(2): 79-92.
96. Toulmin S (1958) *The Uses of Arguments*, Cambridge University Press.
97. Turtle H (1995) Text Retrieval in the Legal World, *Artificial Intelligence and Law*, 3: 5-54.
98. Twining WL and Miers D (1982) *How To Do Things With Rules (2nd ed)*, Weidenfield and Nicolson, London.
99. Tyree AL, Grenleaf G and Mowbray A (1989) Generating Legal Arguments. *Knowledge Based Systems*, 2(1): 46-51.
100. van Engers TM and Glasee E (2001) Facilitating the Legislation Process using a Shared Conceptual Model, *IEEE Intelligent Systems*: 50-57.
101. van Engers TM and Kodelaar P (1998) POWER: Program for an Ontology-based working environment for rules and regulations, *Proc ISMICK'99*.
102. van Kralingen RW (1995) *Frame based Conceptual Models of Statute Law*, Kluwer Law International, The Hague, The Netherlands.
103. Valente A (1995) *Legal Knowledge Engineering; A modeling approach*, IOS Press, Amsterdam, The Netherlands.
104. Visser PRS (1995) *Knowledge Specification for Multiple Legal Tasks: a Case Study of the Interaction Problem in the Legal Domain*, Kluwer Computer/Law Series, 17.

105. Voermans W and Verharen E (1993) A Semi Intelligent Drafting Support System, *Proc 6th Intl Conf Legal Knowledge Based Systems*, Koniklijke Vermand, Tilburg, the Netherlands, November 18-19: 81-94.
106. Vossos G, Zeleznikow J, Moore A and Hunter D (1993) The Credit Act Advisory System (CAAS): Conversion from an expert system prototype to a C++ commercial system, *Proc 4th Intl Conf Artificial Intelligence and Law*, Amsterdam, The Netherlands, June 15-18, ACM Press, New York: 180-183.
107. Waismann F (1951) Verifiability, in Flew A (ed) *Logic and Language*, Blackwell, Cambridge, UK.
108. Walker RF, Oskamp A, Schrickx JA, Opdorp GJ, and van den Berg PH (1991) PROLEXS: Creating law and order in a heterogeneous domain, *Intl J Man Machine Studies*, 35(1): 35-68.
109. Walton D (1998) *The New Dialectic*, Pennsylvania State University Press, PA.
110. Walton D (2002) *Legal Argumentation and Evidence*, Pennsylvania State University Press, PA.
111. Walton DN and Krabbe ECW (1995) *Commitment in Dialogue: Basic Concepts of Interpersonal Reasoning*, SUNY Series in Logic and Language, State University of New York Press, Albany, NY.
112. Waterman DA and Peterson M (1984) Evaluating civil claims: an expert systems approach, *Expert Systems*, 1.
113. Waterman DA, Paul J and Peterson M (1986) Expert systems for legal decision making, *Expert Systems*, 3 (4): 212-226.
114. Wigmore JH (1913) *The Principles of Judicial Proof*, Little Brown & Co, Boston, MA.
115. Wilkenfeld, J, Kraus S, Holley KM and Harris MA (1995) GENIE: A decision support system for crisis negotiations, *Decision Support Systems*, 14: 369-391.
116. Wilkins D and Pillaipakkamnatt K (1997) The Effectiveness of Machine Learning Techniques for Predicting Time to Case Disposition, *Proc 6th Intl Conf Artificial Intelligence and Law*, Melbourne, Australia, June 30 – July 4, ACM Press, New York: 39-46.
117. Williams GR (1983) *Legal Negotiation and Settlement*, West Publishing Co, St Paul, Minnesota.
118. Zeleznikow J (1991) Building intelligent legal tools - The IKBALS project, *J Law and Information Science*, 2(2): 165-184.
119. Zeleznikow J (2002) Using Web-based Legal Decision Support Systems to Improve Access to Justice, *Information and Communications Technology Law*, 11(1): 15-33.
120. Zeleznikow J (2002) Risk, Negotiation and Argumentation – a decision support system based approach, *Law, Probability and Risk*, 1: 37-48.
121. Zeleznikow J (2003) An Australian Perspective on Rsearch and Development required for the construction of applied Legal Decision Support Systems, *Artificial Intelligence and Law*, 10: 237-260.
122. Zeleznikow J (2003) The Split-Up project: Induction, context and knowledge discovery in law, *Law, Probability and Risk* (in press).

123. Zeleznikow J and Hunter D (1994) *Building Intelligent Legal Information Systems: Knowledge Representation and Reasoning in Law*, Kluwer Computer/Law Series, 13.
124. Zeleznikow J, Hunter D and Stranieri A (1997) Using cases to build intelligent decision support systems, *Proc IFIP Working Group 2.6 – Database Applications Semantics*, Stone Mountain, GA, May 30 - June 2, 1995, (eds) Meersman R and Mark L Chapman & Hall, London, UK: 443-460.
125. Zeleznikow J and Stranieri A (1995) The Split-Up system: Integrating neural networks and rule based reasoning in the legal domain, *Proc 5th Intl Conf Artificial Intelligence and Law*, College Park, Maryland, May 21-25, ACM Press, New York: 185-194.
126. Zeleznikow J and Stranieri A (1998) Split Up: The use of an argument based knowledge representation to meet expectations of different users for discretionary decision making, *Proc IAAI'98 – 10th Annual Conf Innovative Applications of Artificial Intelligence*, Madison, Wisconsin, July 27-30, AAAI/MIT Press, Cambridge, MA: 1146-1151.
127. Zeleznikow J, Meersman R, Hunter D and van Helvoort E (1995) Computer tools for aiding legal negotiation, *Proc ACIS95 – 6th Australasian Conf Information Systems*, Curtin University of Technology, Perth, WA, September 26-28: 231-251.
128. Zeleznikow J, Vossos G and Hunter D (1994) The IKBALS project: Multimodal reasoning in legal knowledge based systems, *Artificial Intelligence and Law* 2(3):169-203.

8 Teaming Humans and Intelligent Agents

Christos Sioutis[1], Pierre Urlings[2], Jeffrey Tweedale[2] and Nikhil Ichalkaranje[1]

1. School of Electrical and Information Engineering, University of South Australia, Mawson Lakes, SA 5095, Australia, {Christos.Sioutis,Nikhil.Ichalkaranje}@unisa.edu.au

2. Air Operations Division, Defense Science and Technology Organization, PO Box 1500, Edinburgh, SA 5111, Australia, {Pierre.Urlings,Jeffrey.Tweedale}@dsto.defence.gov.au

8.1 Introduction

This chapter describes initial research into intelligent agents using the Beliefs-Desires-Intentions (BDI) architecture in a human-machine teaming environment. The potential for teaming applications of intelligent agent technologies based on cognitive principles will be examined

Intelligent agents using the BDI-reasoning model can be used to provide a situation awareness capability to a human-machine team dealing with a military hostile environment. The implementation described here uses JACK agents and Unreal Tournament (UT). JACK is an intelligent agent platform while UT is a fast paced interactive game within a 3D-graphical environment. Each game is scenario based and displays the actions of a number of opponents engaged in adversarial roles. Opponents can be humans or agents interconnected via the UT games server. To support research goals, JACK extends the *Bot* class of UT and its agents apply the BDI architecture to build situational awareness.

This chapter provides the background for the use of intelligent agents and their cognitive potential. The research is described in terms of the operational environment and the corresponding implementation that is suitable for intelligent agents to exhibit BDI behavior. The JACK application will be described and specific requirements will be addressed to implement learning in intelligent agents. Other cognitive agent behavior such as communication and teaming are aims of this research.

8.2 Background

During recent decades we have witnessed not only the introduction of automation into the work environment but we have also seen a dramatic change in how automation has influenced the conditions of work. Advances in automation and Artificial Intelligence (AI), and especially in the area of intelligent (machine) agents, have enabled the formation of rather unique teams with human and machine members. The team is still supervised by the human with the machine as a subordinate associate or assistant, sharing responsibility, authority and autonomy over many tasks. The requirement for teaming human and machine in a highly dynamic and unpredictable task environment has led to impressive achievements in many supporting technologies. These include methods for system analysis, design and engineering, and, in particular, for information processing, and for cognitive and complex knowledge engineering [32].

Several major research programs in military aviation have proposed approaches for analysis and design of complex human-machine applications [4, 7, 13]. Although these programs go by different names, their goals are similar and a multitude of traditional Advanced Information Processing (AIP) techniques are employed: knowledge based (rule- and case-based) systems, planning (script based and plan-goal graphs), fuzzy logic and (distributed) blackboard systems. The `Crew Assistant` program introduced a conceptual framework based on a strict modular decomposition of the tasks allocated to the human and machine member of the team. In contrast to previous research, Distributed Problem Solving (DPS) and in particular multi-intelligent-agent systems were proposed as emerging and promising technologies for realization of the `Crew Assistant` architecture [40]. DPS provides a natural transition from the functional framework of a human-machine team into a multi-agent system architecture where the inherent distribution and modularity of functions is preserved and allocated to distributed problem solving agent members of the team.

A multi-agent system consists of heterogeneous agents that have a range of expertise or functionality, normally attributed to humans and their intelligent behavior. These intelligent agents have the potential to function stand-alone but are also able to cooperate with other agents. This provides the basis for a human-machine teaming architecture, based on multiple cooperative agents, where each agent implements an assistant function. Each agent has its own local data, belief sets, knowledge, operations and control that are relevant for the problem or task domain of the function. BDI-agents, which implement a Beliefs-Desires-Intentions architecture, are of particular interest for teaming applications. The beliefs of these agents are its model of the world, and its desires contain the goals it wants to achieve. The intentions are the actions that the BDI-agent has decided to implement in order to fulfill its desires [23].

8.3 Cognitive Engineering

The multi-agent architecture of `Crew Assistant` can be seen as a data processing architecture that processes low-level sensor data in several stages to easy-to-access information to be accessed by the human operator. When analysing candidate operator tasks to be allocated to the machine assistant, it is attractive to classify these tasks into several levels of priority, complexity or workload. The hierarchy between these categories should be that the lowest level represents routine and skill- (control-) based tasks, and that their execution is almost instantaneous. The highest level requires specific knowledge and expertise from the operator and their execution is less time critical [33, 36]. This follows very closely the well-known classification proposed by Rasmussen, who segregates a human's cognitive behaviour into skill-based, rule-based and knowledge-based categories [24].

The cognitive classification has been proposed by Rasmussen as a promising automation approach to overcome the deficiencies of traditional automation. Traditional automation is considered not capable of compensating for the natural human cognitive deficiencies, which results in an automation concept in which the human either becomes overloaded or is kept out-of-the-loop. The core element of the cognitive approach is the natural human information-processing loop, which has the following elements: monitoring of the situation (perception and interpretation), diagnosis of the situation (including goal activation and evaluation), planning and decision-making, and plan execution.

Most current automation research, in particular in military aviation, has adopted the cognitive approach. Examples are the `Cognitive Cockpit` [31], the cognitive man-machine cooperative approach of the `Crew Assistant Military Aircraft` [38] and the cognitive concepts as applied in the `Tactical Mission Management` system [28]. A major contributor to this cognitive success is the fact that the human problem solving strategy and the cognitive process as initially proposed by Rasmussen have been translated into mature and practical engineering frameworks [25, 37]. The approach of Cognitive Work

Analysis (CWA) is of particular interest and has already been successfully introduced into the design of intelligent (BDI) agent applications [39].

8.4 Research Challenge

This chapter describes initial research into using intelligent agents to enhance situational awareness of a human-machine team environment in a military scenario. The research is a collaborative project between the Defence Science and Technology Organisation (DSTO) and the University of South Australia (UniSA) [29, 34, 35]. Although there are several different methods of achieving situational awareness, this project will concentrate on intelligent agents using a BDI architecture and reasoning approach. The BDI-architecture allows the implementation of the cognitive loop within the agent, which is required to stay in line with the cyclic cognitive process of the human member(s) of the team [22]. The project will demonstrate how BDI agents can be used to build situational awareness in a hostile environment. An existing application environment (namely UT) will be used for visualisation and adversarial role-play together with its tools and implementation architecture.

Intelligent agents, in particular agents of the BDI-architecture, have been the subject of many developments in the past decade. These developments aimed at either refinements of the agent shell itself, or at applications of BDI. However, there are still important questions to be answered:
1. How to establish realistic communications between agents within a team and between agents and humans like those in the real world?
2. How robust are intelligent agents in terms of adjusting to complex scenarios and of dealing with non-repeatable (or unforeseeable) events?
3. How do agents learn from past experiences and create knowledge based learning capability?

This project initially aims to research the architecture and requirements for agent oriented software to coordinate and to maintain communication between the agents themselves, and between the agents and human operator(s).

8.4.1 Human-Agent Teaming

A major research aim is to improve the team process of gathering situational awareness by an order of magnitude greater than that of a human acting alone. The proposed increase should result from reorganising the human-machine interaction, by imposing a paradigm shift as shown in Fig. 8.1.

The development of intelligent agents has on one hand been driven in the past by simulation applications where the modelling and inclusion of human-like intelligence and decision-making behaviour is required. This resulted in the development of intelligent agents that act as stand-alone substitutes for relevant human behaviour (such as in operations analysis), and that allow scenarios to be executed much faster than real-time. Research in teaming of intelligent agents to include command and control behaviour has also been initiated in this domain [3]. Stand-alone simulation intelligent agents are further employed to provide non-cooperative opponents to human player(s) in real-time training scenarios.

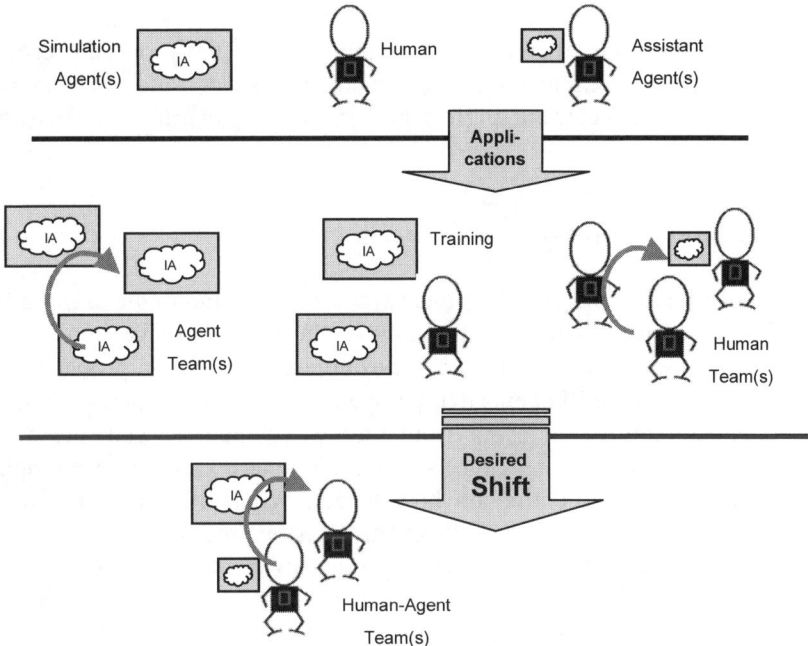

Fig. 8.1. Human-Agent Teaming

The development of intelligent agents on the other hand has been driven by the need for automation and decision support to human operators in critical or high-workload situations. This resulted in the development of machine assistants that can offload human tasks but that remain under supervision and control of the human. Intelligent agents and in particular BDI-agents are attractive when undertaking assistant roles since their structure is well aligned with the cognitive and cyclic character of the knowledge based information process of the human [22].

The aim of the proposed research is to achieve a paradigm shift that realizes an improved teaming environment in which humans, assistant agents and cooperative agents are complementary team members. In formulating such teams, the main change required from a simulation point of view is that humans and machines are not interchangeable, they are complementary. The main issue from an automation point of view is a move away from its traditional design philosophy, which is based on a comparison of tasks in which a human or machine is superior, resulting in fixed allocation of functions and system design. Humans and machines are not comparable, rather they are complementary [30]. It is expected that the human will still be in charge of the team but intensive management and coordination between the different team members will be required. Agent communication will therefore be a main focus of the proposed research. Aspects of communication will include control and management, coordination, self-learning, performance monitoring, warning, and assertive behaviour.

8.4.2 Agent Learning

Another research aim is to introduce aspects of learning to the existing BDI philosophy. Most of the functionality of the agents can be implemented using the existing BDI-syntax of, for example, JACK, but a problem arises from the fact that BDI agents exhibit open-loop decision-making. That is, a BDI agent requires an event from the environment to generate a goal (either directly or by changing its beliefs) so that it can search through its plans to form an intention on how to act. When the plan has finished executing (by succeeding or failing), the decision process finishes.

What happens afterwards depends on the environment because the agent can have several plans executing concurrently corresponding to different goals. In addition, BDI-agents are optimised for a specific scenario, which means that an agent written for a particular environment will not easily translate to another [20]. These shortcomings can be removed by introducing a learning component into the agent. The feedback caused by learning introduces a closed loop decision process analogous to Boyd's Observe-Orient-Decide-Act (OODA) loop [19] as shown in Fig. 8.2. This means that the decision process can proceed more rapidly, and decisions can be evaluated before they are committed. The side with the faster OODA loop triumphs in a battle [5].

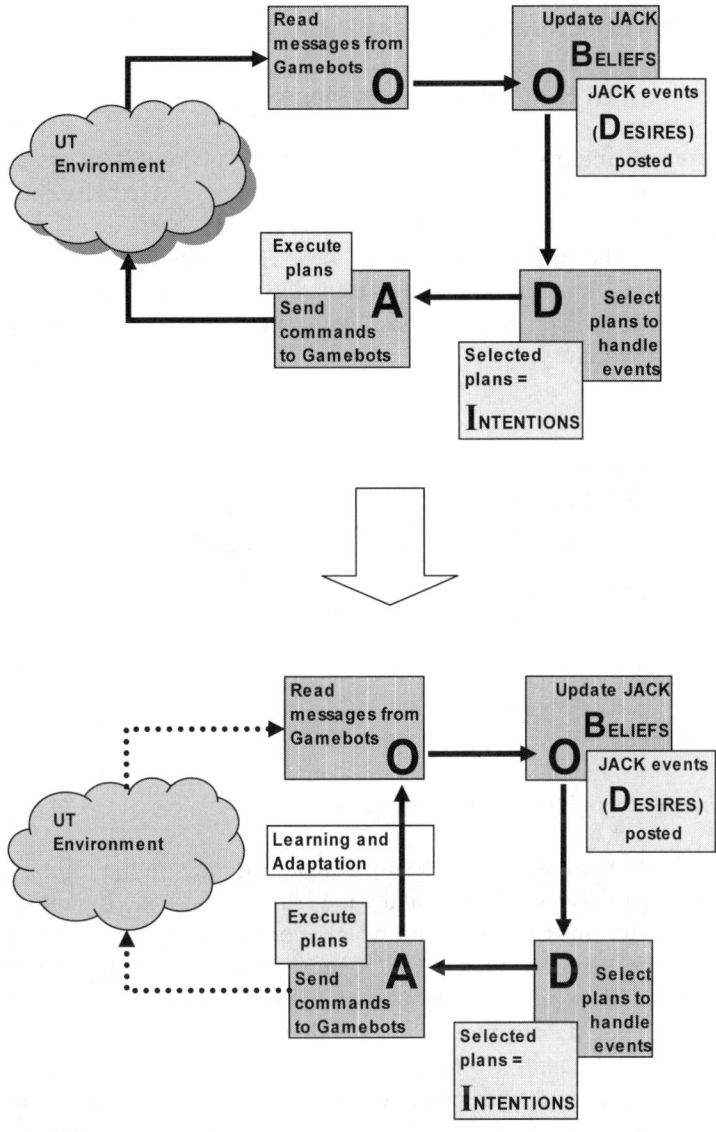

Fig. 8.2. Learning, BDI reasoning and the OODA loop

Introducing learning into a BDI agent is not a trivial problem because current learning techniques tend to require large computational power while agents are small and time-critical processes. It involves analysing the different AI learning techniques, determining which are the most useful and incorporating them into the existing architectures.

8.5 The Research Environment

8.5.1 The Concept of Situational Awareness

Significant research interest in automated situational awareness exists at DSTO with the initial incentive to embody elements of cognitive behaviour in simulation environments. Two systems have already been developed towards this end: the Operator System Integration Environment for airborne applications, and the OneSAF Testbed Baseline for land-based applications [10, 15]. The collaboration with the UniSA is a further extension of this research.

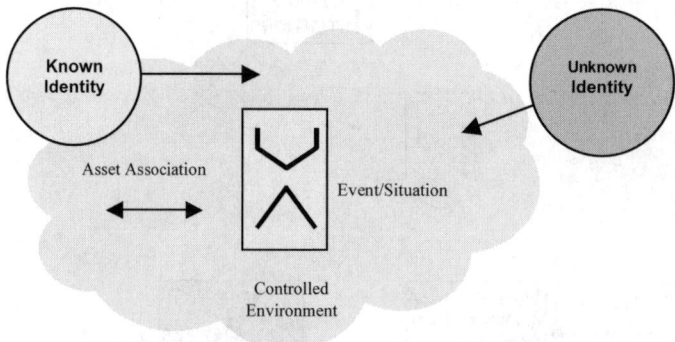

Fig. 8.3. Generic concept of Situational Awareness

The problem with building accurate situational awareness in the military context applies equally to both land and air scenarios. In either environment, an asset must be protected when approached by an unknown entity such as an unidentified plane (Fig. 8.3). A response is instigated to gather information on which to formulate a plan that can react to the perceived threat. This scenario can be extended into the cockpit of an airplane or the turret of a tank. Upon gathering information, the intent of an unknown identity is derived with other assets generally being managed in responses of the perceived threat. As the scenario intensifies, the operator usually needs to contend with an increased volume of information. This scenario can be related to non-military examples, such as: accident sites, bush fires, civil emergencies, or even to games like RoboCup and RoboCup Rescue. Some of these games have already been widely used for research into intelligent agent technology. In addition they provide a publicly available research environment with a challenge in situational awareness.

8.5.2 The `Unreal Tournament` Game Platform

The potential use of several publicly available game platforms was considered – with `RoboCup Rescue`, `Quake` and `Unreal Tournament` being found to be most suitable. `RoboCup Rescue` focuses on the social issues of very large numbers of heterogeneous agents in a hostile environment [27]. However, due to the fact that the rescue simulator was in preliminary stages of development it could not provide the required robust research environment. `Quake` is a popular 3D first-person shooter game developed by ID Software. `Quake` games enjoy a large interest for research into intelligent agent technology [6, 11, 12, 14, 21]. However, `Quake` does not provide a suitable interface for allowing external software to control characters within the game and would require a significant amount of resources to develop such an interface.

The `Unreal Tournament` (UT), a game co-developed by Epic Games, InfoGrames and Digital Entertainment, was selected. The UT game engine has great flexibility for development or customization using a C++ based scripting language called Unreal-Script. This fact was noticed by a team of researchers who developed a UT modification called `Gamebots` [1] that allows bots in the game to be controlled via external network socket connections. The 'thinking' of the bots can therefore be implemented on any machine and in any language that supports TCP/IP networking [1]. A spin-off of the `Gamebots` is `Javabots` [17], which focus on building a low-level Java-based platform for developing `Gamebots` clients.

8.5.3 The `JACK` Agent

Currently a large number and variety of agent-development environments and toolkits exist, however many lack maturity, documentation or support and are therefore difficult to use [1]. The `JACK` agent shell was chosen because of its mature implementation of the desired BDI reasoning philosophy. `JACK` is commercially available and support for the development of research applications is equally available. `JACK` is written in Java and extends the Java language syntax by providing a number of libraries encapsulating much of the framework required to code BDI solutions [2].

8.6 The Research Application

8.6.1 The Human Agent Team

Humans are able to directly interact with the simulation and communicate with others through text messages. A human can join a team comprised of any combination of agents/humans and collaborate with them in order for their team to achieve a victory.

The human is able to assume two types of roles in the team. Firstly, when inexperienced with the game, the human can take on the role of a troop. In this instance, the human is given orders through text messages that he/she must follow if the team is to succeed. The orders begin with simpler and easier objectives and become increasingly more complex and difficult as the human gains more experience. Secondly, after becoming comfortable with the game, the human can assume command and give orders to the rest of the team. While in command, the human is responsible for managing the team assets in a way that will lead the team to victory.

There are three types of agents being developed [8]. Firstly, the *Commander* agent keeps close track of the troops under its command, makes decisions about what each member of the team should be doing and orders its troops accordingly. Additionally, it informs the human player about what is happening and suggests what to do. The human can also give orders to the Commander agent, which in turn controls the team directly. Secondly, the *Communication* agent does not connect to UT; it facilitates in the communication and learning between the Commander and a human by analyzing and understanding the text messages sent to/from the human. Thirdly, the *Troop* agent explores the environment, provides environmental feedback and carries out orders received from the Commander agent. Fig. 8.4 illustrates the relationships between the three agents, human players, and UT.

Agents within UT can observe and act in real-time within a simulated world. Humans are able to interact with the simulation; this includes observing the simulated world, being able to move/act within it, and manipulating simulated entities. Additionally, UT includes team-based scenarios that facilitate communication and collaboration between agents and/or humans.

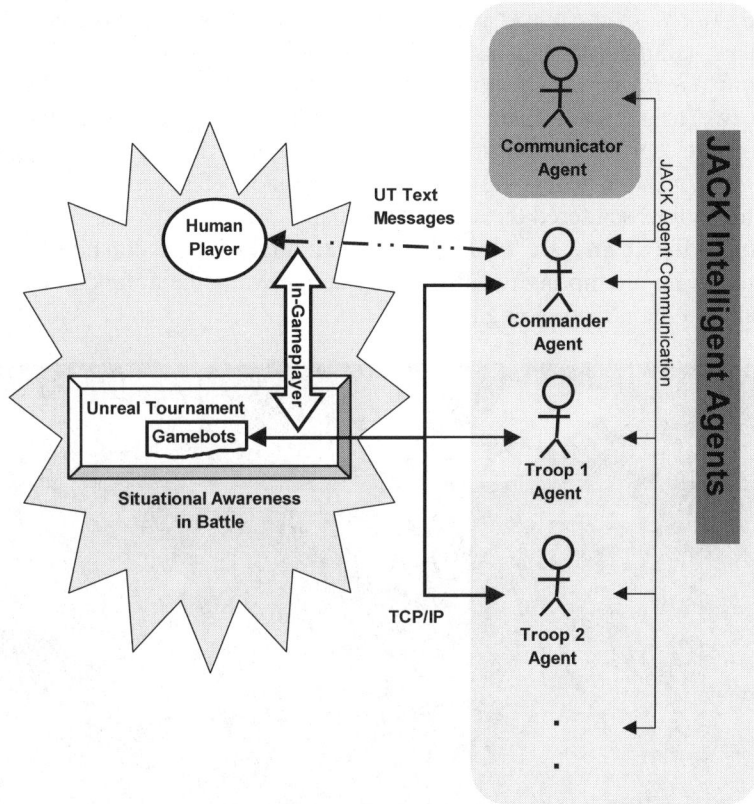

Fig. 8.4. Relationships between the commander agent, communicator agent, troop agents, human player and Unreal Tournament

8.6.2 The Simulated World within Unreal Tournament

Unreal Tournament provides a simulated world within a complete 3D environment. This environment is formed via a pre-designed 3D map and a physics engine. A map can be customized in many different ways. For example, it can have buildings, doors, elevators and water. The physics engine can also be customized, such as by varying the gravity to simulate a game in outer space. Fig. 8.5 shows a screen capture of UT.

UT also offers a powerful multi-player mode, which allows two or more humans to enter the same game and play in the same team or in opposing teams. There are four ways to play a game of UT, called *GameTypes* as described below [26]:

1. *DeathMatch,* every player is acting by themselves and the goal is to kill as many competitors as possible while avoiding being killed.
2. *CaptureTheFlag,* which introduces team play. The players are divided into two teams, each team has a base with a flag that it must defend. Points are scored by capturing the opposing team's flag.
3. *Domination,* two teams fight for the possession of several domination control points scattered throughout the map.
4. *Assault,* the teams are classed as either attackers or defenders. The attackers are attempting to fulfill an objective while the defender's job is to prevent them from doing this.

Fig. 8.5. The Unreal Tournament Environment

An important feature of UT is that it allows humans and agents to interact from within the same virtual world. It is possible to form teams comprised of any desired combination, which include: Human only teams, Agent only teams and Human-Agent combined teams. The latter allows researching and testing issues related to Human-Agent collaboration.

8.6.3 Interacting with Unreal Tournament

Interfacing agents to UT is performed using contributions made with the Gamebots [1] and Javabots [17] projects. Gamebots allows the UT characters in the game to be controlled via network sockets connected to clients [26]. It also provides information to the clients through a specific protocol. Based on the information acquired, the client (agent) can decide for itself what actions it needs to take (in other words, move, shoot, talk, and so on).

Some useful tools have been designed for use with Gamebots. VizClients are visualization tools for observing and analyzing the behavior of agents in the game. There are two types of VizClients available: global and agent-specific. The global VizClient identifies the location of all humans, agents and objects in the environment as well as team and individual agent scores. The information presented can be overlaid over a birds-eye view of the UT map. The agent-specific VizClient describes the movement of an individual agent; it can draw the path of the agent as it moves in the environment. Also it is possible to use the VizClient streams in order to generate logs which can be analyzed to determine decision patterns of the agents and compare their performance [1].

Javabots provides a selection of Java packages that are designed for handling the low level communication to the Gamebots server [17]. The Javabots developers intended to create a toolkit whereby, if used, the agent developer would not require much knowledge of the specifics of the Gamebots protocol, only the ability to implement the *Bot* interface in the Javabots package [17].

The Gamebots-Javabots combination allows any Java-based software to interact with UT. JACK is built on top of Java and it provides all the functionality of Java along with agent-oriented extensions [16]. JACK agents, with this in mind, should be able to interact with UT. This is true, however a more complex custom interface is required in order to provide an agent with information that it can use.

8.6.4 The Java Extension

The Java portion of the interface provides three enhancements to the standard Javabot classes. Firstly, it has an improved message parser that supports parsing property lists and paths of the Gamebots protocol format, since the parser included with Javabots does not parse them properly. Secondly, it has specific classes for capturing the information from messages. For example, objects created using the *UtPlayer* class have the ability to capture Gamebots messages of the Player format and extract the data into values an agent can use.

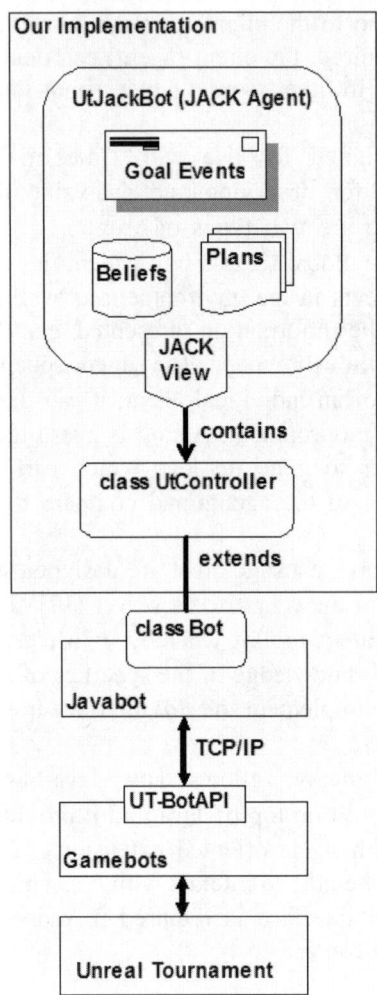

Fig. 8.6. Research Implementation

Thirdly, it has a class called *UtController* which extends the Javabots's *Bot* class (and through inheritance can do anything the *Bot* class can do), along with the ability to obtain a reference to a JACK view.

Once that reference has been initialized, *UtController* uses it to pass information over to JACK.

The custom interface proposed here is called *UtJackInterface*; its general architecture is shown in Fig. 8.6. The interface works by encapsulating a *UtController* object within JACK View, which generates specific events based on the messages that arrive. The events originating from the View are used to update agent beliefs accordingly. When the agent beliefs are updated by the view, more events are posted by the agent internally depending on how the beliefs were changed. The posted events are handled by selecting an appropriate plan (which becomes an intention), and the agent acts by executing the instructions of the selected plan.

It is important to note that the *UtJackInterface* only provides the underlying framework for interfacing to UT; it is up to the individual JACK developer to implement any behaviors required for an agent to survive and ultimately win a UT game session.

8.6.5 The JACK Component

The JACK portion of the interface allows an agent to interact with the UT environment. It consists of a number of standard calls that are used to populate belief-sets and respond to UT events within the UT environment. This interface handles all type and protocol conversions corresponding to different types of messages generated by the Gamebots. An agent can sense the environment in real-time by having some of its beliefs continuously updated using the Gamebots synchronous message queue. For example, when another player is in field-of-sight, the agent's *Players* belief-set is updated with information about this player. In other words, the agent believes that it is looking at another player when its belief-set says so. Synchronous messages are transmitted in blocks at a specified interval (the default is every 10 milliseconds).

The belief-sets can also be configured to generate events under some conditions. For example, a *FindHealthPack* goal event can be fired when the value of *Health* in the *Self* belief-set drops below 10%. An agent senses the environment through JACK events corresponding to Gamebots asynchronous messages. Generally, sensory events are fired when something important happens in UT that the agent should pay special attention to. To understand this better, an agent 'hears' when it receives an *UtJackAsyncHearNoise* event. Furthermore, it 'feels' that is has been shot when it receives an *UtJackAsyncDamage* event. In most cases sensory events stop an agent in its tracks and force it to assume a different behaviour.

Finally, agents take action in the world by transmitting commands back to the server. Some of the commands are Jump, RunTo and TurnTo. According to the Gamebots documentation [18] some commands such as RunTo are persistent; this means that when a Gamebots client issues a RunTo command, it will keep on running until it reaches its destination. Persistent commands can however be interrupted when some important events occur (such as being killed[1]). The agent therefore needs to keep track of its status while in the middle of executing persistent commands.

8.7 Demonstration System

As a first implementation for the project a demonstration system is required to verify that a JACK-based agent can reason about and interact with UT. A series of fundamental actions were chosen and implemented for two missions with the following criteria:

- Autonomously explore the UT environment.
- Defend an asset for a short amount of time.

The implemented *DomDefender* agent attempts to explore five NAV points and then attempts to defend a DOM point for ten seconds, where 'Defending' this point means holding ground on that location. The agent does not use weapons in this demonstration. The missions are treated sequentially – in other words, if mission A succeeds the agent then attempts to achieve mission B. If any of the two objectives fail then the agent reports 'Mission Failed', otherwise when both objectives have been successful the agent says 'Mission Accomplished'. Sample JACK code of the *exploreAndDefend* reasoning method used to implement this behavior is shown in Fig. 8.7 (the important words are highlighted in bold).

---

[1] An illustration of this problem is that an agent has issued a command to move to location-A and it starts to move towards that location. Does it then assume that it got there after a few seconds? No, it may have been killed, or perhaps slammed into a wall that sits in between where it started and where its destination was before the event.

```
if(!postEventAndWait(expEv.explore(eNm)))
{
    System.out.print("Mission Failed!");
    System.exit(1);
}

if(!postEventAndWait(ddEv.domDefend(dNm)))
{
    System.out.print("Mission Failed!");
    System.exit(1);
}

System.out.print("Mission Accomplished");
System.exit(0);
```

Fig. 8.7. ExploreAndDefend Behaviour Sample Code

JACK reasoning methods are special methods that can be defined within an agent and can be executed by a program external to the agent. They are primarily used to post goal events to the agent in order to force it to assume some behavior. Inspection of the *exploreAndDefend* sample code reveals that it uses a key JACK method called *postEventAndWait*. An event posted using the *postEventAndWait* method is handled synchronously. That is, the agent executes the event as a separate task and the calling method must wait until this task has been completed before continuing. The *postEventAndWait* method also returns a Boolean value indicating whether the event was successfully handled [16]. The sample code therefore posts an 'explore' event synchronously, at which point it waits for the result. If the result is a Boolean false, it prints 'Mission Failed' and exits. This is repeated for a 'defend' event. If both events are successfully handled it follows on to print 'Mission Accomplished' and exits.

8.7.1 Wrapping Behaviours in Capabilities

The capabilities of an agent encapsulate some of the things that it can do in intuitive groupings. The implemented *DomDefender* agent uses four capabilities, these being *SyncHandler*, *AsyncHandler*, *UtExploreCap*, *DomDefendCap* and *UtMoveCap*. The relationships between the capabilities can be illustrated using a JACK design diagram as shown in Fig. 8.8. Relationships are automatically generated by the JACK design tool using the developer's code.

The diagram illustrates that the *DomDefender* agent contains a private instance of *UtJackInterface*. In JACK, 'private' scope means that the agent has its own copy of that data type, which it can read and modify as required independently of all other agents in the system [16]. In addition the agent 'has' instances of the *SyncHandler, AsyncHandler, UtExploreCap* and *DomDefendCap* capabilities, each capability providing the agent with different behaviors. The *SyncHandler* capability contains the agent's beliefs and plans for seeing the environment as explained earlier. Similarly, the *AsyncHandler* capability contains the beliefs and plans that the agent uses whenever a sensory event occurs. It is also shown that the agent has the capability to move but only indirectly through the *UtExploreCap* and the *DomDefendCap* capabilities. This means that the agent does not continuously move unless it is trying to achieve an exploring or defending goal.

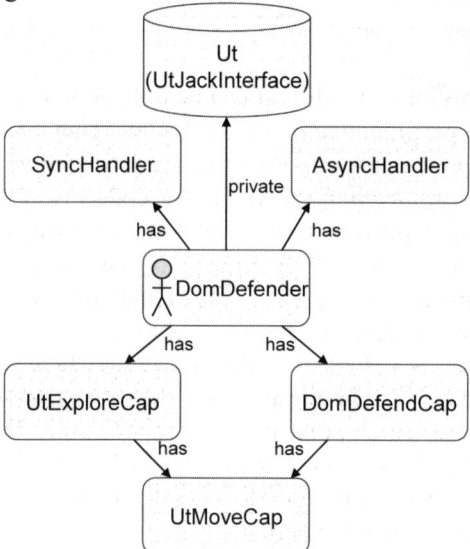

Fig. 8.8. The DomDefender Agent Behaviours

8.7.2 The Exploring Behaviours

The *UtExploreCap* algorithm (Fig. 8.9) implements the behavior of exploring a specified number of Navigation (NAV) points. It is started by posting a *UtExploreEvent* goal event which is then handled by the *UtExplorePlan* plan. The capability uses a java-based algorithm called *NavsToExplore* to tell the agent which NAV to choose in addition to a customized counter that counts how many NAV points have already been explored.

As shown in Fig. 8.10, the relationship between *UtExploreEvent* and *UtExplorePlan* shows both 'post' and 'handle'. This is because the design of the algorithm within *UtExplorePlan* recursively posts a new *UtExploreEvent* for each unexplored NAV. This is shown by the sample pseudo code for the *UtExplorePlan* plan given in Fig. 8.9. Note that for a whole plan to succeed all instructions in the plan must succeed. That is, if for example the statement 'Move to next NAV' fails then the whole plan fails and consequently the entire attempt to explore also fails.

```
IF there are unexplored nAVs
   Move to next NAV
   Increment count
   IF count <5
      Synchronously post an UtExploreEvent goal
   ENDIF
ENDIF
```

Fig. 8.9. Exploring Behaviour Algorithm

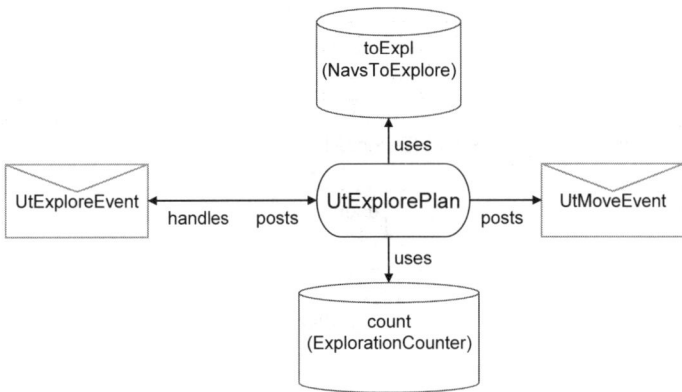

Fig. 8.10. Exploring Behaviour Relationship Diagram

8.7.3 The Defending Behaviour

DomDefendCap implements the behavior of defending a domination point for a specified number of seconds. It starts by receiving a *DomDefendEvent* goal event. When the agent executes this behavior it moves to one of the

DOMs that is has seen while previously exploring. When it arrives it starts counting (1 count for for every second passed). If control of the DOM point is lost (in other words, by shooting or pushing the agent away) it will attempt to regain control and restart the counting process. The algorithm behind this behavior is shown in Fig. 8.11.

```
Select DOM to defend
Move to DOM
DOWHILE count < 10
     IF I am not at the DOM
          Move to defending DOM
          Reset count to 0
     ENDIF
     Wait for 1 second
     Increment count
ENDDO
```

Fig. 8.11. Defending Behaviour Algorithm

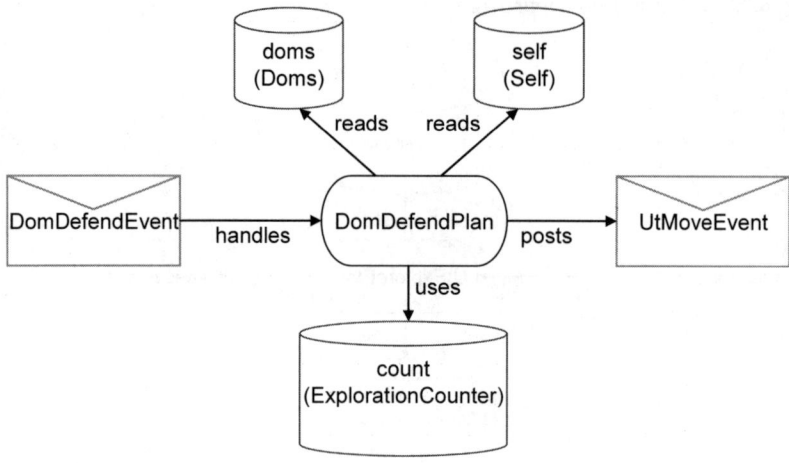

Fig. 8.12. DomDefend Relationship Diagram

The relationship diagram for *DomDefendCap* is shown in Fig. 8.12. The DOMs belief-set identifies any DOMs that the agent has seen. The Self belief-set can specify the current location of the agent and is used to determine whether the agent has been pushed away from the DOM that it is defending. In this plan, the 'select DOM to defend' statement fails if the agent has not seen any DOM points. The 'Move to defending DOM' statement fails if the agent is not able to reach the DOM point it wants to defend.

Finally, the *UtMoveCap* capability provides the agent with the ability to move around the map in a controlled fashion. Upon receiving a *UtMoveEvent*, *UtMovePlan* sends the appropriate command to UT and an internal *UtMoveTraceEvent* is posted synchronously. *UtMoveTracePlan* then monitors the location of the agent through the Self belief-set. It succeeds when the agent has reached the target location before a specified time-out period. If the time-out occurs then it fails. Fig. 8.13 shows this process in graphical form via the relationship diagram.

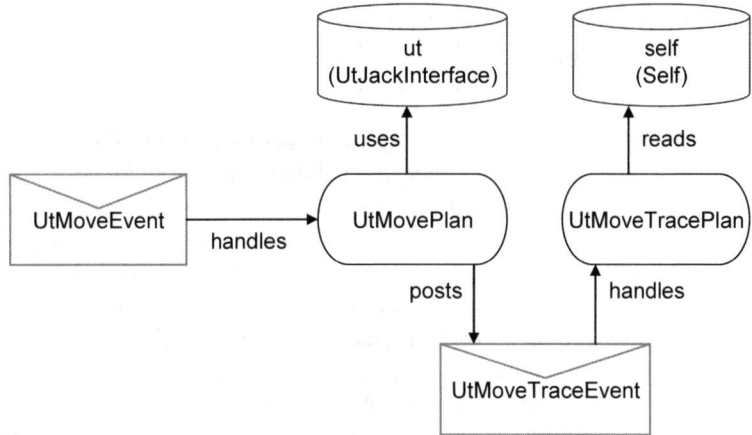

Fig. 8.13. Moving behavior relationship diagram

8.8 Conclusions

This chapter argues the case for the use of intelligent agents as cognitive team members in a human-machine team environment. This team will still need to be supervised by human(s) but will be further composed of machine members that provide complementary capabilities. The BDI type of agent is expected to exhibit the cognitive behavior that is required to align with the cyclic character of the knowledge based information process of the human.

Research is proposed to examine the use of intelligent agents that are able to enhance situational awareness in a hostile environment. To this purpose the environment of Unreal Tournament has been selected because of its mature development tools, its large research following, and its well-developed library of scenarios. The JACK intelligent agent shell was chosen because of its BDI-reasoning philosophy and its compatibility with UT. Furthermore, JACK is commercially available and is well supported in the development of research applications.

UtJackInterface uses previous work available from the Gamebots and Javabots projects. It extends the previous work and incorporates a layer of JACK constructs that can be used by a JACK agent. An example agent has been implemented in order to demonstrate the interface in action.

Further research includes investigating how teams communicate, behave in teams and learn in their environment.

Acknowledgements

The authors would like to acknowledge that papers with brief versions of the work presented in this Chapter have been published in conference proceedings – namely the *10th International Conference on Human-Computer Interaction* [35], the *7th International Conference on Knowledge-Based Intelligent Information & Engineering Systems* [34], and the *3rd International Conference on Hybrid Intelligent Systems* [29]. The authors extend their appreciation to the conference proceedings publishers: Lawrence Erlbaum Associates Inc, Springer-Verlag and IOS Press, respectively for their permission to combine and publish the work here. The authors also thank Ms. Irene Sofianos from UniSA for her invaluable help and comments when proof-reading the Chapter text.

References

1. Adobbati R, Marshall AN, Scholer A, Tejada S, Kaminka G, Shaffer S and Sollitto C (2001) Gamebots: A 3D Virtual World Test-Bed for Multi-Agent Research, *Proc 2nd Intl Workshop on Infrastructure for Agents, MAS, and Scalable MAS,* Montreal, Canada, AAAI, June 2001.
2. AOS (2002) JACK Intelligent Agents, *Agent Oriented Software Pty Ltd*, http://www.agent-software.com.au (accessed: 9/11/2003).
3. Appla D, Heinze C, Goss S and Connell R (2002) Teamed Intelligent Agents Software, *Proc Defence Operations Analysis Symposium, DOAS 2002,* Edinburgh, Australia.
4. Champigneux G (1993) In-flight mission planning in the Copilote Electronique, *AGARD Lecture Series on New Advances in mission planning and rehearsal systems,* October : 701-712.

5   Clemens SM (1997) The one with the most information wins? The quest for information superiority, Master of Science in Information Resource Management Thesis, Air Force Institute of Technology: 128.
6   Cronin E, Filstrup B and Kurc A (2001) A distributed multiplayer game system, *Electrical Engineering and Computer Science Department, University of Michigan,* Course Project Report UM EECS589, May 2001.
7   Gibson CP and Garret AJ (1989) Towards a Future Cockpit - The prototyping and Integration of the Mission Management Aid (MMA), *Proc AGARD Symposium on Situation Awareness in Aerospace Operations*, AGARD, AGARD-CP-478.
8   Ichalkaranje N, Urlings P, Tweedale J and Jain L (2002) Intelligent Agents for Airborne Mission Systems, *University of South Australia*, Adelaide, Research Report EIE-KES-IAAMS-2002-LJNI-01, May 2002.
9   Kaminka GA, Veloso MM, Schaffer S, Sollitto C, Adobbati R, Marshall AN, Scholer A and Tejada S (2002) Gamebots: a flexible test bed for multi-agent team research, *Communications ACM*, 45: 43-5.
10 KESEM (2003) The Operator System Integration Environment (OSIE) Project, http://www.kesem.com.au/siteprojects/projects_osie.htm (accessed: 16/1/2003).
11 Laird JE (2001) It Knows What You're Going To Do: Adding Anticipation to a Quakebot, *Proc 5th Intl Conf Autonomous Agents,* Montreal, Canada, May, ACM Press, New York: 385-392.
12 Laird JE (2001) Using a computer game to develop Advanced AI, *Computer Magazine*, July, 34: 70-75.
13 LaPuma AA and Marlin CA (1993) Pilot's Associate: A synergistic System reaches Maturity, *AIAA-93-4665-CP*: 1131-1141.
14 Lent MV, Laird J, Buckman J, Joe Hartford, Houchard S, Steinkraus K and Tedrake R (1999) Intelligent Agents in Computer Games, *Proc 16th National Conf Artificial Intelligence,* Orlando, Florida, AAAI Press, Menlo Park, CA: 929-930.
15 Lui F (2002) Mapping Cognitive Work Analysis (CWA) Towards An Intelligent Agents Software Architecture: Command Agents, *Proc Defence Human Factors Special Interest Group (DHFSIG) 2002,* DSTO Melbourne, Australia, November: 21-22.
16 Maisano P (2003) JACK Intelligent Agents™ JACK Manual, Agent Oriented Software Pty Ltd.
17 Marchall AN, Vaglia J, Sims JM and Rozich R (2003) JavaBot for Unreal Tournament, http://utbot.sourceforge.net (accessed: 10/11/2003).
18 Marshall AN, Gamard S, Kaminka G, Manojlovich J and Tejada S (2003) Gamebots: Network API, http://planetunreal.com/Gamebots/docapi.html (accessed: 10/11/2003).
19 MindSim (2000) Decision Making: OODA Loop, http://www.mindsim.com/MindSim/Corporate/OODA.html (accessed: 17/2/2003).
20 Norling E (2000) Flexible, Reusable Agents for Modelling Human Operators, *Proc Defence Human Factors SIG Workshop,* Melbourne, October.

21 Norling E (2002) Capturing the Quake Player: Using a BDI Agent to Model Human Behaviour, in Rosenschein JS, Sandholm T, Wooldridge M and Yokoo M (eds), *Proc 2nd Intl Joint Conf Autonomous Agents and Multi-Agent Systems*, Melbourne, Australia, July, ACM Press, New York: 1080-1081.
22 Onken R and Walsdorf A (2001) Assistant Systems for Vehicle Guidance: Cognitive Man-Machine Cooperation, *Aerospace Science Technology*, 5: 511-520.
23 Rao AS and Georgeff MP (1995) BDI Agents: From Theory to Practice, *Proc 1st Intl Conf Multi-Agent Systems (ICMAS-95)*, San Francisco, CA, June, AAAI Press: 312-319.
24 Rasmussen J (1986) *Information Processing and Human-Machine Interaction: An Approach to Cognitive Engineering*, Elsevier Science Ltd, North Holland.
25 Rasmussen J, Pejtersen AM and Goodstein LP (1994) *Cognitive Systems Engineering*, John Wiley & Sons, New York.
26 Robinson WD, Harlick B and Fish R (2000), *Unreal Tournament Manual*, InfoGrames, Epic Games and Digital Extremes, Woodinville.
27 RoboCup (2003) RoboCup Rescue Official Webpage, http://www.r.cs.kobe-u.ac.jp/RoboCup-rescue/ (accessed: 17/1/2003).
28 Schulte A (2002) Mission Management and Crew Assistance for Military Aircraft - Cognitive Concepts and Prototype Evaluation, in *Tactical Decision Aids and Situational Awareness*, System Concepts and Integration Panel, Amsterdam, Sofia, Madrid, Maryland, NATO Research and Technology Organisation, Canada, January, RTO-EN-019, SCI-113: 4(1)-4(19).
29 Sioutis C, Ichalkaranje N and Jain L (2003) A Framework for Interfacing BDI agents to a Real-time Simulated Environment, *Proc 3rd Intl Conf Hybrid Intelligent Systems,* Melbourne Australia, December, IOS Press: 24-30.
30 Statler IC (1991) Voyage to Mars: A Challenge to Collaboration between Man and Machines, *Symposium on the occasion of the farewell of Prof Dr. O.Gerlach IR as Chairman of the Board of NLR,* National Aerospace Laboratory (NLR), Amsterdam, The Netherlands, October: 5-29.
31 Taylor RM, Bonner MC, Dickson B, Howells H, Miller CA, Milton N, Pleydell-Pearce K, Shadbolt N, Tennison J and Whitecross S (2002) Cognitive Cockpit Engineering: Coupling Functional State Assessment, Task Knowledge Management and Decision Support for Context Sensitive Aiding, in McNeese M & Vidulich M (eds) *Cognitive Systems Engineering in Military Aviation Environments: Avoiding Cogminutia Fragmentosa!: A report produced under the auspices of The Technical Cooperation Programme Technical Panel HUM TP-7 Human Factors in Aircraft Environments (HSIAC-SOAR-2002-01),* Wright Patterson Air Force Base, OH: Human Systems Information Analysis Center, April: 253-314.
32 Taylor RM and Reising J (1998) The Human-Electronic Crew; Human-Computer collaborative Team working, in *Collaborative Crew Performance in Complex Operational Systems,* RTO Human Factors and Medicine Panel (HFM), Edinburgh, RTO MP-4, NATO Research Technology Organisation, Canada, December, RTO-MP-4: 22(1)-22(17).

33. Urlings P and Jain L (2002) Teaming Human and Machine: a Conceptual Framework, in Abraham A and Köppen M (eds) *Proc Hybrid Information Systems – 1st Intl Workshop Hybrid Intelligent Systems,* Adelaide, Australia, Advances in Soft Computing, Physica-Verlag, Berlin, December, 711:721.
34. Urlings P, Tweedale J, Sioutis C and Ichalkaranje N (2003) Intelligent Agents and Situation Awareness, in Palade V, Howlett RJ, Jain L (eds.), *Proc 7th Intl Conf Knowledge-Based Intelligent Information & Engineering Systems,* United Kingdom, Springer-Verlag, Berlin, September, 2: 723-733.
35. Urlings P, Tweedale J, Sioutis C, Ichalkaranje N and Jain L (2003) Intelligent Agents as Cognitive Team Members, *Proc 10th Intl Conf Human-Computer Interaction,* Greece, June, Lawrence Erlbaum Associates: 14-19.
36. Urlings PJM and Zuidgeest RG (1997) A generic Architecture for Crew Assistant Systems, *Proc AGARD MSP Symposium on Advanced Architectures for Aerospace Mission Systems,* Istanbul, July, AGARD, CP-581: 12.
37. Vicente KJ (1999) *Cognitive Work Analysis: Toward Safe, Productive, and Healthy Computer-based Work,* Lawrence Erlbaum Associates.
38. Walsdorf A and Onken R (2000) Cognitive Man-Machine Cooperation: Modelling Operator's General Objectives and its Role within a Cockpit Assistant System, *Proc IEEE Intl Conf Systems, Man and Cybernetics,* October, IEEE Press, Piscataway, NJ, 2: 906-913.
39. Watson M and Lui F (2003) Knowledge Elicitation and Decision-Modelling for Command Agents, *Proc 7th Intl Conf Knowledge-Based Intelligent Information and Engineering Systems (KES 2003),* Oxford, UK, September, Springer-Verlag, Berlin, 2: 704-713.
40. Zuidgeest RG and Urlings PJM (1996) A generic Architecture for in-flight Crew Assistant Systems based on Advanced Information Processing Technology, *Proc 20th ICAS Conf,* Sorrento, Italy: 14-25.

9 Fuzzy Multivariate Auto-Regression Method and its Application

N Arzu Sisman-Yilmaz[1], Ferda N Alpaslan[2] and Lakhmi C Jain[3]

1   Central Bank of the Republic of Turkey, 06100 Ankara, Turkey, Arzu.sisman@tcmb.gov.tr.

2   Computer Engineering Department, Middle Eeast Technical University, 06531 Ankara, Turkey, alpaslan@ceng.metu.edu.tr.

3   Knowledge-Based Intelligent Engineering Systems Centre, University of South Australia, Mawson Lakes, SA 5095, Australia, Lakhmi.Jain@unisa.edu.au

9.1 Introduction

In this Chapter, a fuzzy multivariate auto-regression method is introduced which is based on fuzzy linear regression. The aim of the study is to model multivariate time series data.

Time series data is tabular data in the form of columns varying in time intervals. In other words, a time series is a collection of observations made sequentially in time. The basic characteristics of time series analysis are that successive observations are usually not independent and that the analysis must take into account the time order of the observations. When successive observations are dependent, future values may be predicted from past observations. Moreover, when there are two or more time series dependent on each other, one of the time series can be expressed in terms of previous values of itself and the other time series. If there is only one time series to predict the future values, this model is called univariate time series. An analysis of several data sets for the same sequence of time intervals is called multivariate time series analysis. Multivariate time series analysis is the focus of this study.

In multivariate time series analysis, it is possible to define each time series in terms of previous values of itself and previous values of other time series in the same system. The definitions of each time series can be represented as a rule which can be used in a rule-based system. These rules can be utilized for forecasting the future values of the system. The model obtained at the end of the processing can be used to forecast the future behavior of the time series data.

By using these principles, the fuzzy linear regression concept is combined with statistical multivariate time series analysis in a hybrid approach. In the following sections, background information on fuzzy regression and fuzzy time series analysis is introduced. The approach proposed in this study is described in detail and the results of two real data experiments are presented.

9.2 Fuzzy Data Analysis

Fuzzy MAR is based on Fuzzy Linear Regression (FLR) and the Multivariate Auto- Regression (MAR) algorithm, which is a non-fuzzy method. In order to explain the relationship between variables having ambiguity, the fuzzy regression method has been applied in the literature. In order to show how fuzzy data changes in time, fuzzy time series analysis is used. Studies related to fuzzy regression and fuzzy time series analysis are presented in the following subsections.

9.2.1 Fuzzy Regression

Tanaka introduced the pioneering work called Fuzzy Linear Regression (FLR) [24]. They designed a fuzzy structure represented as a fuzzy linear function whose parameters are given by fuzzy sets. Unlike the conventional linear regression function where the differences between the observed and the expected values are treated as observational errors, the differences are treated as the fuzziness in the system. In the fuzzy model, each fuzzy parameter is represented by a triangular fuzzy number (a number having a center and a width equal at both sides of the center), and which are coefficients in the fuzzy linear function.

The vagueness in the system is represented by the total width of the parameters (in other words coefficients). The problem is to find the parameters in a linear programming problem where the vagueness is to be minimized. The Fuzzy MAR model developed in this study uses fuzzy FLR. Fuzzy MAR accepts non-fuzzy data as input and gives fuzzy values as output. For that reason, FLR will be described in detail in Sect. 9.2.3.

It is also possible to apply possibilistic linear systems to fuzzy data analysis [22]. Possibilistic linear systems are based on the extension principle. The resulting approach is called fuzzy interval analysis. The research on possibilistic linear regression is described in [23, 26]. They claim that the probabilistic models take central tendency into account and ignore some observations, whereas all data are assumed to occur in possibilistic models. Another study based on fuzzy linear regression principles is a two-step procedure named Fuzzy Least Squares Linear Regression (FLSLR), which provides an enhancement of the minimal vagueness criterion of Tanaka's fuzzy linear regression model [15]. Fuzzy Membership Least Squares Regression (FMLS) is described in [12]. Their method uses minimization of the difference of fuzzy membership values between observed and estimated fuzzy numbers. Interval numbers are defined by means of two multilayer feedforward neural networks in [11]; the BackPropagation algorithm (BP) is used for training. The approach supplies a nonlinear approximation ability because neural networks can be used for modeling more complex systems. Non-symmetric fuzzy number coefficients for the fuzzy regression problem are suggested in [10]. They claim that symmetric triangular fuzzy numbers are not flexible enough to represent all data types. They modified the model to handle non-symmetric trapezoidal fuzzy number coefficients. They used a fuzzified neural network with fuzzy number coefficient weights.

Statistical and fuzzy regression models are compared in [13] by using criteria which contains the number of data points and quality of data (whether the variance is small or large). In their study, they reported that fuzzy regression is appropriate when the observations are fuzzy numbers or some data are collected by measurements, yet others are estimated subjectively (qualitative human expert knowledge) – in other words, there is not enough experiments for deriving a valid statistical model. They claim that a fuzzy model is less effective in forecasting when data quality is bad and some outliers exist in the model.

An objective function, which aims to minimize the total width of the estimated fuzzy number is used in [29]. This model accepts fuzzy values as expected output, which is a variation on the original fuzzy regression model. An evolutionary model for finding a fuzzy regression function is suggested in [4]. The regression function can be linear or nonlinear. An Evolutionary Algorithm searches a library of fuzzy functions, which can be linear, polynomial, exponential or logarithmic. The best function is the one with the minimum error value. The error function minimizes the distance between the function value and the observed value. This is another approach to fuzzy regression, and fits the best model to the data. The set of models is restricted where a univariate model is found.

The Fuzzy MAR Algorithm assumes that the data is defined by a linear model, in which a variable is defined by previous values of itself, as well as the previous values of other variables.

9.2.2 Fuzzy Time Series Analysis

There are different approaches to fuzzy time series analysis. Some of the models apply classical time series analysis methods directly to data. Some others apply fuzzy regression theory to fuzzy time series analysis, and another group of researchers uses fuzzy relations in order to describe temporal relationships.

The fuzzy time series analysis by [2] deals with the fuzzy representation of a classical time series. The model relates gas flow with carbon dioxide concentration for gas furnace data. The model aims to construct an expert system having fuzzy rules. It uses classical time series analysis methods to obtain a set of fuzzy rules.

A fuzzy time series model is presented in [30], where fuzzy time series is formulated by using a fuzzy function of time whose parameters are fuzzy numbers. The fuzzy numbers recorded at different time instances are possibilistic observation data and represented by fuzzy sets. They are used for forecasting future behavior. Linear trend and seasonal cycles are also described in the model.

Fuzzy trend and fuzzy seasonality concepts were introduced to fuzzy time series in [5]. A fuzzy ARIMA (Auto-Regression Integrated Moving Average) model was presented in [28]. The algorithm has two phases. They utilize the basic concept of ARIMA for finding a model. The model uses classical time series analysis methods and applies fuzzy regression theory on it.

Unlike conventional time series models, linguistic values (fuzzy sets) are studied as observations in the fuzzy time series model introduced in [15 - 17]. Fuzzy relational equations are employed in order to describe the dynamic process. The fuzzy time series model aims to forecast future values by using fuzzy relational equations. Song suggested seasonality forecast for fuzzy time series in [19], based on [17].

Modifications to the fuzzy time series model were made in [9, 17]. The new method is based on a time-variant fuzzy time series model. In [21], fuzzy time series modeling is compared to Markov modeling. The analogy between the two models is achieved by replacing the membership functions in the fuzzy approach with probability density functions.

There are various approaches to time series analysis dealing with fuzzy data. In our fuzzy multivariate analysis method, the fuzzy linear regression concept is utilized. For that reason, in the following Section, the fuzzy linear regression concept is described.

9.2.3 Fuzzy Linear Regression (FLR)

The FLR model was first introduced by [24]. In this Section, definitions related to FLR are presented, then the problem is defined as described in [27]. In the scope of the study only non-fuzzy data processing is described.

Basic Definitions

Conventional regression methods treat the difference between the observed data and the obtained output as the error. In the FLR method, this difference is accepted as the ambiguity in the data [27]. This ambiguity is represented as the fuzzy coefficients in the fuzzy linear equations. These fuzzy coefficients are fuzzy numbers.

A fuzzy number A is represented by a center α and a width c. The membership value for a fuzzy number is given by $A(x) = L((x-\alpha)/c)$. The membership function must be symmetric.

Linear Programming Problem

Fuzzy linear regression finds an equation for a dependent variable in which fuzzy numbers are used as coefficients of independent variables. The data handled can be either standard (with non-fuzzy output) or fuzzy data. In the scope of the study only FLR for non-fuzzy data is investigated.

The input data is given as $x_i = (x_{i1}, x_{i2}, \ldots, x_{in})$ for non-fuzzy output y_i where i denotes the sample index. The fuzzy linear regression model produces the equation:

$$Y_i = A_0 + A_1 * x_{i1} + A_2 * x_{i2} + \ldots + A_n * x_{in} \tag{9.1}$$

where Y_i is the fuzzy output, x_i is the input vector, and * denotes the scalar multiplication operation. The coefficients A_j, where $j=1,\ldots,n$ are triangular fuzzy numbers with (α_j, c_j).

The coefficients in the model are computed by using the following linear programming (LP) problem, namely minimization of the following objective function:

$$\text{Minimize } J(c) = \sum_{i=1}^{T} c^t |x_i| \tag{9.2}$$

with the constraints being:

$$y_i \leq x_i^t \alpha + |L^{-1}(h)| \, c^t \, |x_i|$$
$$y_i \geq x_i^t \alpha - |L^{-1}(h)| \, c^t \, |x_i| \tag{9.3}$$

where T is the size of data set, $|x_i| = (|x_1|,...,|x_n|)^t$, $L(x) = 1 - |x|$, $c > 0$ and $i = 1,...,n$. h denotes the degree to which the given data (y_i, x_i) is included in the output fuzzy number Y_i [27]. If a coefficient has a width c closer to 0, then the variable has less ambiguity, since the coefficient is similar to a non-fuzzy value.

9.3 Fuzzy Multivariate Auto-Regression Algorithm

The Fuzzy MAR algorithm is used for obtaining fuzzy relationships between the variables of a multivariate time series data. Fuzzy MAR is based on the non-fuzzy Multivariate Auto-Regression (MAR) algorithm, which is used to model multivariate time series data and obtain the rules necessary for predicting the values of variables in future time intervals. The method is a statistical analysis one for understanding the dynamics and temporal structure of several data sets for the same sequence of time intervals.

MAR was introduced in [1]; it is a stepwise procedure producing IF-THEN rules for a rule-based system. Each of the resulting IF-THEN rules is an explanation of a variable in terms of possibly all variables in the system, and also having temporal relationships. The rules are extracted from time series data which is in the form of a table of numerical values defined over different time intervals. There are n variables and T time instances in the multivariate temporal data, as shown in Table 9.1.

Table 9.1. Multivariate Temporal Data for n Variables and T Time Instances

Time	Variable		
	x_1	...	x_n
1	$x_{1,1}$	...	$x_{n,1}$
2	$x_{1,2}$	...	$x_{n,2}$
...	...	...	...
T	$x_{1,T}$	...	$x_{n,T}$

In multivariate time series data, the dependent variables must be defined in terms of previous values of itself and the other variables in the system. For that purpose, previous time instances of the other variables must also be considered. An information criterion must be used in order to find the best model describing the variables or to choose the best variable set among candidates in order to describe a dependent variable. The information criterion compares the models containing different combinations of the variables and different lags (time intervals). The information criterion used in testing the MAR Algorithm is BIC (Bayesian Information Criterion) which will be described in the next Section. For each iteration of the algorithm, a linear equation is found for a dependent variable by using a regression method. The BIC value of the equation is computed by using the RSS (Residual Sum-of-Squares), the number of regressors (predictor variables) and the number of observations. The model having the minimum BIC is determined to be the model needed at the end of processing. The final output of the system is the set of rules having the variables and their corresponding time intervals in the antecedent part and the variable to be described in the consequent part of the IF-THEN rule, as in the following example:

Example – Gas Furnace Data Processed by MAR Algorithm

The Gas Furnace data contains two multivariate time series data, namely X and Y; X is *GasFlowRate* and Y is *CO2Concentration*. There is one dependent series *CO2Concentration* and one independent series *GasFlowRate*.

The MAR algorithm processes the time series *CO2Concentration* as follows:
- An equation set containing previous intervals of CO2Concentration is found,
- The equation set containing previous instances of GasFlowRate is also found,
- BIC information criterion is used to compare the equations and to find the best equation.

The resulting equation found is:

$$CO2Concentration(t) = \quad (9.4)$$
$$f(GasFlowRate(t-4), CO2Concentration(t-1))$$

where t represents the present time instance. This result shows that the variable *CO2Concentration* is most appropriately explained by using *GasFlowRate* at time instance (t-4) and *CO2Concentration* at time (t-1).

The result of the process is described by means of an IF-THEN rule:

IF GasFlowRate(t-4) is a AND
    CO2Concentration(t-1) is b
THEN CO2Concentration(t) =
    f(GasFlowRate(t-4),CO2Concentration(t-1))

A dependent variable can be represented by using many different combinations of dependent variables and different lags of these variables. The Bayesian Information Criterion (BIC) is utilized to choose the best variable set among the candidates. The variable set providing the minimum BIC value is chosen among the candidates and set as the antecedent part.

In the example above, *CO2Concentration* can be described by a linear function of previous time instances of itself and *GasFlowRate*. For each variable-lag combination a regression equation is found. The BIC values of different regression equations are compared and the one with smallest BIC value is defined as the linear equation describing *CO2Concentration*. *CO2Concentration* is defined in terms of its value at the time instance (t-1) and the value of the other variable *GasFlowRate* at time instance (t-4); a and b are numerical values.

9.3.1 Model Selection

Model selection is the process of choosing the best set of regressors (predictor variables) among a set of candidate variables. In forecasting the linear regression models, some criteria are used to compare competing models and to choose the best among them [7]. In-sample forecasting and out-of-sample forecasting are concerned with determining a model which forecasts future values of the regressand (dependent variable) given the values of the regressors. In-sample forecasting shows how the chosen model fits the data in a given sample. Out-of-sample forecasting is concerned with determining how a fitted model forecasts future values of the regressand, given the values of the regressors.

There are various criteria that are used for this purpose. R^2, Adjusted R^2, Akaike Information Criterion (AIC), and Schwarz or Bayesian Information Criterion (BIC) are some examples. The aim of the model selection criteria is to minimize RSS (Residual Sum-of-Squares). All criteria except R^2 aim to minimize the number of regressors by putting a penalty on the inclusion of an increasing number of regressors. The resulting regression model includes the least number of predictor variables (as well as minimum RSS) as possible. R^2 is an in-sample criterion, which does not take into account adding more variables in the model. It depends on the RSS of the regression model for a regressand variable. Adjusted R^2 is the adaptation of R^2 to a criterion which puts a penalty on choosing a larger number of variables. Akaike Information Criterion (AIC) applies putting a penalty on the increasing number of regressors. The mathematical formula is:

$$AIC = e^{2k/n} RSS/n \qquad (9.5)$$

where k is the number of regressors and n is the number of observations. For mathematical convenience, the *ln* (natural logarithm) transform of *AIC*

$(ln(AIC) = (2k/n + ln(RSS/n)))$ is used. AIC is preferable in both in-sample and out-of-sample forecasting.

Schwarz (or Bayesian) Information Criterion (BIC) is similar to AIC. The mathematical formula is:

$$BIC = n^{k/n} RSS/n \tag{9.6}$$

and its logarithmic transform is $ln(BIC) = (k/n) \ ln(n) + ln \ (RSS/n)$, where the penalty factor is $[(k/n)ln(n)]$. The model with the smallest BIC is chosen as the best model. It is also convenient to use BIC to compare the in-sample or out-of-sample forecasting performance of a model.

In multivariate time series data, too many sets of variables and time instances can be used in defining a regressand variable. For that reason, it becomes important to decide on the best model that describes the data set. Moreover, use of redundant variables must be prevented. In order to choose the best set of variables and time instances in the data set, an information criterion is used which prevents use of redundant variables by placing a heavier penalty on them. The information criterion used is BIC (Schwarz or Bayesian Information Criterion). The mathematical formula of BIC is as follows:

$$BIC = T * ln(RSS) + K*ln(T) \tag{9.7}$$

In this formula, K stands for the number of regressors (in other words, the predictor variables on the right hand side of the model equation), T for the number of observations and RSS for residual sum-of-squares. As K or T increases, BIC puts a heavier penalty on the model. The variable set producing the minimum information criterion is accepted as the definitive model.

9.3.2 Motivation for FLR in Fuzzy MAR

The FLR model is applied to a set of input-output data pairs in order to obtain a relationship between input and output data. The relationship is an equation in which fuzzy numbers are used as coefficients. In this model, there exists an output variable and one or more input variables. The output equation of the model contains all the variables defining the regressand (dependents variable). The coefficients $A_i = (\alpha_i, c_i)$ can have a value greater than or equal to 0 (zero) for α (center) and c (width).

In our problem domain, we assume the availability of a multivariate model where there is more than one output variable. The output can be defined by using more than one variable changing over time. Each variable in the model is processed by the MAR algorithm to obtain a function of regressors (predictor variables) that defines the variable concerned. Moreover, more than one variable in the multivariate system can be described in terms of other variables in the system.

In order to find the linear function for each variable in the multivariate system, a variant of Least Squares Estimation (LSE) is used. Least squares approximation uses the error between the observed and computed output variables. FLR treats the computational error as the ambiguity in the data observed. The total width of the fuzzy output is the total ambiguity in the output of the system. So, FLR is similar to LSE where total observational error is replaced by total width of the fuzzy coefficients.

The reasons why FLR is applied to the MAR algorithm are as follows:
- Since MAR is a non-fuzzy method, its application to fuzzy systems is not possible; therefore adaptation of the method is required,
- There exists a parallel between obtaining a linear function for a dependent variable in LSE and obtaining a fuzzy linear function in FLR, and
- Ambiguity in the temporal data can be handled by fuzzy regression.

Consequently, the FLR method is used in the MAR algorithm to handle ambiguity in multivariate data. Compared to FLR, Fuzzy MAR is a new approach for processing multivariate autoregressive time series data. When considering from a fuzzy modeling perspective, the MAR algorithm:
- deals with standard (non-fuzzy input and output) data and yields fuzzy rules in order to cover the ambiguity in the temporal data,
- produces fuzzy rules that can be used in any fuzzy system, and
- provides fuzzy rules for forecasting applications.

9.3.3 Fuzzification of Multivariate Auto-Regression

The MAR algorithm is a non-fuzzy method which produces linear functions that define the variables in a multivariate system. Fuzzification is used to obtain fuzzy rules that can be stored in fuzzy rule-bases where FLR is used to obtain fuzzy linear functions for each variable. These functions can then be used as fuzzy rules in fuzzy expert systems.

The resulting equations contain fuzzy coefficients as in a conventional fuzzy linear regression equation, such as:

$$Y = A_1 x_1 + A_2 x_2 + ... + A_p x_p \qquad (9.8)$$

where Y is the output, x is the input vector and each A_i is a fuzzy number, as stated previously. The method of finding a fuzzy linear regression is based on minimization of total width of fuzzy coefficients in the equation.

The model is the solution of an LP (linear programming) problem in which the objective function is to minimize total width of fuzzy coefficients.

Before considering multivariate temporal data, the modifications needed for the auto-regressive model will be explained. Given the linear function to define the value of the variable x_i at time instance t is:

$$x_{i,t} = f(x_{i,t-1},...,x_{i,t-p}) \qquad (9.9)$$

which indicates that $x_{i,t}$ is dependent on p previous time instances of itself which is explained as:

$$x_{i,t} = A_1 x_{i,t-1} + A_2 x_{2,t-2} + ... + A_p x_{i,t-p} \qquad (9.10)$$

In the above equation, the value A_i is a fuzzy number with center α and width c. The objective function of the LP problem is also modified as follows:

$$\text{minimize } J(c) = \sum c^t |x_i| \qquad (9.11)$$

where $|x_i| = (|x_{i,t-1}|,...,|x_{i,t-p}|)$ and the constraints are:

$$x_{i,t} \leq x_i^t \alpha + |L^{-1}(h)| c^t |x_i| \qquad (9.12)$$

$$x_{i,t} \geq x_i^t \alpha - |L^{-1}(h)| c^t |x_i|$$

where $L(x)=1 - |x|$, $c>0$ and $i=1,...,n$.

Each of the variables in the fuzzy linear regression equation is a previous value of the variable to be described for all other possible variables. For multivariate time series analysis, the Multivariate Auto-Regression algorithm is adapted to the fuzzy linear regression analysis.

9.3.4 Bayesian Information Criterion in Fuzzy MAR

A *modified_BIC* is used as the criteria to evaluate the models obtained by using different sets of variables; originally *RSS* was used as the main item in model selection criteria. Since the difference between the expected and obtained output is treated as the ambiguity in FLR, the measure of ambiguity is used instead of RSS in *modified_BIC*. As a result, *modified_BIC* is as follows:

$$modified\_BIC = T*ln(J(c)) + K*ln(T) \qquad (9.13)$$

where T is the number of observations used in the FLR process in order to find a linear function containing fuzzy coefficients, K is the number of regressors in the model function and $J(c) = \sum c_i |x_i|$ is the total width of fuzzy coefficients in the function. *Modified_BIC* is used as the criterion in order to compare the models obtained for a variable; the best model is the one with the minimum *modified_BIC* value.

9.3.5 Obtaining a Linear Function for a Variable

In order to obtain a model for a variable x_i in a multivariate temporal data, all the variables (including their p previous time instances) are tested. It should be noted that p is always less than the total number of time instances in the time interval. At each step, the obtained model (a temporary model) is compared with the previous best model (the model with minimum *modified_BIC*).

The auto-regressive model was presented in the previous Sections. In a multivariate auto-regressive model, all variables must be considered. If there is any input-output relationship between the regressand variables and the regressors, these relationships must be embedded within the functions.

The model for a variable x_i in a vector of multivariate data $\vec{x}_t$ of size n for T time instances (as shown in Table 9.1) is defined as follows:

$$x_{i,t} = f(x_{1,t-1}, \cdots, x_{1,t-p_1}, \cdots, x_{j,t-1}, \cdots, x_{j,t-p_j}, \cdots, x_{n,t-1}, \cdots, x_{n,t-p_n}) \quad (9.14)$$

$$= A_{11} x_{1,t-1} + \cdots + A_{1p_1} x_{1,t-p_1} + \cdots + A_{n1} x_{n,t-1} + \cdots + A_{np_n} x_{n,t-p_n}$$

which indicates that $x_{i,j}$ is dependent on p_j previous time instances of x_j.

The objective function of the LP problem by using the Eq. (9.12) is formulated as follows:

$$\text{minimize } J(c) = \sum c^t |x_i| \quad (9.15)$$

where $|x_i| = |x_{1,1}|, \cdots, |x_{1,p_1}|, \cdots, |x_{j,1}|, \cdots, |x_{j,p_j}|, \cdots, |x_{n,1}|, \cdots, |x_{n,p_n}|$ and $x_{j,p_j} = x_{j,t-p_j}$, and the constraints are:

$$x_{i,t} \leq x_i^t \alpha + |L^{-1}(h)| c^t |x_i| \quad (9.16)$$

$$x_{i,t} \geq x_i^t \alpha - |L^{-1}(h)| c^t |x_i|$$

where $L(x) = 1 - |x|$, $c \geq 0$ and $i = 1, \cdots, n$.

It should be noted that the above LP problem is a general multivariate auto-regressive model which is formulated for a variable x_i and for a specific set of input variables $x_{1,1}, \ldots, x_{1,p1}, \ldots, x_{j,i}, \ldots, x_{j,pj}, \ldots, x_{n,1}, \ldots, x_{n,pn}$. In the set of input variables, $j=1,...,n$ and $p_j <= T$, where n is the number of variables and T is the number of time instances in the time interval.

Since for a variable all possible time intervals of all the variables are tested, the general LP problem described in this Section will be repeated for each variable.

9.3.6 Processing of Multivariate Data

The Fuzzy MAR algorithm gets multivariate data and outputs fuzzy linear functions stored as fuzzy rules. The multivariate data is input to the algorithm in tabular format as in Fig.9.1. The algorithm given runs for each variable x_i in a vector of multivariate data x_t. The crucial point in the process is that the variables are processed one by one by the algorithm. At the end, all the variables in the system are defined by a linear function with fuzzy coefficients. From the viewpoint of complexity, it can be thought that the algorithm is exponential, where all variables are tested for all time instances. On the other hand, the model search is terminated when a predefined p time instances are reached for a variable. Moreover, the algorithm does not compare all possible models.

Another limitation is applied to the number of variables. An auto-regressive model is found first. Then the models containing other variables are searched. All the models are compared as they are obtained with the previous ideal model (the model with minimum *modified_BIC*). As a model of one variable (input) is found, this specific variable is not processed any more. FLR is formulated for all possible sets of variables for all possible p time instances. Whenever a model is found, FLR is called – in other words, the LP problem is solved.

By using the algorithm, a regression equation is found for each variable in the multivariate auto-regressive system. Since the variable depends on its values at previous time instances, the algorithm first finds an auto-regressive equation for that variable. The function $f(x_{i,t-1}, ..., x_{i,t-k})$ found in Step 2 is an auto-regressive function such that $k=1, ..., p$ and p is a predetermined limit of previous time instances to be used in the algorithm.

Whenever a function is found, *modified_BIC* is computed and the function with the minimum criterion is assigned to f_i (the multivariate auto-regressive function for the variable x_i), where i is the index of the variable x_i.

Since in the multivariate auto-regressive model the variable can depend on the previous values of the other variables, the auto-regressive functions containing other variables as regressors are found, and their *modified_BIC* values are computed. These values are compared with the f_i's and the function with the minimum *modified_BIC* is assigned as the auto-regressive equation.

---

1. Input x_t of size n for T time intervals such that each variable is x_i where $i=1,...,n$.
2. For each x_i, find a linear function $f_{ii}(x_{i,t-1},...,x_{i,t-k})$ for $x_{i,t}$ such that the function uses p previous time instances of x_i. Note that k changes from 1 to p.
   (a) For each k, compute *modified_BIC*.
   (b) Find the minimum *modified_BIC* value, choose the function f_{ii} for previous time instance k as f_{ii}. f_{ii} is an auto-regressive function such that
   $$x_{i,t} = A_1 * x_{i,t-1} + A_2 * x_{i,t-2} + ... + A_k * x_{i,t-k}$$
   The equation above describes $x_{i,t}$ in terms of its p previous values.
3. Repeat Step 2 for all variables, $j=1,...,n$ such that
   (a) Find a function for $x_{i,t}$ like $f_{ij}(x_{j,t-1},...,x_{j,t-k})$ where $k=1,...,p$.
   (b) Compute *modified_BIC* for all the variables $j=1,...,n$.
   (c) Choose the function f_{ij} with minimum *modified_BIC*.
4. Compare f_{ii} with f_{ij}.
   (a) Choose the one with minimum *modified_BIC* and assign it to f_i.
   (b) Note the index *ind* for which f_i is found.
5. Find a function f_{ij} as follows:
   (a) Concatenate $x_{j,k}$ to f_i where $j=1,...,n$ and $j \neq ind$ and $k=1,...,p$.
   (b) Compute *modified_BIC* for f_{ij}.
   (c) Compare f_i with $\bar{f}_{ij}$.
   (d) Choose the one with minimum *modified_BIC* and assign it to f_i.
   (e) Continue for all $j=1,...,n$. Find the function containing all possible variables for all possible time instances.
6. Find equations for all variables in the multivariate data by using Steps 2-5.

Fig. 9.1. Fuzzy MAR Algorithm.

In order to supply other variables in addition to the auto-regressive model, the variables other than the auto-regressive model are added as in Step 5.(a) and *modified_BIC* computed. The ultimate goal is to obtain the equation having all the variables possible with the minimum *modified_BIC*. The above steps are repeated for all variables. The number of variables in the regression equation, total number of observations, and total width of fuzzy coefficients ($J(c)$) are the three values used in *modified_BIC* in order to compare the equations.

The output of the Fuzzy MAR Algorithm is the set of equations which describe all of the dependent variables in the system. The following is an example of a rule:

$$\text{IF } x_{t-4} \text{ is } B_1 \text{ AND } y_{t-1} \text{ is } B_2$$
$$\text{THEN } y_t = A_0 + A_1 \, x_{t-4} + A_2 \, y_{t-1}$$

It can be seen that y_t is described by means of a function of previous instances of itself and other variables, namely x_{t-4} and y_{t-1}.

9.4 Experimental Results

Experiments were performed using two real data sets. The first data set is Gas Furnace data [3]; the second data set is taken from [14]. Each of the series variables in the system is described by a fuzzy equation by using a number of lags of a number of variables. Fuzzy outputs are obtained by using fuzzy equations and comparing with the expected output.

9.4.1 Experiments with Gas Furnace Data

Gas Furnace data contains two variables which are gas flow rate (X_0) and CO_2 concentration (X_1). The fuzzy equations for the variables X_0 and X_1 are shown in Table 9.2. $X_{0,t-1}$ is the largest coefficient center value for the fuzzy equation for X_0, and $X_{1,t-1}$ is the largest coefficient value for X_1. The accuracy rate of the model for X_0 (0.482877) is worse than the rate of the model for X_1 (0.708904).

In Fig. 9.2, linearity between the output variables and the defining variables can be seen. The defining variables with largest coefficient center values in the fuzzy equations have apparent linear relationship with the respective variables. The widths of the obtained results for the variable X_0 are large.

9.4.2 Experiment with Interest Rate Data

Monthly Interest Rate data contains three series, which are Federal Funds Rate X_0, 90-Day Treasury Bill Rate X_1, and the One-Year Treasury Bill Rate X_2.

Table 9.2. Fuzzy Equations for Gas Furnace Data

Variable	Fuzzy Equation
X_0	$(1.121, 0.1912)\, X_{0,t-1} + (0, 0.1117)\, X_{0,t-4} + (0, 0.004617)\, X_{1,t-1}$
X_1	$(0.9662, 0.01582)\, X_{1,t-1} + (0.03636, 0)\, X_{1,t-4} + (0, 0.1352)\, X_{0,t-1}$

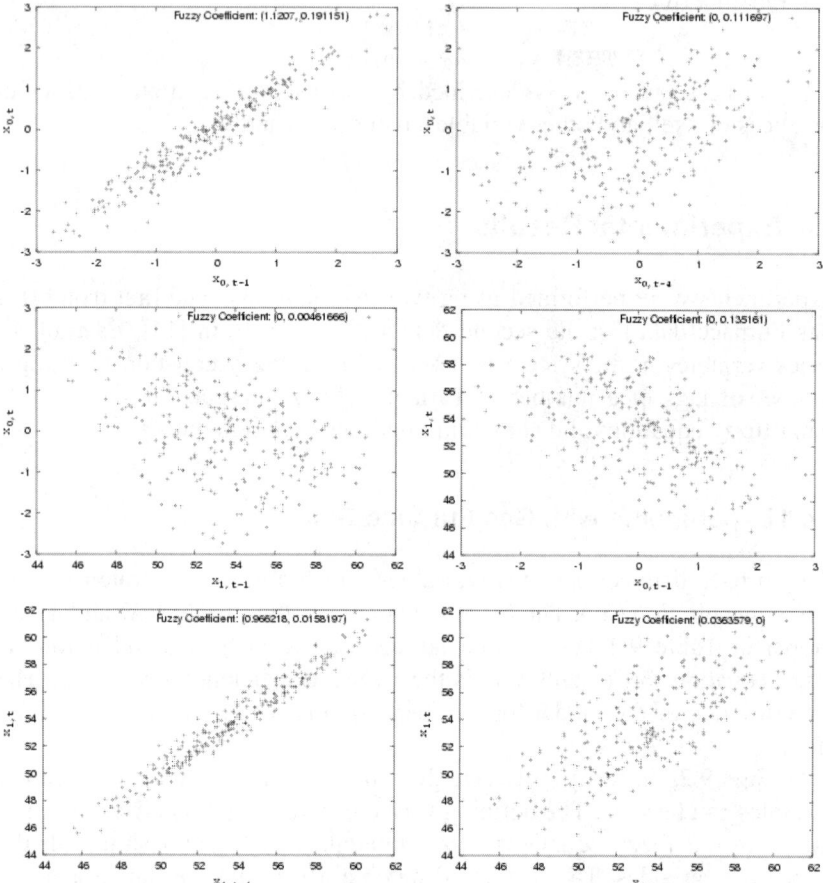

Fig. 9.2. Linearity Figures for Gas Furnace Data

Table 9.3. Fuzzy Equations for Interest Rate Data

Variable	Fuzzy Equation
X_0	$(1.024, 0)\, X_{0,t-1} + (0, 0.01188)\, X_{1,t-1}$
X_1	$(0.7448, 0)\, X_{1,t-1} + (0.2335, 0)\, X_{1,t-2} + (0, 0.09282)\, X_{2,t-1}$
X_2	$(0.534, 0.09717)\, X_{2,t-1} + (0.5283, 0)\, X_{1,t-1}$

In Fig. 9.3, the linear relationship between dependent and independent variables is evident.

Fuzzy MAR yields the model of the variable X_0 with the maximum accuracy – which is 0.852941. The models for both X_1 and X_2 provide 0.844538 accuracy.

9.4.3 Discussion of Experimental Results

Experimental results for two real data sets are given for the Fuzzy MAR Algorithm. The aim of using the algorithm is to extract the fuzzy equations defining each time series variable in a multivariate system. The results show that:
- The algorithm yields approximately linear relationships between variables, as seen in the related figures,
- The variable with the largest center value and smallest width value has the most linear relationship with the variable to be defined,
- The Fuzzy MAR Algorithm aims to find the equation that involves as many lagged variables as possible by using the *modified_BIC* value; the obtained equation may contain excess variables,
- A refinement method can be used to limit the number of variables, and
- Another criterion which is based on individual coefficient center and width values can be developed for that purpose.

9.5 Conclusions

A time series analysis approach, based on fuzzy linear regression has been designed in order to construct rules for defining variables in multivariate time series data. Fuzzy multivariate time series analysis is used to extract fuzzy temporal rules defining the time series data in order to predict the future values. FLR is applied to multivariate time series analysis in order to find a linear function in multivariate time series data. The algorithm is based on finding fuzzy linear functions and comparing them by using BIC. The function containing the optimum number of variables and lags is found as a result of executing the algorithm. The functions provide the necessary information for any fuzzy rule-based system. The Fuzzy MAR algorithm produces the best number of previous instances and number of variables by means of a predetermined information criterion. This algorithm is based on Fuzzy Linear Regression Analysis. In Fuzzy Linear Regression, the difference between the actual and the expected values is treated as the

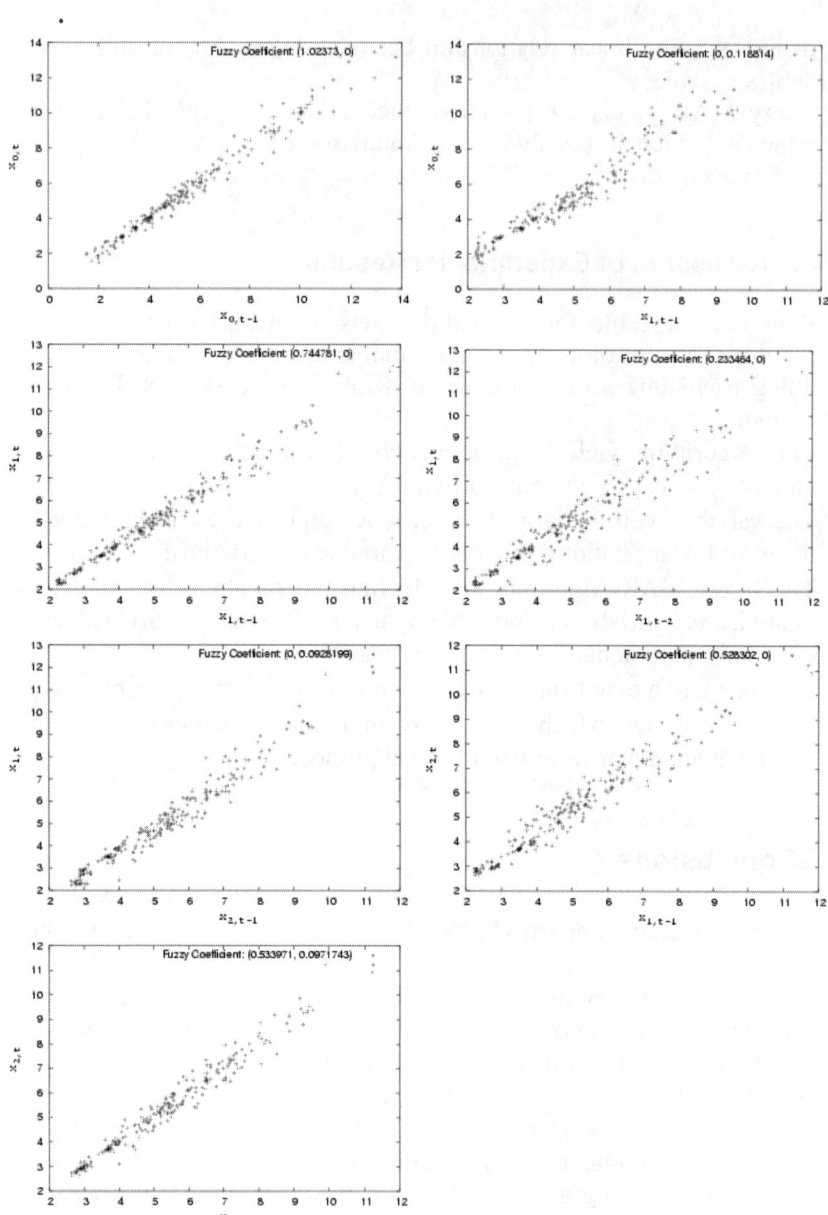

Fig. 9.3. Linearity Figures for Interest Rates for the Federal Funds Rate

vagueness of the system. The aim of fuzzy linear regression is to find the fuzzy parameters which minimizes the cost function and finds the best parameters.

As a future direction, the model can be enhanced in order to improve accuracy and robustness. The equation obtained at the end of execution of the Fuzzy MAR Algorithm may be further processed in order to extract the surplus variables. Since the algorithm is applied to the raw time series data, sometimes excess information is obtained for a variable.

References

1. Akleman D and Alpaslan FN (1997) Temporal Rule Extraction for Rule-Based Systems Using Time Series Approach, *Proc ISCA CAINE-97*, San Antonio, Texas.
2. Bintley H (1987) Time series analysis with reveal, *Fuzzy Sets and Systems*, 23: 97-118.
3. Box GE and Jenkins GM (1970) *Time Series Analysis: Forecasting and Control*, Holden Day, San Francisco, CA.
4. Buckley J and Feuring T (2000) Linear and non-linear fuzzy regression: Evolutionary Algorithm Solutions, *Fuzzy Sets and Systems*, 112: 381-394.
5. Chang PT (1997) Fuzzy seasonality forecasting, *Fuzzy Sets and Systems*, 90: 1-10.
6. Chatfield C (1996) *The Analysis of Time Series: An Introduction*, Chapman and Hall, London.
7. Gujarati DN (2003) *Basic Econometrics*, McGraw Hill, Boston, MA.
8. Heshmaty B and Kandel A (1985) Fuzzy Linear Regression and Its Applications to Forecasting in Uncertain Environment, *Fuzzy Sets and Systems*, 15: 159-191.
9. Hwang JR, Chen SM and Lee CH (1998) Handling Forecasting problems using fuzzy time series, *Fuzzy Sets and Systems*, 100: 217-228.
10. Ishibuchi H and Nii M (1996) Fuzzy Regression Analysis by Neural Networks with Non-Symmetric Fuzzy Number Weights, *Proc Intl Conf Neural Networks*, Washington, DC.
11. Ishibuchi H and Tanaka H (1992) Fuzzy regression analysis using neural networks, *Fuzzy Sets and Systems*, 50: 257-265.
12. Kim B and Bishu RR (1998) Evaluation of fuzzy linear regression models by comparing membership functions, *Fuzzy Sets and Systems*, 100: 343-352.
13. Kim KJ, Moskowitz H and Koksalan M (1996) Fuzzy versus statistical regression, *European J Operational Research*, 92: 417-434.
14. Reinsel GC (1997) *Elements of Multivariate Time Series Analysis*, Springer-Verlag, New York.
15. Savic DA and Pedrycz W (1991) Evaluation of fuzzy linear regression models, *Fuzzy Sets and Systems*, 39: 51-63.
16. Song Q and Chissom BS (1993) Forecasting enrollments with fuzzy time series-Part I, *Fuzzy Sets and Systems*, 54: 1-9.

17. Song Q and Chissom BS (1993) Fuzzy time series, *Fuzzy Sets and Systems*, 54: 269-277.
18. Song Q and Chissom BS (1994) Forecasting enrollments with fuzzy time series-Part II, *Fuzzy Sets and Systems*, 62: 1-8.
19. Song Q (1999) Seasonal forecasting in fuzzy time series, *Fuzzy Sets and Systems*, 107: 235-236.
20. Sugeno M and Kang GT (1988) Structure Identification of Fuzzy Model, *Fuzzy Sets and Systems*, 28: 15-33.
21. Sullivan J and Woodall WH (1994) A comparison of fuzzy forecasting and Markov modeling, *Fuzzy Sets and Systems*, 64: 279-293.
22. Tanaka H (1987) Fuzzy Data Analysis by possibilistic Linear Models, *Fuzzy Sets and Systems*, 24: 363-375.
23. Tanaka H, Hayashi I and Watada J (1989) Possibilistic linear regression analysis for fuzzy data, *European J Operational Research*, 40: 389-396.
24. Tanaka H, Uejima S and Asai K (1982) Linear Regression Analysis with Fuzzy Model, *IEEE Trans Systems, Man and Cybernetics*, SMC-12(6): 903-907.
25. Tanaka H and Lee H (1998) Interval Regression Analysis by Quadratic Programming Approach, *IEEE Trans Fuzzy Systems*, 6(4): 473-481.
26. Tanaka H and Watada J (1988) Possibilistic linear systems and their application to the linear regression model, *Fuzzy Sets and Systems*, 27: 275-289.
27. Terano T, Asai K and Sugeno M (1992) *Fuzzy Systems Theory and Its Applications*, Academic Press Inc., San Diego, CA.
28. Tseng FM, Tzeng GH, Yu HC and Yuan BJC (2001) Fuzzy ARIMA model for forecasting the foreign exchange market, *Fuzzy Sets and Systems*, 118: 9-19.
29. Wang HF and Tsaur RC (2000) Insight of a fuzzy regression model, *Fuzzy Sets and Systems*, 112: 355-369.
30. Watada J (1992) Fuzzy Time Series Analysis and Forecasting of Sales Volume, in Kacprzyk J and Fedrizzi M (Eds) *Studies in Fuzziness Volume I: Fuzzy Regression Analysis*.

10 Selective Attention Adaptive Resonance Theory and Object Recognition

Peter Lozo[1], Jason Westmacott[2], Quoc V Do[3], Lakhmi C Jain[3] and Lai Wu[3]

1. Weapons Systems Division, Defence Science and Technology Organisation, PO Box 1500, Edinburgh, SA. 5111, Australia, peter.lozo@dsto.defence.gov.au

2. Tenix Defence Systems Pty Ltd, Second Avenue, Technology Park, Mawson Lakes, SA, 5095, Australia, jason.westmacott@tenix.com

3. Knowledge-Based Intelligent Engineering Systems Centre, University of South Australia, Mawson Lakes, SA 5095, Australia, Lakhmi.Jain@unisa.edu.au

10.1 Introduction

The concept of *selective attention* as a useful mechanism in Artificial Neural Network models of visual pattern recognition has received a lot of attention recently, particularly since it was found that such a mechanism influences the receptive field profiles of cells in the primate visual pathway by filtering out non-relevant stimuli [28, 29]. It is believed that the massive feedback pathways in the brain play a role in attentional mechanisms by biasing the competition amongst the neural populations that are activated by different parts of a scene [8, 14, 15].

Inspired by the neurophysiological data on selective visual attention from the primate visual cortex one of the authors [26, 27] has, in the context of higher levels of vision, proposed a neural network model called Selective Attention Adaptive Resonance Theory (SAART). The SAART network is an extension of the Adaptive Resonance Theory (ART) of Grossberg [21, 22] and the ART-based neural networks of Carpenter & Grossberg [3, 4, 6]. The new model embeds novel top-down synaptic modulatory feedback pathways (*top-down presynaptic facilitation*) which, together with ART's top-down memory pathways, provides stronger top-down attentional mechanisms for object recognition in cluttered images than is available in the conventional ART models. The top-down presynaptic facilitation pathways enable the network to use its active short term memory (whose contents may be transferred from the long term top-down memory upon its activation by the input) to selectively bias competitive processing at the input of the network. This enables object recognition when a familiar object is presented simultaneously with other objects and clutter, with differential illumination across the object of interest.

In this chapter we provide a brief conceptual background of the ART model and discuss its limitations in the context of difficult pattern recognition problems. In these problems the ART neural network model fails primarily because its attentional subsystem is limited in scope. We will then describe the Selective Attention ART model, its neural network implementation, and object recognition capabilities in cluttered images. This will be followed by a description of some recent extensions that allow the network to recognize distorted 2D shapes. Recent extensions to the network [30] demonstrate that, in the context of 2D shape recognition, one can embed a form of fuzziness into the network to handle object distortions. This is accomplished by designing the input layer of the network to have cells that sample their inputs via 2D receptive fields, whose signal transmission gain is presynaptically modulated by the 2D pattern carried in the top-down pathways.

10.2 Adaptive Resonance Theory (ART)

ART was originally introduced as a physical theory of cognitive information processing in the brain [21, 22]. The theory was derived from a simple feedforward real-time competitive learning system called Instar [20], in response to a problem that real-time competitive learning systems face the *plasticity-stability dilemma*. The dilemma is that a real-time competitive learning system must be plastic to learn significant novel events, while at the same time it must be stable to prevent the corruption of previously learned memories by erroneous activations. The adaptive resonance concept suggests that only the resonant state, in which the reverberation between feedforward (bottom-up) and feedback (top-down) computations within the system are consonant, can lead to adaptive changes.

Since the introduction of ART in the late 1970s and early 1980s, a large family of ART based artificial neural network architectures have been proposed. These include: ART-1 for binary inputs [3], ART-2 for binary and analog inputs [4], ART-3 for hierarchical neural architectures [6], ARTMAP for supervised self-organization of memory codes [7], and various other versions.

As shown in Fig.10.1, the ART model embeds bottom-up and top-down adaptive pathways in a competitive network that contains two subsystems that regulate learning: (i) an attentional subsystem where top-down expectancies (recalled memories) interact with the bottom-up information; and (ii) an arousal (orienting or vigilance) subsystem that is sensitive to the mismatch between the two. Interactions between these two subsystems ensure that memory modification occurs under exceptional circumstances. Memory can only be modified when an approximate match has occurred between the neural pattern at the input and the resultant pattern across F1. This state is called *adaptive resonance*. F0, F1 and F2 are competitive neural Fields whose neural pattern of activity is represented by $x_1^{(0)}$, $x_1^{(1)}$ and $x_1^{(2)}$ respectively. If $x_1^{(0)}$ and $x_1^{(1)}$ match above the preset threshold level then resonance is established between F1 and F2, and this leads to learning in the memory pathways between the active cells in F1 and F2.

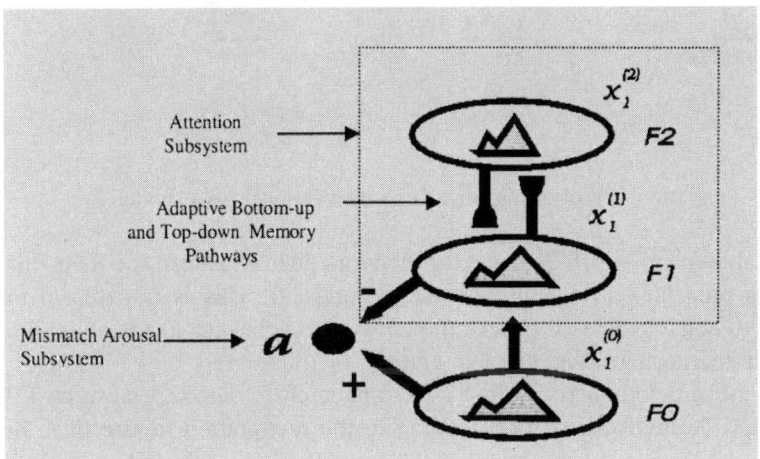

Fig. 10.1. Schematic of the ART model

10.2.1 Limitation of ART's Attentional Subsystem with Cluttered Inputs

Below we illustrate a class of difficult 2D pattern recognition problems that expose a limitation of ART's attentional subsystem. The problem that we wish to consider is that of recognizing a shape in a cluttered input. It can be demonstrated via computer simulations [27] that if a conventional ART model (such as ART-2 or ART-3 neural network) has previously learned a 2D pattern, it will not be able to recognize that pattern when it is presented in a complex background of other patterns and clutter, as illustrated in Fig.10.2.

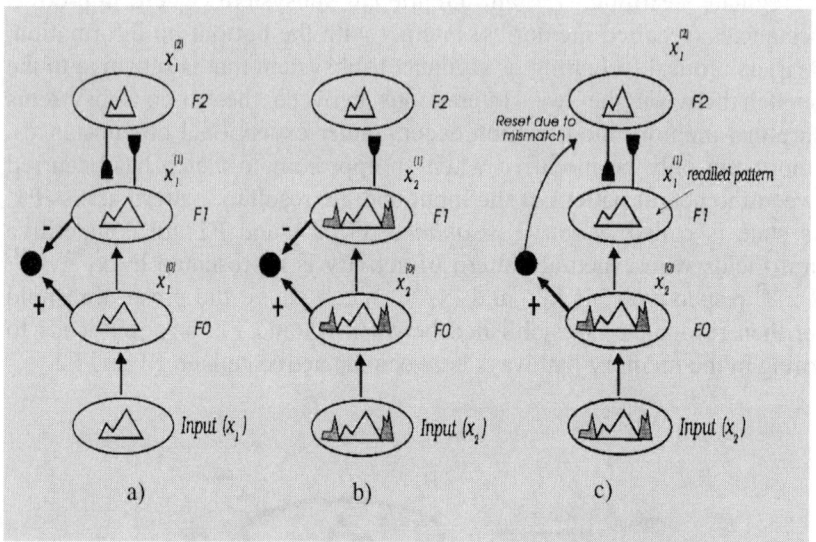

Fig. 10.2. Illustration of processing stages in a conventional ART model

If, as shown in Fig.10.2, the ART network learns pattern x_1, then this pattern is presented in a cluttered background (b). This is transferred to Field F1 to activate F2 (c). Before the network reaches its steady state, the top-down memory of the learned pattern is transferred to F1 where it replaces the previous activity in F1. A mismatch is detected between F0 and F1 and the network is reset leading to the recognition failure of x_1 in input x_2.

Fig.10.2 illustrates the processing stages of the conventional ART model. As shown, whenever Field F0 is activated by an input pattern in which the known pattern is embedded, the network does not have the capability to selectively pay attention to the known portion of the input. Thus, even if the correct top-down memory is activated, the ART model fails to match it with the input because it compares the whole pattern across F0 with the whole pattern across F1. In the next Section we show how the model can be extended to enable selective attention to known portions of the input pattern.

10.3 Selective Attention Adaptive Resonance Theory (SAART)

Rather then resetting the ART network as in Fig.10.2, Lozo has proposed that object recognition in cluttered backgrounds can be solved by extending the capability of ART's attentional subsystem with an additional but functionally different set of top-down feedback pathways [26, 27]. These new pathways run from F1 to F0, allowing the recalled memory across F1 to selectively focus on the portions of the input which it can recognize. This is illustrated in Fig.10.3.

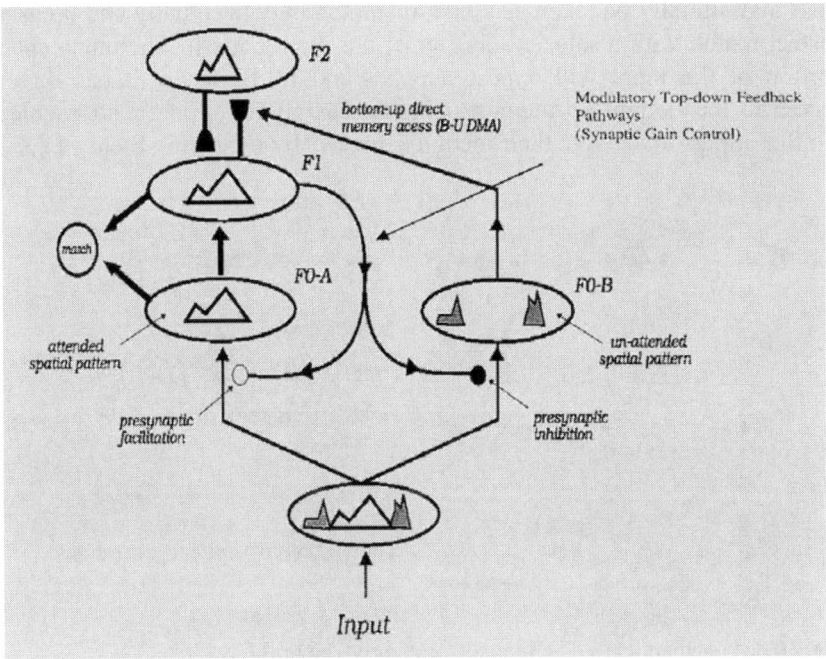

Fig. 10.3. Schematic of the Selective Attention ART model

With reference to Fig.10.3, the feedback pathway from F1 to the input Field F0-A is modulatory, and acts on the signal transmission gain of the bottom-up input synapses of F0-A (and F0-B). Thus, each cell in F1 sends a top-down facilitatory gain control signal to the input synapse of the corresponding cell in F0-A. Lateral competition across F0-A will suppress the activity of all neurons whose input gain is not enhanced by the corresponding neurons in F1. These feedforward-feedback interactions enable *selective resonance* to occur between the recalled memory at F1 and a selected portion of the input at F0-A. Because the facilitatory signals and the competitive interactions in F0-A do not act instantaneously, the resonant steady state develops over a period of time during which the network may be in a highly dynamic state.

To follow the progress of these interactions, it becomes necessary to measure the degree of match between the spatial patterns across the Fields F0-A and F1 as well as the time rate of change of this match. To protect the long term memory from unwarranted modification by non-matching inputs during these rapid changes, long term memory is updated when the system is in a stable resonant state. Similarly, the certainty of the network's pattern recognition response increases as the steady state is approached. Thus, what may initially be taken as a bad mismatch may eventually end up as a perfect match with a selected portion of the input pattern. The unmatched portion of the input will appear across Field F0-B, where it has direct access to the bottom-up memory (this feature of the model also enables familiar inputs to activate their memory directly by bypassing Field F1).

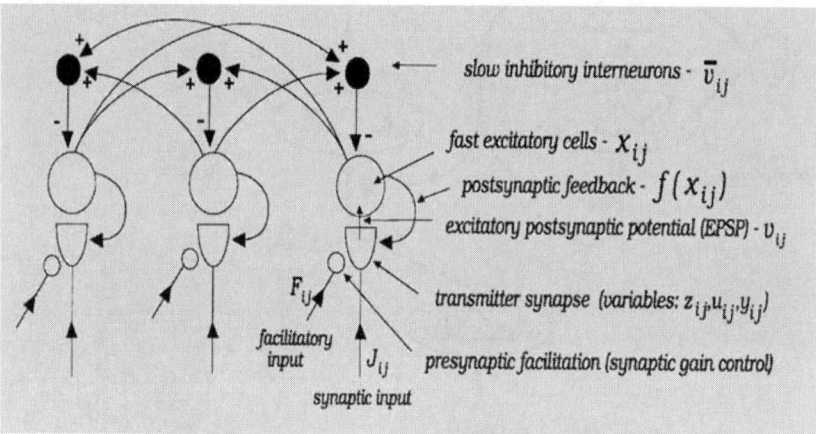

Fig. 10.4. Presynaptically Facilitated Competitive Neural Layer (PFCNL)

10.3.1 Neural Network Implementation of SAART

Fig.10.5 shows the architecture of the SAART neural network. Note that the network has been further modified from the model shown in Fig.10.3 by splitting the Field F1 into three parts. The purpose of this is to prepare the network for more advanced extensions, such as memory guided search [27], where it is desirable to retain the short-term memory of the input in a reverberatory loop (not shown in Fig.10.5) without activating the long-term memory cells (cells in SAART's processing Field C1).

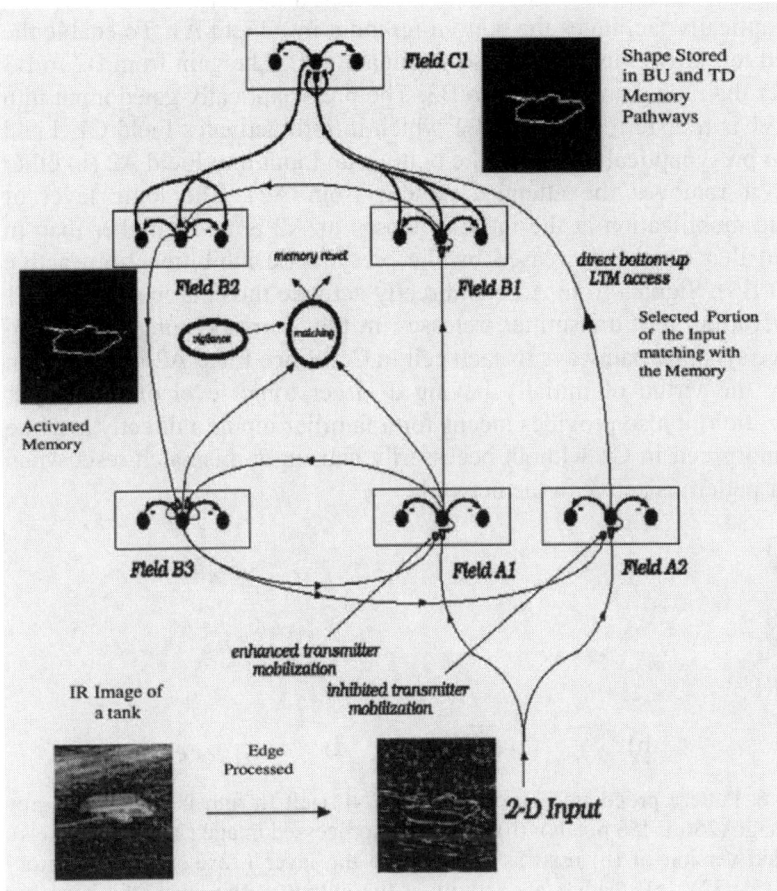

Fig. 10.5. SAART Neural Network

Superimposed are example images to illustrate SAART's Processing Capability. The basic building block of the SAART neural network is a Presynaptically Modulated Competitive Neural Layer, a version of which is shown in Fig.10.4. All layers in the SAART network are two-dimensional, with the exception of Field C1. Each cell in Field C1 sums its inputs from active neurons in Field B1, whose signals are gated by the adaptive bottom-up memory pathways. Cells in Field C1 compete in a winner take all fashion. The winning cell in C1 projects its signals via the top-down adaptive memory pathways to Field B2.

The top-down gain from C1 to B2 is larger than the gain from B1 to B3. This enables the top-down memory to eventually dominate in B2. Field B2 projects to Field B3 which, in addition to receiving an input from Field A1, presynaptically facilitates the bottom-up input into Field A1. To enable the recalled top-down memory to also dominate in B3 the gain from B2 to B3 is larger than the gain from A1 to B3. The presynaptically gated input into Field A1 is transferred to Field B1 which in turn activates Field C1. Field B3 also presynaptically inhibits the bottom-up input into Field A2 (in other words, it removes the attended pattern from A2). The tonic level of synaptic mobilization at the input synapses of A2 is much higher than in A1 (but that can be decreased by the presynaptic inhibition from active cells in B3). Signals from A2 can directly activate the neurons in Field C1 by activating the transmitter release in the corresponding bottom-up adaptive synaptic pathways to each cell in C1. Since Field A2 is faster than A1 (by the virtue of initially having a larger tonic level of transmitter mobilization) it also provides means for a familiar input to directly activate its memory cell in C1 without necessarily causing a mismatch reset when another pattern is active in the network.

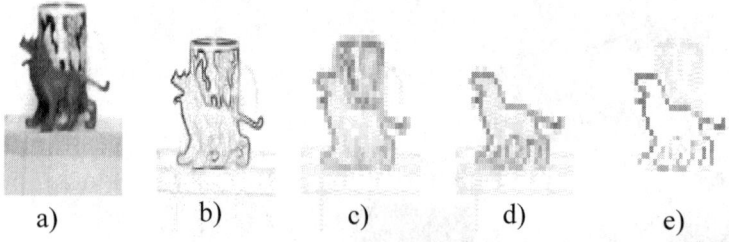

Fig. 10.6. Pattern processing property of PFCNL (left to right): (a) original grey level image (256 x 256 pixels); (b) Sobel edge processed image (256 x 256 pixels); (c) scaled version of (b) used as the input to the layer (32x32); (d) facilitatory pattern (32x32); (e) steady state activity at the output of the layer (the layer has 32x32 neurons)

The Presynaptically Facilitated Competitive Neural Layer (PFCNL) is implemented in two layers: a layer of fast excitatory cells and a layer of slow inhibitory interneurons (large black circles). The inhibitory neural layer mediates the shunted competition in the fast excitatory layer. Each neuron in the excitatory layer receives its excitatory input through a facilitated synaptic pathway whose internal dynamics is slower than the postsynaptic dynamics. The pattern processing property of the layer is shown in Fig.10.6. Since all the layers in the SAART network are based on a version of the presynaptically modulated competitive neural layer we will describe the mathematics of only one layer.

Postsynaptic Cellular Activity

The postsynaptic cellular activity of the ij^{th} neuron in the layer is described by the following shunting competitive equation:

$$\frac{dx_{ij}}{dt} = -Ax_{ij} + (B - x_{ij})Gv_{ij} - x_{ij}\bar{G}\bar{v}_{ij} \quad (10.1)$$

where Gv_{ij} is the amplified transmitter gated excitatory postsynaptic potential (EPSP) presented in [3], G is the amplification factor, $\bar{G}\bar{v}_{ij}$ is the amplified lateral feedback inhibition that is mediated by slow inhibitory interneurons, A is the passive decay rate, and B is the upper saturation. Eq.(10.1) thus represents shunted competition of a layer of neurons whose cellular activity is restricted to range $(0, B)$. It is solved at the equilibrium to give:

$$x_{ij} = \frac{BGv_{ij}}{A + Gv_{ij} + \bar{G}\bar{v}_{ij}} \quad (10.2)$$

Excitatory Postsynaptic Potential

The Excitatory PostSynaptic Potential (EPSP) acting on a cell is due to the gating of the synaptic input signal by the level of the transmitter that is in the mobilised state (and hence available for release). The EPSP is given by the following equation

$$\frac{dv_{ij}}{dt} = -Dv_{ij} + J_{ij}y_{ij}(\rho_v + K_v f(x_{ij})) \quad (10.3)$$

where D is the passive decay rate, J_{ij} is the synaptic input signal, y_{ij} is the level of the mobilised transmitter, ρ_v and K_v are constants, and $f(x_{ij})$ is the thresholded postsynaptic cellular feedback signal that interacts multiplicatively with the input signal and the mobilised transmitter to further increase the EPSP under the condition of correlated pre-postsynaptic activity.

Lateral Competition

The competition in the layer is mediated by slow inhibitory neurons whose dynamics is given by Eq.(10.4):

$$\frac{d\bar{v}_{ij}}{dt} = -\bar{A}\bar{v}_{ij} + \bar{B}\frac{(n-1)}{n}\sum_{(p,q)\neq(i,j)} f(x_{pq}) \quad (10.4)$$

where $\bar{A}$ and $\bar{B}$ are constants, n is the number of neurons in the layer. Thus only those neurons whose activation $f(x_{pq})$ is above a threshold contribute to the competition.

Transmitter Dynamics

The following two interacting equations represent the dynamics of the transmitter in the model of a facilitated chemical synapse.

$$\frac{du_{ij}}{dt} = \alpha_u(z_{ij} - u_{ij}) - (\beta_u + K_u J_{ij} f(x_{ij}))(u_{ij} - y_{ij}) \quad (10.5)$$

This first equation represents the dynamics of the stored transmitter. The first term is the accumulation of the stored transmitter (where z_{ij} is the transmitter production rate held constant (= 1) in all of SAART's layers, except in the adaptive bottom-up and top-down memory pathways where it is subject to learning).

$$\frac{dy_{ij}}{dt} = -\gamma y_{ij} - J_{ij} y_{ij}(\rho_y + K_y f(x_{ij}))(u_{ij} - y_{ij}) + (\beta_y + F_{ij})(u_{ij} - y_{ij}) \quad (10.6)$$

This second equation represents the dynamics of the mobilized transmitter. It says that the transmitter mobilization rate is increased by the facilitatory signal and that the transmitter is released (and hence depleted) by the correlated activity of the input signal and the postsynaptic feedback. The postsynaptic cellular activity must be above a pre-set threshold before it can influence the synapse.

The pre-postsynaptic interactions in the model of the facilitated chemical synapse as implied by the above equations is schematized in Fig.10.7. Although there is no direct neuroscientific evidence from the visual cortex that selective attention affects the dynamics of a chemical synapse as proposed above, there is evidence from the study of the neural circuit controlling the gill withdrawal reflex in the sea slug *Aplysia* [1, 16, 17, 23-25] that facilitatory signals interact with the internal dynamics of a chemical synapse by increasing the pool of the releasable neurotransmitter (in other words, it enhances the mobilization rate of the stored transmitter). Note that the model of a chemical synapse as presented here is an extension of a previously introduced model which did not include the facilitatory synaptic gain control mechanism [2, 18]. A simpler version of the Carpenter & Grossberg model has been used in the adaptive bottom-up and top-down pathways of the ART-3 neural network [5].

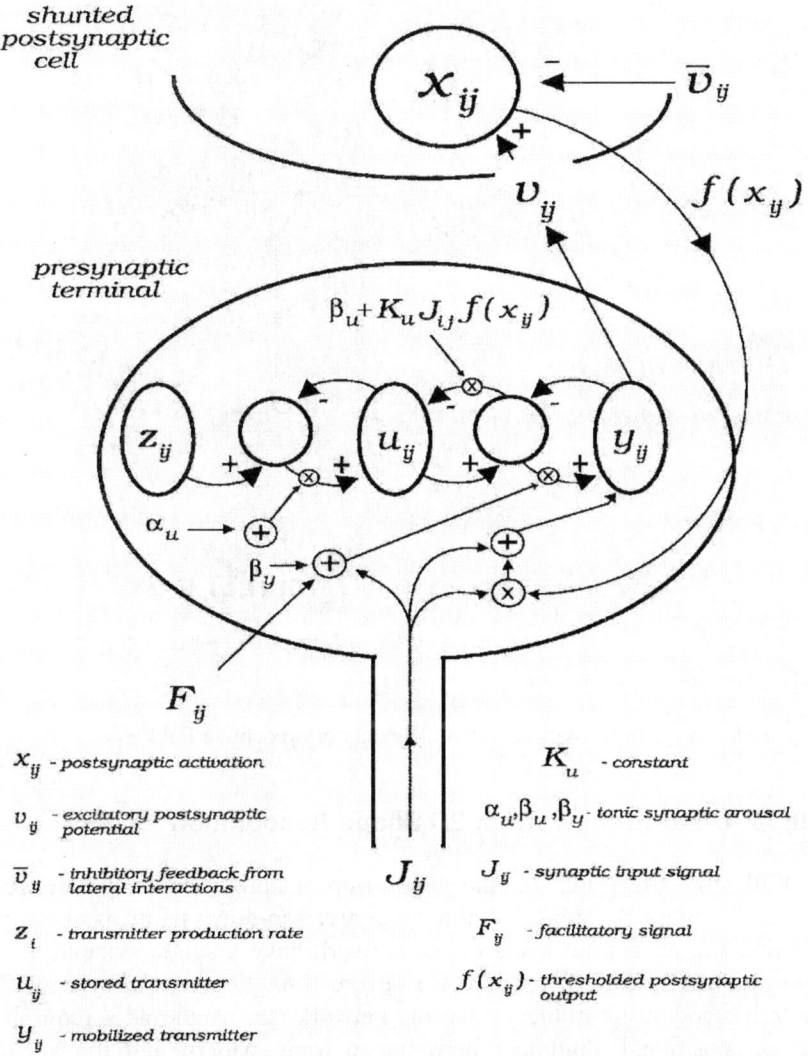

Fig. 10.7. Schematic of pre-postsynaptic interactions in a model of a facilitated chemical synapse

In the next Section we describe a recent extension to the SAART network that enables it to recognise distorted 2D shapes embedded in clutter.

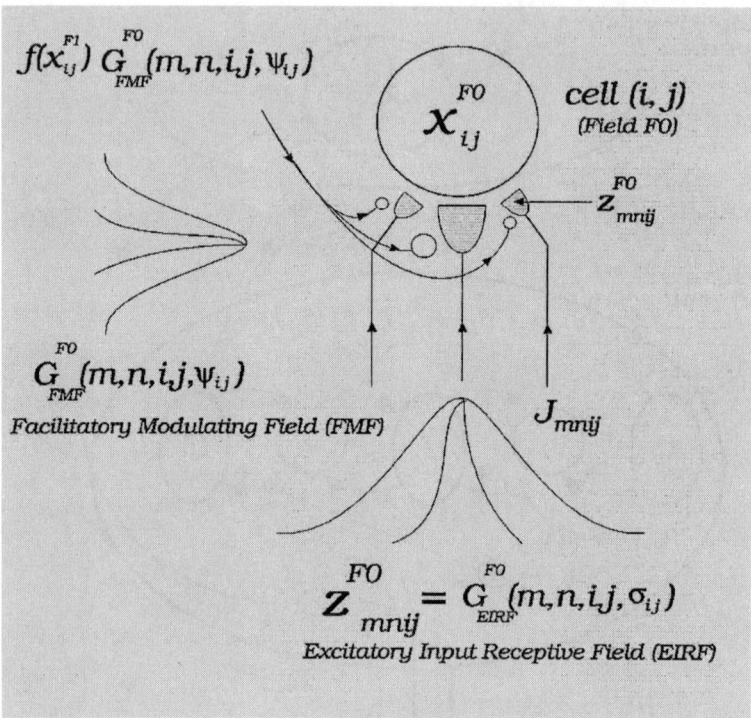

Fig. 10.8. Presynaptically modulated 2D excitatory receptive field

10.3.2 Distortion-invariant 2D Shape Recognition

With the exception of the bottom-up adaptive pathways between SAART's Field B1 and C1 and the adaptive top-down pathways between C1 and B2, all neural layers of the network have a single synaptic input and output pathway. This places a severe limitation on the type of 2D pattern recognition problems that the network can handle as it requires a perfect positional alignment between an input pattern and the version stored in the memory. Thus even a positional offset by one cell position between the stored memory and the input shape will lead to recognition failure and often this misalignment may be such that the input pattern will fail to activate the memory via the bottom-up adaptive pathways. In order to solve this problem the input layer of the network (Field A1) can be extended such that each cell samples over a local 2D region of the input space (as was proposed in [27], but initially implemented and simulated on a 2-layered neural circuit comprising the Fields B3 and A1).

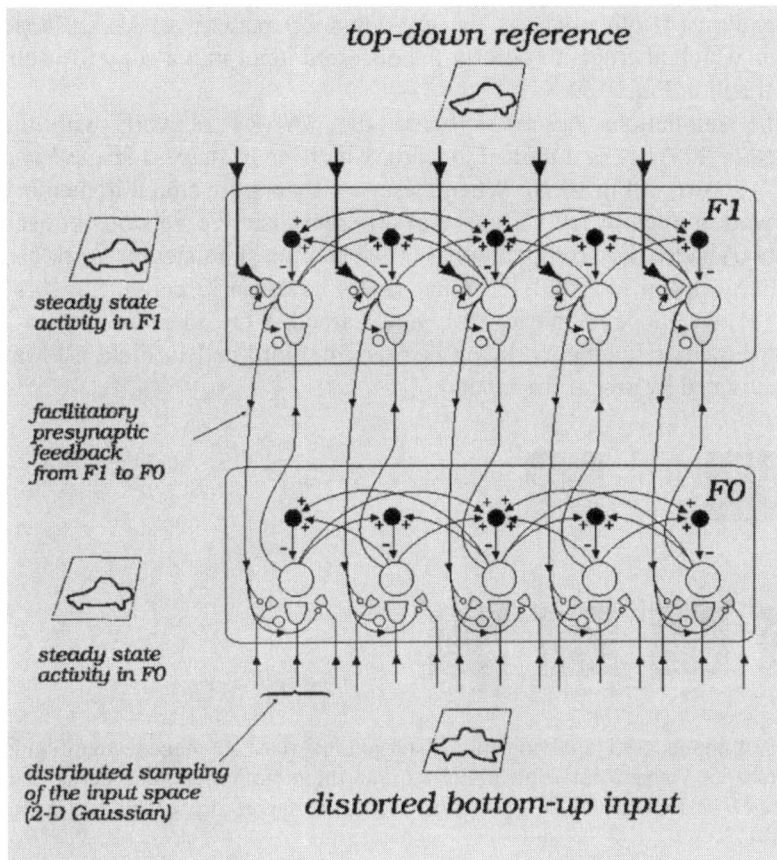

Fig. 10.9. Feedforward-Feedback interactions between SAART's Fields B3 and A1 (each active neuron in B3 projects a 2D Gaussian weighted facilitatory field to the corresponding neuron in Field A1)

Fig.10.8 shows a modified cell in Field A1 whose input is sampled via a distributed 2D Gaussian weighted excitatory input receptive field that is presynaptically facilitated by another 2D Gaussian (referred to as the Facilitatory Modulating Field in the figure) from neurons in Field B3. Fig.10.9 shows the interactions between Field B3 and the modified Field A1. When each cell in Field A1 (and A2) is modified to sample via 2D receptive fields then the network can deal with small positional displacements and distortions as is demonstrated in the simulation results described below. Top-down facilitation from Field B3 and the competitive interactions across B3 and A1 ensure that the pattern transferred into Field A1 does not appear fuzzy but is deformed into a clean copy of the top-down pattern (in other words, the top-down 2D pattern acts as a *shape attractor* which attempts to deform the distorted input into a copy of itself, as illustrated in Fig.10.9).

In the simulations described below the SAART network with the modified Field A1 is first trained to learn two different shapes (Shape 1 and Shape 2) shown in Fig.10.10. When these two shapes are stored in memory the network is then tested with several different distorted versions of each of the two shapes. For computational efficiency the simulated network had only three cells in Field C1 (all other layers have 32x32 cells). Fig.10.11 shows the two shape memories that were learned in the adaptive pathways (the third memory remained blank because the third cell in Field C1 was never activated by any of the inputs).

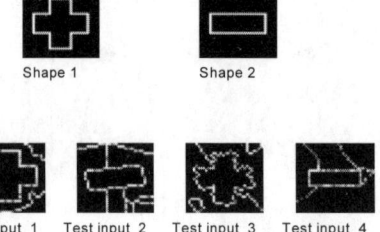

Fig. 10. 10. Inputs used to demonstrate distortion invariant 2D shape recognition in clutter (shapes 1 and 2 are sequentially fed into the modified SAART network to be learned. Various distorted versions of these two shapes are embedded in test inputs 1-4)

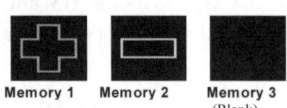

Fig. 10.11. SAART's memory after training (and testing) – Note that there are two copies of each one in the bottom-up and the other in the top-down memory pathways

Each of the four test inputs shown in Fig.10.10 contain either a displaced, distorted, size changed or a slightly rotated version of the original two shapes but embedded in a background of non-relevant clutter.

Thus, test input 1 embeds a displaced version of Shape 1 (1 pixel right, 1 pixel up); test input 2 embeds a slightly rotated version of Shape 2; test input 3 embeds a distorted version of Shape 1; test input 4 embeds a smaller version of Shape 2 (reduced on all sides by 1 pixel). Simulation results on the four test inputs are shown in Figs.12-15 (shown are the resultant steady state patterns in SAART's various processing Fields).

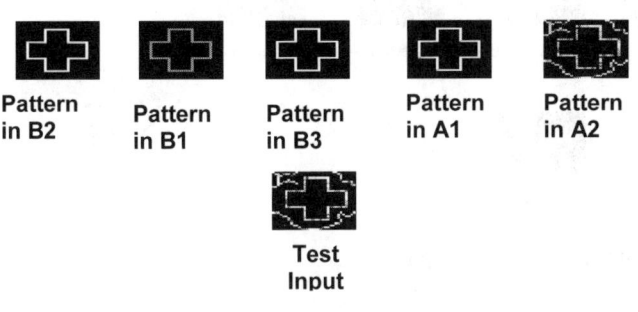

Fig. 10.12. Steady state results on test input 1

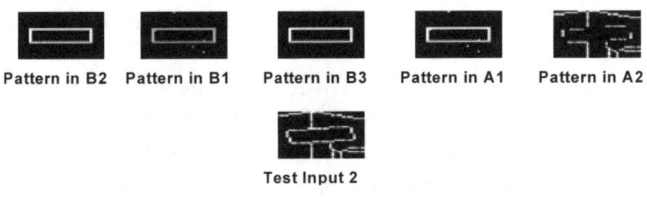

Fig. 10.13. Steady state results on test input 2

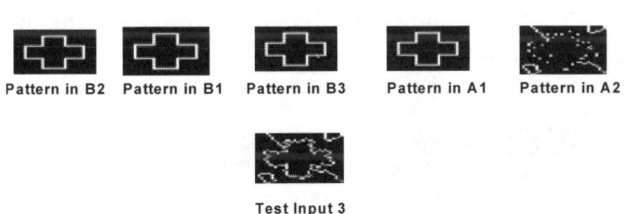

Fig. 10.14. Steady state results on test input 3

The above simulation data demonstrates that the SAART network whose input layer is designed such that the neighbouring cells in the 2D array have overlapping 2D excitatory input receptive fields that presynaptically facilitated by the top-down 2D modulating field can recognize familiar but distorted 2D shapes in a cluttered bottom-up input. Because the extension that was made to Field A1 has not yet been implemented in Field, A2 the steady state pattern across A2 still shows some residual parts of the attended and recognised shape (ideally this should be removed, leaving only the unattended portion of the input across A2).

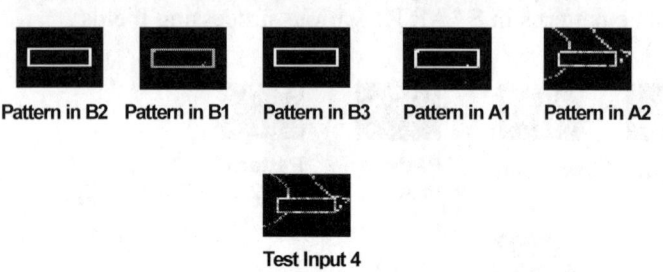

Pattern in B2   Pattern in B1   Pattern in B3   Pattern in A1   Pattern in A2

Test Input 4

Fig. 10.15. Steady state results on test input 4

To demonstrate what happens when the currently active pattern in the network does not match a portion of the input we show in Fig.10.16 the steady state results at the time when the test input 1 (containing a version of Shape 1) is replaced by the test input 2 (containing a version of Shape 2).

The steady state results shown in Fig.10.16 demonstrate how the top-down active pattern in the network (pattern in B3) attempts to deform a selected portion of the input pattern across A1 to match itself. However, because the input receptive fields of A1 are small (5x5 input pathways) in comparison to the difference between the pattern across B3 and A1, the degree of match between B3 and A1 is below the preset threshold level. This causes the network to reset. Upon reset, the test input 2 then activates the memory of Shape 2 and, as shown in Fig.10.13, this leads to the recognition of Shape 2 in the test input 2.

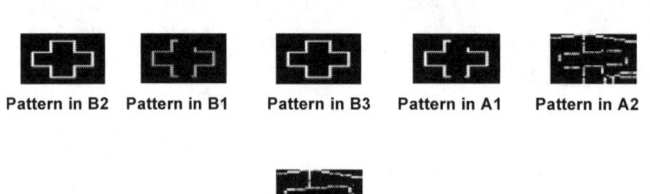

Pattern in B2   Pattern in B1   Pattern in B3   Pattern in A1   Pattern in A2

Test Input 2

Fig. 10.16. Mismatched steady state results when test input 1 is followed by test input 2

Fig.10.17 shows the degree of match between patterns across B3 and A1 throughout the complete simulation of the network. Also shown are the instants at which the network is reset by the mismatch. Since Field A2 in this instance still samples via single input pathways it cannot always provide a direct route to the memory for distorted versions of familiar shapes because of the positional misalignment between the pattern across A2 and the bottom-up memory pathways. This problem can easily be solved by allowing A2 to also sample via 2D receptive fields.

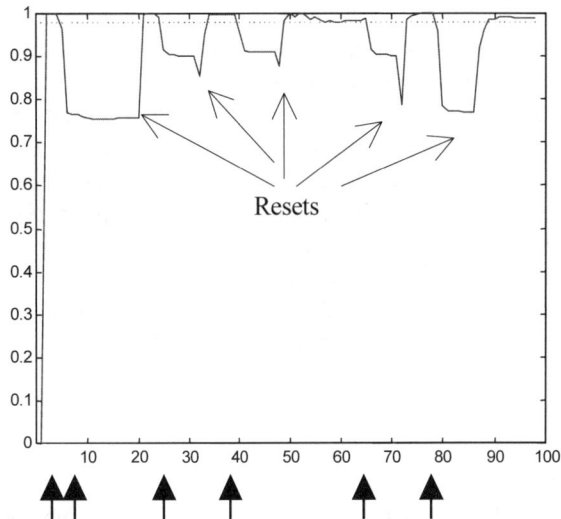

Fig. 10.17. Degree of match throughout the first 100 iterations of the network. The required match for recognition is 0.98 (indicated by the dotted horizontal line). Large arrows indicate the times when a new input pattern is presented to the network (this occurs after each resonance)

10.4 Conclusions

We have described how the Adaptive Resonance Theory (ART) model may be extended to provide a real-time neural network that has more powerful attentional mechanisms. The Selective Attention Adaptive Resonance Theory (SAART) neural network uses the established memory to selectively bias the competitive processing at the input to enable, under the current design constraints, object recognition in cluttered real-world visual and InfraRed images. The SAART network as presented here is still very limited in scope for it to be applicable in general to a wider range of object recognition problems. For example, there is a need to address the problem of larger positional displacements than can be catered for by the 2D receptive fields of the input layer. This is a problem called translational invariance. Lozo has proposed how the above model can be extended by an additional three neural layers at the input to provide for memory guided search and translation invariant object recognition in a wide and cluttered bottom-up edge processed input image [27]. Some of these extensions have been partially implemented and demonstrated recently [10].

In addition to the problem of translation invariant recognition, it has also been proposed, on the basis of the available psychophysical data [11-13], that the SAART network can be extended to handle size and rotation invariant recognition by spatial transformations of both the input and the active top-down memory via a set of parallel and competing neural layers in the bottom-up and the top-down pathways [27]. Bottom-up spatial transformations are required to activate the stored memory while the top-down spatial transformations of the recalled memory are required to selectively match it with the input. Some of these extensions and other related problems are currently being investigated.

References

1. Carew TJ (1987) Cellular and molecular advances in the study of learning in Aplysia, in Changeux JP and Konishi M (Eds) *The Neural and Molecular Bases of Learning*, John Wiley & Sons, New York: 177-204.
2. Carpenter GA and Grossberg S (1981) Adaptation and transmitter gating in vertebrate photoreceptors, *J Theoretical Neurobiology*, 1: 1-42.
3. Carpenter GA and Grossberg S (1987) A massively parallel architecture for a self-organizing neural pattern recognition machine, *Computer Vision, Graphics, and Image Processing*, 37: 54-115.
4. Carpenter GA and Grossberg S (1987) ART-2: Self-organization of stable pattern recognition codes for analog input patterns, *Applied Optics*, 26: 4919-4930.
5. Carpenter GA and Grossberg S (1989) Search mechanisms for Adaptive Resonance Theory (ART) architectures, *Proc Intl Joint Conf Neural Networks*, June 18-22, Washington, DC, I: 201-205.
6. Carpenter GA and Grossberg S (1990) ART-3: Hierarchical search using chemical transmitters in self-organizing pattern recognition architectures, *Neural Networks*, 3: 129-152.

7. Carpenter GA, Grossberg S and Reynolds JH (1991) ARTMAP: Supervised real-time learning and classification of nonstationary data by a self-organizing neural network, *Neural Networks*, 4: 565-588.
8. Chelazzi L, Miller EK, Duncan J and Desimone R (1993) A neural basis for visual search in inferior temporal cortex, *Nature*, 363: 345-347.
9. Chelazzi L, Duncan J, Miller EK and Desimone R (1998) Responses of neurons in inferior temporal cortex during memory-guided visual search, *J Neurophysiology*, 80: 2918-2940.
10. Chong EW, Lim CC and Lozo P (1999) Neural model of visual selective attention for automatic translation invariant object recognition in cluttered images, *Proc 3rd Intl Conf Knowledge-Based Intelligent Information Engineering Systems,* Adelaide, August 31 – September 1, IEEE Press, Piscataway, NJ: 373-376.
11. Corbalis M (1988) Recognition of Disoriented Shapes, *Psychological Review*, 95(1): 115-123.
12. Corbalis M and McLaren R (1982) Interaction Between Perceived and Imagined Rotation, *J Experimental Psychology: Human Perception and Performance*, 8(2): 215-224.
13. Cooper LA and Shepard RN (1973) Chronometric studies of the rotation of mental images, in Chase WG (Ed), *Visual information processing*, Academic Press, New York: 75-176.
14. Desimone R, Wessinger M, Thomas L and Schneider W (1990) Attentional control of visual perception: Cortical, and subcortical mechanisms, *Proc Cold Spring Harbour Symposium in Quantitative Biology*, 55: 963-971.
15. Desimone R (1996) Neural mechanisms for visual memory and their role in attention, *Proc National Academy of Sciences, USA*, 93: 13494-13499.
16. Gingrich KJ, Baxter DA and Byrne JH (1988) Mathematical model of cellular mechanisms contributing to presynaptic facilitation, *Brain Research Bulletin*, 21: 513.
17. Gingrich KJ and Byrne JH (1985) Simulation of synaptic depression, posttetanic potentiation and presynaptic facilitation of synaptic potentials from sensory neurons mediating gill-withdrawal reflex in Aplysia, *J Neurophysiology*, 53: 652-669.
18. Grossberg S (1968) Some physiological and biochemical consequences of psychological postulates, *Proc National Academy of Sciences, USA*, 60: 758-765.
19. Grossberg S (1969) On the production and release of chemical transmitters and related topics in cellular control, *J Theoretical Biology*, 22: 325-364.
20. Grossberg S (1972) Neural expectation: Cerebellar and retinal analogs of cells fired by learnable or unlearned pattern classes, *Kybernetik*, 10: 49-57.
21. Grossberg S (1976) Adaptive pattern classification and universal recoding, II: Feedback, expectation, olfaction, and illusions, *Biological Cybernetics*, 23: 187-202.
22. Grossberg S (1980) How does a brain build a cognitive code?, *Psychological Review*, 87: 1-51.

23. Hawkins RD, Abrams TW, Carew TJ and Kandel ER (1983) A cellular mechanism of classical conditioning in Aplysia: Activity-dependent amplification of presynaptic facilitation, *Science*, 91: 400-405.
24. Kandel ER (1979) *Behavioral Biology of Aplysia*, Freeman, San Franciso, CA.
25. Kandel ER and Schwartz JH (1982) Molecular biology of learning: Modulation of transmitter release, *Science*, 218: 433-443.
26. Lozo P (1995) SAART: Selective Attention Adaptive Resonance Theory Neural Network for Neuroengineering of Robust ATR Systems, *Proc IEEE Intl Conf Neural Networks*, Perth, November 26 – December 1: 2461-2466.
27. Lozo P (1997) Neural Theory and Model of Selective Visual Attention and 2D Shape Recognition in Visual Clutter, PhD Thesis, Department of Electrical & Electronic Engineering, University of Adelaide.
28. Moran J and Desimone R (1985) Selective attention gates visual processing in the extrastriate cortex, *Science*, 229: 782-784.
29. Motter BC (1993) Focal attention produces spatially selective processing in visual cortical areas V1, V2, and V4 in the presence of competing stimuli, *J Neurophysiology*, 70: 909-919.
30. Westmacott J, Lozo P and Jain L (1999) Distortion Invariant Selective Attention Adaptive Resonance Theory Neural Network, *Proc 3rd Intl Conf Knowledge-Based Intelligent Information Engineering Systems*, Adelaide, August 31 – September 1, IEEE Press, Piscataway, NJ: 13-16.

Index

accumulation index 4
accuracy 112
adaptive mutual information 66
Adaptive Resonance Theory –
 ART 301, 302, 317
Adaptive Technical Analysis –
 ATA 12
Advanced Information Processing –
 AIP 256
agent communication 260
Aibo 194
Akaike Information Criterion –
 AIC 288
Alternative Dispute Resolution –
 ADR 242
Amari-Cichocki-Yang algorithm 67
anomaly 109, 113
anomaly detection 104, 109, 115, 121
Ant Colony Optimisation – ACO
 141, 157, 162, 170, 176
Ant Colony System – ACS 144
Ant Multi-Tour System –
 AMTS 145
Ant System – AS 143
anti-Hebbian 65
A-PHONN 37
Aquabot pool cleaner 189
Aquavac pool cleaner 189
Artificial Intelligence – AI
 99, 102, 181, 197, 204, 256
Artificial Neural Network – ANN
 12, 17, 98, 104, 109, 111, 301
Artificial Neural Network ensemble 121
Artificial Neural Network Expert System –
 ANSER 39
Artificial Neural Network Group 33
Assault game type 26

Auto-Regression Integrated Moving
 Average Model – ARIMA 283
Attentional subsystem 302, 303, 317
Australian All Ordinaries Index
 7, 9, 12, 13
Automated Guided Vehicle –
 AGV 196
Automatically Defined Function –
 ADF 8, 14
automation 256
auto regression 281

BackPropagation algorithm – BP
 10, 17, 104, 120, 130, 283
banknote fraud detection 80
Bayesian Information Criterion –
 BIC 287, 288, 291
Bayesian network 206
Bayesian Ying Yang – BYY 82
BDI agent 257, 258, 260, 261, 275
BDI architecture 258
BDI reasoning 261, 263, 276
Beliefs-Desires-Intentions –
 BDI 255
Best Alternative To a Negotiated
 Agreement – BATNA 240
biomedical signal processing 76
Blind Source Separation – BSS 60
Bot class 255
Bot interface 267
bottom-up 302
bounded rationality 2
Boyd's Observe-Orient-Decide-Act
 loop – OODA 260
Burma data set 146, 148, 152
buy-and-hold strategy 9, 11

Index

CaptureTheFlag game type 266
Case Based Reasoning –
    CBR 211, 219, 221, 246
Chebyshev-Hermite polynomial 67
classification 104, 105
cleaning robot 183
clustering 118, 129
cluttered (noisy) image 303, 314, 317
Cognitive Cockpit 257
cognitive engineering 257
Cognitive Work Analysis – CWA 257
collective action 136
collective intelligence 133, 136
commander agent 264
commonplace cases 212
communication agent 264
Comon algorithm 71
competitive learning – CL 302
compliance 103
computational finance 81
Contextual ICA algorithm 73
contrast function 62, 68
cost function 64
Crew Assistant program 256, 257
Criteria, Context, Contingency guideline
    Framework – CCCF 229

Data Mining – DM 98, 99, 102, 222
DeathMatch game type 266
Decision Support Systems – DSS 202
Decision Tree – DT 40, 43, 223
decorrelation method 69
Defence Science and Technology
    Organisation, Australia –
    DSTO 258
defending behaviour 273
dense data 106
dependant variable 289
discriminant 10
distortion-invariant shape recognition 312
Distributed Problem Solving – DPS 256
dividend 4
Dolphin pool cleaner 189
domain expert 129
DomDefender agent 270, 274
Domination game type 266

DOM(ination) point 270
e-Commerce 207
Efficient Market Hypothesis –
    EMH 1, 14
ElectroCardioGraph – ECG 76
ElectroEncephaloGraph – EEG 76
ensemble of ANNs 121
entropy 64
Evolutionary Algorithm – EA
    8, 101, 109, 118, 133, 283
exclusive or – XOR 19, 107
Expectation Maximization – EM 78
Expert System – ES
    98, 111, 121, 127, 130, 183
ExploreAndDefend reasoning 270
exploring behaviour 272
Extended InfoMax algorithm 66

F0-field (layer) 303
F1-field (layer) 303
F2-field (layer) 303
facilitatory gain control 305
Facilitatory Modulating Field
    (2D Gaussian) 313
facilitatory signal 305
fallback bargaining 241
FastICA algorithm 68
feature selection 109
feature space 10
Feedforward-feedback interaction
    305, 313
Feedforward neural network
    12, 109, 283
financial market 1
financial time series analysis 82
forecasting 282
Frequency domain ICA 77
function optimization 140
fuzzification 290
fuzzy auto regression 281, 290
Fuzzy Auto-Regression Integrated
    Moving Average Model
    – Fuzzy ARIMA 284
Fuzzy Least Squares Linear
    Regression – FLSLR 283

Index

Fuzzy Linear Regression – FLR 281, 282, 285, 289, 290, 297
Fuzzy Membership Least Squares Regression – FMLSR 283
Fuzzy Multivariate Auto-Regression – Fuzzy MAR 282, 283, 286, 289, 290, 291, 294, 297, 299
fuzzy neural network 283
fuzzy rule 283, 290
fuzzy rule base 290
fuzzy set 284
fuzzy time series analysis 282

GameBots 263, 267, 276
GameTypes 265
gas furnace 287, 295
Gaussian function 3, 4, 61, 68, 313
Gaussian Mixture Model – GMM 78
Gaussian noise 74
Gaussianity 75
Genetic Algorithm – GA 8
Genetic Programming – GP 7, 14
Givens rotation matrix 71
gradient method 65
Gram-Charlier expansion 66

Health Insurance Commission, Australia – HIC 97, 120, 129
HelpMate 182
Herault-Jutten algorithm 75
heuristic, local 155
Higher-Order Neural Network – HONN 18
human-agent collaboration 266
human-agent teaming 258, 264
human-machine interaction 259
human player 259

if-then rule 287
image clutter 303, 314, 317
image noise 303, 314, 317
Independent Component Analysis– ICA 60
InfoMax ICA algorithm 64, 69
Information Technology – IT 201
InStar 302
intelligent (machine) agent 256, 258, 259
Intelligent Knowledge-Based Systems 208

intelligent systems 201
intelligent tutoring systems 207
Interactive Flash Flood Analyzer – IFFA 45
interest rate data 295
iris data set 162
iRobot 184, 195

Jack intelligent agent shell 255, 260, 263, 269, 276
Java 264, 267
JavaBots 263, 267, 276
Joint Approximate Diagonalization of Eigenmatrices – JADE 74, 81

Kalman filter 82
k-means clustering 118
knowledge-based systems 202
Knowledge Discovery in Databases – KDD 210, 229, 233, 246
Knowledge worker Desktop Environment – KDE 236
Kullback-Leibler divergence 63
kurtosis 3, 79, 118

landmark cases 212
lateral competition 305, 309
Least Squares Estimation – LSE 290
Legal Knowledge-Based Systems – LKBS 202, 208, 228, 231
Legal Knowledge Discovery in Databases – KDD 222, 229, 233, 246
legal ontologies 234
legal reasoning 209
Legal Support Systems – LSS 202
Lego Mindstorms 193
Lego Robolab 193
linear programming – LP 285, 291
linear regression 281
local heuristic 155

MagnetoEncephaloGraph – MEG 77
Markov model 284
Mars rover 180

Index   324

maximum likelihood 65, 74
Max-Min Ant System 145
McCulloch and Pitts neuron model 18
Medicare 97, 126, 127
Micro-Net 121
Mindstorms 193
minimum path length 140
mobile robot 179
model selection 288
Molgedey-Schuster algorithm 72
Moving Windows paradigm 2
M-PHONN 38
multi-agent system 256
Multi-Layer Perceptron – MLP 12, 17, 19
Multivariate Auto-Regression – MAR 282, 284, 286, 291
multivariate temporal data 292
multivariate time series 281, 293
mutual information methods 62

natural law 204
NAV(igation) point 270, 272
negentropy 68
negotiation support systems 240
Neuron-Adaptive Higher-Order Neural Network – NAHONN 30
noise removal 82
nonlinear decorrelation 75
Nuclear Magnetic Resonance (NMR) spectroscopy 83

objective (cost) function 64
Observe-Orient-Decide-Act loop – OODA 260
oil flow monitoring 79
OneSAF Testbed Baseline 262
On-line Dispute Resolution – ODR 242
Operator System Integration Environment 262
optimization 138, 140
outlier 113

Particle Swarm Optimisation – PSO 165, 172, 176
pattern recognition 17, 302, 312
Pharmaceutical Benefits Scheme – PBS 97

pheromone 134
plasticity-stability dilemma 302
Polynomial Higher-Order Neural Network – PHONN 20
PostEventAndWait method 271
postsynaptic 308, 309
precision 112
prediction 104, 106
predictor variable 288
presynaptic 306
Presynaptically Facilitated Competitive Neural Layer – PFCNL 306, 308
Principal Component Analysis – PCA 75
pro se litigant 203
probabilistic linear systems 282
Probability Distribution (Density) Function – PDF 61, 284
PT-HONN 36

Quake 263
QuickProp 17

Radial Basis Function – RBF 10
rainfall estimation 45, 46
rainfall prediction 36
Rasmussen Cognitive Classification 257
reasoning network 43
regressand (dependant variable) 289
regression 281
regressor (predictor variable) 288
relative entropy 68
Residual Sum-of-Squares – RSS 287, 288
Resilient BackPropagation – Rprop 17
reward-to-variability ratio 4
RoboCup 262
RoboCup Rescue 262
RoboLab 193
RoboMower 187
robot courier 182
robot lawn mower 187
robot people transporter 191

robot pool cleaner 189
robot toy 193
robot vacuum cleaner 183
Roomba 184, 195
rotation method 69
rule-based reasoning 214, 246
rule-based systems 206, 212, 281

SAART neural network 306
satellite rainfall estimation 38
scatterplot 117
Schwartz (Bayesian) Information Criterion – BIC 289
scripting language 263
Segway human transporter 191
selective attention 301
Selective Attention Adaptive Resonance Theory – SAART 301, 305, 317
semantic web 234
sensor noise 84
shape (pattern) recognition 302, 312
Share Price Index – SPI 5, 9, 12, 13
Sharpe Ratio 4, 11, 13
situational awareness 262
skewness 3
sockets 263, 267
SoftLaw 202, 204, 217, 234
Sojourner 180
SomulAnt algorithm 140, 147, 163
sonar sensor 183
Sony Aibo 194
Sorting 140, 158, 162
sparse data 105
speech extraction 77
spatio-temporal ICA 73
Split-Up 205, 224, 227, 231, 232, 243
Stationarity 76
Statistical Learning Theory 10
statistical multivariate time series 282
statistical regression 283
stigmergy 133, 134
Support Vector Machine – SVM 10, 14

Tactical Mission Management System 257
TCP/IP network 263
team 259

Technical Analysis – TA 1
temporal correlation 76
temporal decorrelation 71
time series 28
top-down 301
Toulmin Argument Structure 226, 232
transaction costs 5
Travelling SalesPerson – TSP 140, 171
Trigonometric Higher-Order Neural Network – THONN 23
Trilobite vacuum cleaner robot 183
troop agent 264
2D Gaussian 313
2D shape (pattern) recognition 302, 312

univariate time series 281
Unreal-Script 263
Unreal Tournament – UT 263, 275
unsupervised classification 78
UtController class 269
UtJackInterface architecture 269, 276
UtPlayer class 268

VizClient tool 267

weather forecasting 35
World Wide Web – WWW 204, 233

XOR 19, 107

Printing: Strauss GmbH, Mörlenbach
Binding: Schäffer, Grünstadt